Lecture Notes in Geoinformation and Cartography

Marco Painho · Maribel Yasmina Santos · Hardy Pundt
Editors

Geospatial Thinking

Springer

Editors
Prof. Marco Painho
Universidade Nova Lisboa
Inst. Superior Estatistica e
Gestao Informacao (ISEGI)
Depto. Estatistica & Economia
Travessa Estevao Pinto
1070-312 Lisboa
Campus de Campolide
Portugal
painho@isegi.unl.pl

Prof. Maribel Yasmina Santos
Universidade do Minho
Depto. Sistemas de Informacao
Campus de Azurem
4800-058 Guimaraes
Portugal
maribel@dsi.uminho.pt

Prof. Hardy Pundt
University of Applied Sciences Harz
Dept. for Automation Computer Science
Friedrichstr. 57-59
38855 Wernigerode
Germany
hpundt@hs-harz.de

ISSN 1863-2246 e-ISSN 1863-2351
ISBN 978-3-642-12325-2 e-ISBN 978-3-642-12326-9
DOI 10.1007/978-3-642-12326-9
Springer Heidelberg Dordrecht London New York

Library of Congress Control Number: 2010924734

Cover design: deblik

Printed on acid-free paper

Springer is part of Springer Science+Business Media (www.springer.com)

Preface

For the fourth consecutive year, the Association of Geographic Information Laboratories for Europe (AGILE) promoted the edition of a book with the collection of the scientific papers that were submitted as full-papers to the AGILE annual international conference. Those papers went through a competitive review process. The 13th AGILE conference call for full-papers of original and unpublished fundamental scientific research resulted in 54 submissions, of which 21 were accepted for publication in this volume (acceptance rate of 39%).

Published in the *Springer Lecture Notes in Geoinformation and Cartography*, this book is associated to the 13th AGILE Conference on Geographic Information Science, held in 2010 in Guimarães, Portugal, under the title "*Geospatial Thinking*".

The efficient use of geospatial information and related technologies assumes the knowledge of concepts that are fundamental components of *Geospatial Thinking,* which is built on reasoning processes, spatial conceptualizations, and representation methods. *Geospatial Thinking* is associated with a set of cognitive skills consisting of several forms of knowledge and cognitive operators used to transform, combine or, in any other way, act on that same knowledge.

The scientific papers published in this volume cover an important set of topics within Geoinformation Science, including: Representation and Visualisation of Geographic Phenomena; Spatiotemporal Data Analysis; Geo-Collaboration, Participation, and Decision Support; Semantics of Geoinformation and Knowledge Discovery; Spatiotemporal Modelling and Reasoning; and Web Services, Geospatial Systems and Real-time Applications.

The organization of this annual conference, either in its scientific and its organizing component, was only possible with the support of a large number of individuals and institutions. We therefore would like to gratefully acknowledge the effort of the authors and reviewers of this book, who made this edition possible in a short schedule. We thank the local organizing chair Adriano Moreira (University of Minho) and his team for giving all local support. We would also like to acknowledge Filipe Meneses (University of Minho) for his great support in the edition of this book and

Agata Oelschläger (Springer) who also helped us to publish this book on time.

Additionally, we would like to thank ESRI Inc. and the Nordic Network in Geographic Information Science for sponsoring the 13th AGILE International Conference on Geographic Information Science.

Marco Painho, Maribel Yasmina Santos, Hardy Pundt

February 2010

Programme Committee

Programme Chair Marco Painho,
New University of Lisbon (Portugal)

Programme Co-Chair Maribel Yasmina Santos,
University of Minho (Portugal)

Programme Co-Chair Hardy Pundt
University of Applied Sciences Harz (Germany)

Local Organising Committee

Adriano Moreira, University of Minho (Chair)
Jorge Gustavo Rocha, University of Minho
Filipe Meneses, University of Minho
Fernando Bação, New University of Lisbon

Scientific Committee

Adriano Moreira, University of Minho (Portugal)
Agnès Voisard, Fraunhofer ISST (Germany)
Alexander Wolff, TU Eindhoven (The Netherlands)
Anders Oestman, University of Gävle (Sweden)
Anders Friis-Christensen, National Survey and Cadastre (Denmark)
Andrea Rodríguez, Universidad de Concepción (Chile)
Antonio Krüger, University of Muenster (Germany)
Arie Croitoru, The University of Alberta (Canada)
Arnold Bregt, Wageningen University (The Netherlands)
Arzu Çöltekin, University of Zurich (Switzerland)
Atsuyuki Okabe, University of Tokyo (Japan)
Bela Markus, University of West Hungary (Hungary)
Bettina Speckmann, TU Eindhoven (The Netherlands)
Bin Jiang, University of Gävle (Sweden)
Bisheng Yang, Wuhan University (China)
Christian Heipke, Leibniz Universität Hannover (Germany)
Christoph Brox, University of Muenster (Germany)

Christophe Claramunt, Naval Academy Research Institute (France)
Claus Brenner, Leibniz Universität Hannover (Germany)
Claus Rinner, Ryerson University (Canada)
Danny Vandenbroucke, Katholieke Universiteit Leuven (Belgium)
Dave Abel, CSIRO Mathematical and Information Sciences (Australia)
Didier Josselin, Université d'Avignon et des Pays du Vaucluse (France)
Dieter Pfoser, Institute for the Management of Information Systems, RC ATHENA (Greece)
Femke Reitsma, University of Canterbury (New Zealand)
Fernando Bação, New University of Lisbon (Portugal)
Filipe Meneses, University of Minho (Portugal)
Florian Probst, SAP Research CEC Darmstadt (Germany)
Francis Harvey, University of Minnesota (USA)
Fred Toppen, University of Utrecht (The Netherlands)
Gerard Heuvelink, Wageningen University (The Netherlands)
Gilberto Camara, National Institute for Space Research (Brazil)
Gábor Mezosi, University of Szeged (Hungary)
Hans-Gerd Maas, Dresden University of Technology (Germany)
Hartwig Hochmair, University of Florida (USA)
Isabel Cruz, University of Illinois (USA)
Itzhak Benenson, Tel Aviv University (Israel)
Javier Nogueras, University of Zaragoza (Spain)
Javier Zarazaga-Soria, University of Zaragoza (Spain)
Jochen Renz, The Australian National University (Australia)
Joep Crompvoets, Katholieke Universiteit Leuven (Belgium)
Johannes Schoening, University of Muenster (Germany)
John Stell, University of Leeds (United Kingdom)
Jorge Rocha, University of Minho (Portugal)
Juan Suárez, Forestry Commission (United Kingdom)
Jürgen Döllner, Universität Potsdam (Germany)
Lars Bernard, Technische Universität Dresden (Germany)
Lars Bodum, Aalborg University (Denmark)
Lars Harrie, Lund University (Sweden)
Lars Kulik, University of Melbourne (Australia)
Leila De Floriani, University of Genova (Italy)
Marc van Kreveld, University of Utrecht (The Netherlands)
Marco Painho, New University of Lisbon (Portugal)
Maribel Yasmina Santos, University of Minho (Portugal)
Marinos Kavouras, National Technical University of Athens (Greece)
Martin Raubal, University of California (USA)
Max Craglia, Joint Research Centre (Italy)
May Yuan, University of Oklahoma (USA)

Menno-Jan Kraak, ITC - International Institute of Geo-Information Science and Earth Observation (The Netherlands)
Michael Gould, ESRI Inc.
Michael Lutz, Joint Research Centre (Italy)
Michela Bertolotto, University College Dublin (Ireland)
Monica Wachowicz, Wageningen University (The Netherlands)
Monika Sester, Leibniz Universität Hannover (Germany)
Pedro Muro Medrano, University of Zaragoza (Spain)
Peter Fisher, University of Leicester (United Kingdom)
Poulicos Prastacos, Foundation for Research and Technology (Greece)
Pragya Agarwal, Lancaster University (United Kingdom)
Ralf Bill, Universität Rostock (Germany)
Rob Lemmens, ITC - International Institute of Geo-Information Science and Earth Observation (The Netherlands)
Rob Weibel, University of Zurich (Switzerland)
Ross Purves, University of Zurich (Switzerland)
Sara Fabrikant, University of Zurich (Switzerland)
Spiros Skiadopoulos, University of Peloponnese (Greece)
Stephan Winter, The University of Melbourne (Australia)
Stephen Hirtle, University of Pittsburgh (USA)
Suchith Anand, University of Nottingham (United Kingdom)
Takeshi Shirabe, Technical University Vienna (Austria)
Tapani Sarjakoski, Finnish Geodetic Institute (Finland)
Thomas Brinkhoff, Institute for Applied Photogrammetry and Geoinformatics (Germany)
Thomas Kolbe, Technical University Berlin (Germany)
Thérèse Steenberghen, Katholieke Universiteit Leuven (Belgium)
Tiina Sarjakoski, Finnish Geodetic Institute (Finland)
Tobias Dahinden, Leibniz Universität Hannover (Germany)
Tumasch Reichenbacher, University of Zurich (Switzerland)
Volker Paelke, Leibniz Universität Hannover (Germany)
Wolfgang Reinhardt, Universität der Bundeswehr München (Germany)

Contributing Authors

Carlos Abargues
Centre for Interactive Visualization, Universitat Jaume I, Castellón, Spain

Ivo Anastácio
Instituto Superior Técnico / INESC-ID Lisboa, Portugal

Marius Austerschulte
Institute for Geoinformatics, University of Münster, Germany

Peter Bak
Visual Analytics Group, Dept. of Computer and Information Science, University of Konstanz, Germany

Bastian Baranski
Institute for Geoinformatics, University of Münster, Germany

Rubén Béjar
Computer Science and Systems Engineering Department
University of Zaragoza, Spain

Arturo Beltran
Centre for Interactive Visualization, Universitat Jaume I, Castellón, Spain

Alberto Belussi
Dipartimento di Informatica, Università degli Studi di Verona, Verona, Italy

Sandro Bimonte
Cemagref, France

Rizwan Bulbul
Vienna University of Technology, Department of Geoinformation and Cartography, Austria

Chunyuan Cai
Institute for Geoinformatics, University of Muenster, Germany

Pável Calado
Instituto Superior Técnico / INESC-ID Lisboa, Portugal

Dave Chapman
University College London, UK

Cláudia Costa
University of Coimbra, Departamento de Arquitectura, Coimbra, Portugal

Jürgen Döllner
Hasso-Plattner-Institut, University of Potsdam, Germany

Vincenzo Del Fatto
Dipartimento di Matematica e Informatica, Università di Salerno, Italy

Aneta J. Florczyk
Department of Computer Science and Systems Engineering
Universidad de Zaragoza, Spain

Theodor Foerster
International Institute for Geo-Information Science and Earth Observation, Enschede, The Netherlands

Andrew U. Frank
Vienna University of Technology, Department of Geoinformation and Cartography, Austria

Carlos Granell
Centre for Interactive Visualization, Universitat Jaume I, Castellón, Spain

Thimmaiah Gudiyangada
Centre for Interactive Visualization, Universitat Jaume I, Castellón, Spain

Dirk Hecker
Fraunhofer IAIS, Germany

Carsten Keßler
Institute for Geoinformatics, University of Münster, Germany

Daniel Keim
University of Konstanz, Konstanz, Germany

Slava Kisilevich
University of Konstanz, Konstanz, Germany

Jan Klimke
Hasso-Plattner-Institute, University of Potsdam, Germany

Christine Körner
Fraunhofer IAIS, Germany

Jan Eric Kyprianidis
Hasso-Plattner-Institut, University of Potsdam, Germany

Thomas Leduc
CERMA laboratory, France

Francisco J. Lopez-Pellicer
Department of Computer Science and Systems Engineering
Universidad de Zaragoza, Spain

Adriana Loureiro
University of Coimbra, Departamento de Arquitectura, Coimbra, Portugal

Bruno Martins
Instituto Superior Técnico / INESC-ID Lisboa, Portugal

Michael May
Fraunhofer IAIS, Germany

Francis Miguet
CERMA laboratory, France

Adriano Moreira
Algoritmi Research Centre, University of Minho, Guimarães, Portugal

Sara Migliorini
Dipartimento di Informatica, Università degli Studi di Verona, Verona, Italy

Pedro R. Muro-Medrano
Department of Computer Science and Systems Engineering
Universidad de Zaragoza, Spain

Mauro Negri
Dipartimento di Elettronica e Informazione, Politecnico di Milano, Milan, Italy

Javier Nogueras-Iso
Department of Computer Science and Systems Engineering
Universidad de Zaragoza, Spain

Itzhak Omer
Urban Space Analysis Laboratory, Dept. of Geography and Human Environment, Tel Aviv University, Israel

Daniel Orellana
Centre for Geo-Information, Wageningen University and Research, The Netherlands

Luca Paolino
Dipartimento di Matematica e Informatica, Università di Salerno, Italy

Giuseppe Pelagatti
Dipartimento di Elettronica e Informazione, Politecnico di Milano, Milan, Italy

Martin Raubal
Department of Geography, University of California, Santa Barbara, USA

Lior Rokach
Department of Information Systems Engineering and The Deutsche Telekom Laboratories, Ben-Gurion University of the Negev, Beer-Sheva, Israel

Paula Santana
University of Coimbra, Departamento de Arquitectura, Coimbra, Portugal

Maribel Yasmina Santos
Algoritmi Research Centre, University of Minho, Guimarães, Portugal

Rita Santos
University of Coimbra, Departamento de Arquitectura, Coimbra, Portugal

Sven Schade
Institute for Environment and Sustainability, European Commission - Joint Research Centre, Ispra, Italy

Bastian Schäffer
Institute for Geoinformatics, University of Münster, Germany

Markus Schneider
Department of Computer & Information Science & Engineering, University of Florida, Gainesville, USA

Tobias Schreck
Interactive Graphics Systems Group, Technische Universität Darmstadt, Germany

Monica Sebillo
Dipartimento di Matematica e Informatica, Università di Salerno, Italy

Amir Semmo
Hasso-Plattner-Institut, University of Potsdam, Germany

Vincent Tourre
CERMA laboratory and Ecole Centrale de Nantes, France

Giuliana Vitiello
Dipartimento di Matematica e Informatica, Università di Salerno, Italy

Monica Wachowicz
Centre for Geo-Information, Wageningen University and Research, The Netherlands

Patrick Weber
University College London, UK

Stephan Winter
Department of Geomatics, The University of Melbourne, Australia

Philippe Woloszyn
RESO laboratory, Université de Haute Bretagne - Rennes II, France

Stefan Wrobel
Fraunhofer IAIS, Germany

Wenjie Yuan
Department of Computer & Information Science & Engineering, University of Florida, Gainesville, USA

F. Javier Zarazaga-Soria
Department of Computer Science and
Systems Engineering
Universidad de Zaragoza, Spain

Table of contents

Geo-Collaboration, Participation, and Decision Support

Semantics of Geoinformation and Knowledge Discovery

Spatiotemporal Modelling and Reasoning

Intersection of Nonconvex Polygons Using the Alternate Hierarchical Decomposition

Rizwan Bulbul, Andrew U. Frank

Vienna University of Technology
Department of Geoinformation and Cartography
Gusshausstrasse 27-29 E127, A-1040 Vienna, Austria
{bulbul, frank}@geoinfo.tuwien.ac.at

Abstract. Intersection computation is one of the fundamental operations of computational geometry. This paper presents an algorithm for intersection computation between two polygons (convex/nonconvex, with nonintersecting edges, and with or without holes). The approach is based on the decomposed representation of polygons, alternate hierarchical decomposition (AHD), that decomposes the nonconvex polygon into its convex components (convex hulls) arranged hierarchically in a tree data structure called convex hull tree (CHT). The overall approach involves three operations (1) intersection between two convex objects (2) intersection between a convex and a CHT (nonconvex object) and, (3) intersection between two CHTs (two nonconvex objects). This gives for (1) the basic operation of intersection computation between two convex hulls, for (2) the CHT traversal with basic operation in (1) and, for (3) the CHT traversal with operation in (2). Only the basic operation of intersection of two convex hulls is geometric (for which well known algorithms exist) and the other operations are repeated application of this by traversing tree structures.

1 Introduction

The intersection operation is of fundamental importance (Shamos and Hoey 1976) as it provides basis for computing other Boolean operations like union and difference etc. Also, it is the most expensive operation

M. Painho et al., (eds.), *Geospatial Thinking*, Lecture Notes in Geoinformation and Cartography,
DOI 10.1007/978-3-642-12326-9_1, © Springer-Verlag Berlin Heidelberg 2010

computationally, roughly taking 80 % of the running time (Greiner and Hormann 1998). The intersection operation has two problems, intersection detection and intersection computation. The intersection detection between two convex objects is a basic geometric operation (Chazelle and Dobkin 1987) and a great account of the topic can be found in (David 1997). We are not focusing on the issue of intersection detection and for simplicity we assume hereafter that the objects under consideration for intersection computation do intersect.

The Problem: Given two polygons (convex or nonconvex, with non-intersecting edges, and with or without holes), compute the intersection region that may be;

a) Empty (Figure 1a)

b) Convex (Figure 1b)

c) Nonconvex (Figure 1c)

d) A set of convex and/or nonconvex disjoint regions (Figure 1d)

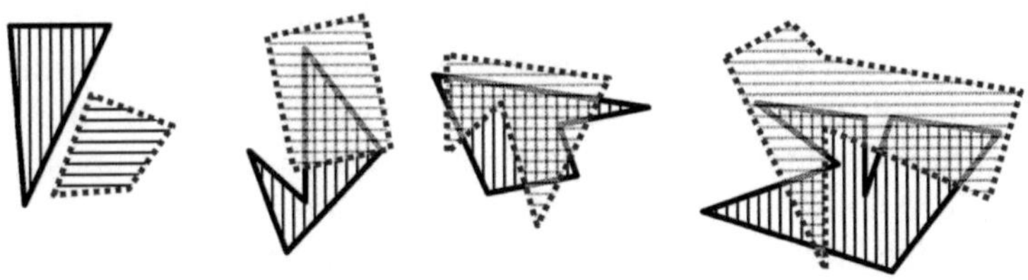

(a) (b) (c) (d)

Fig. 1. Different polygon intersection scenarios

In GIS domain the set-theoretic Boolean operations (intersection, union and symmetric difference) are extensively used for extracting useful spatial information out of spatial data modeled as polygons (Margalit and Knott 1989). For example, polygon clipping is a frequent operation in GIS (Liu et al. 2007). Other Boolean operations in GIS include overlay, windowing, join and merge etc (Rigaux et al. 2001) (FranciscoMartınez et al. 2009). The map overlay operations are key operations in the GIS domain.

For convex polygons, optimal algorithms for intersection computation are known. Although, variety of solutions also exist for Boolean operations on complex polygons (nonconvex polygons with or without holes), these solutions are intricate having complex data structures leading to difficult implementations.

Our approach for the intersection computation is a simple algorithm based on the decomposed representation of polygons, AHD (Bulbul and Frank 2009), that decomposes the nonconvex polygon into its convex components (convex hulls) arranged hierarchically in a tree data structure called CHT (more in section 3). The intersection computation involves three operations;

1) Intersection between two convex objects.
2) Intersection between a convex and a CHT (nonconvex object).
3) Intersection between two CHTs (two nonconvex objects).

This gives for (1) the basic and simple operation of intersection computation between two convex hulls. For (2), the CHT traversal with basic operation in (1) and for (3), the CHT traversal with operation in (2). The approach is further discussed in detail in section 4. Only the basic operation of intersection of two convex hulls is geometric (for which well known algorithms exist) and the other operations are repeated application of this by traversing tree structures.

2 Previous Work

Many algorithms for Boolean operations on polygons have been reported in literature. The preliminary work by Shamos and Hoey (1976) provided a basis for geometric intersection problems. They have shown that the intersection of two simple plane *n*-gons can be detected in $O(n \log n)$. The intersection of two convex *n*-gons and two nonconvex *n*-gons can be computed in $O(n)$ and $O(n^2)$ respectively. Bentley and Ottmann (1979) gave the classical sweep line algorithm for counting and reporting all intersections by extending the work by Shamos and Hoey. They provided an algorithm for reporting all *k* intersections between two general planar objects in $O(n \log n + k \log n)$. The work by Bentley and Ottmann was further extended by Lauther (1981), and the reported algorithm has the expected time complexity of $O(n \log n)$. Another $O(n)$ time algorithm was presented by O'Rourke etal (1998). The algorithm is simple but is limited to convex polygons only.

The two plane sweep algorithms by Nievergelt and Preparata (1982) compute the geometric intersection of two nonconvex polygons in $O((n+k) \log n)$ and two convex polygons in $O(n \log n + k)$. The polygons can have self intersecting edges but degenerate cases are not tackled. The data structure is complex and implementation details are not given.

The work by Chazelle and Dobkin (1987) provides lower bounds on algorithms for intersection of convex objects in two and three dimensions. Their work is based on the assumption that the intersecting objects are available in random access memory, eliminating reliance on linear input reading time. The time bounds for two convex polygons in 2D and two convex polyhedra in 3D cases are $O(\log n)$ and $O(\log^3 n)$ respectively (in 3D case an additional multiplicative factor of *log n* for data structure preprocessing for standardization).

The work by Margalit and Knott (1989) presents an algorithm for set operations on polygon pairs having worst time complexity of $O(n^2)$. They give partial correctness proof of their solution and implementation is discussed but still complex and not easily understandable.

The work by Rappoport (1991), extended convex difference tree (*ECDT*) for representing two n-dimensional polygons (polytopes) and performing intersection and union operations, is similar to our approach. The differences between our approach and their approach both at data structure and operations level are given in Table 1.

Table 1. Difference between our and Rappoport's Approach

Our Approach	**Approach by Rappoport**
A single CHT data structure with build and process operations	Implementation details of how data structure is actually built and processed is not mentioned
Can easily handle holes	Nothing said about
Our approach is robust, no special topological handling	Needs special topological handling for robustness
Can handle multiple polygons with slight modifications	Two polygons
Simple structure as the children convex hulls of a parent node are contained within the convex region of that parent node.	Primitives in a right sub-tree of a difference node are contained within the primitive on the left side of that node.

The state of the art for finding all intersections among segments is given by (Bernard and Herbert 1992). The algorithm by Vatti (1992) is for clipping arbitrary polygons against arbitrary polygons. The polygons may be convex, concave or self intersecting. However, the self intersecting polygon is converted to a nonintersecting polygon by inserting the points of intersection during the clipping process. The algorithm also supports polygon decomposition by allowing the output in the form of trapezoids if required. The solution is complex and implementation is not easy. Its performance has not been proved asymptotically, rather a comparison with traditional clipping methods is provided.

The algorithm proposed by Greiner and Hormann (1998) also deals clipping arbitrary polygons like Vatti's algorithm. However, it is a simple algorithm based on the boundary segment manipulation and performs better than Vatti's algorithm over randomly generated general polygons. The data structure is a doubly link list or lists in case of multiple polygons. Only few degenerate cases are mentioned and it can not treat overlap degeneracies, and no complete complexity analysis provided. The solution is limited for 2D polygons and robustness issues related to the fixed precision floating point arithmetic are not catered.

The algorithm by Rivero and Feito (2000) to calculate Boolean operations for general planar polygons (manifold and non manifold, with and without holes) is based on simplicial chains and their operations. The strategy has been demonstrated for 2D and claimed to be valid for 3D polyhedra. The algorithm does not need special treatment of degenerate cases, and it is shown that its time is similar to Greiner's algorithm. The slight modifications in the work of Rivero and Feito by Peng, Yong et al. (2005) resulted in an algorithm which has been shown to be more efficient (execution time less than one third of that by Rivero and Feito).

CGAL (Fogel et al. 2006) provides Boolean operations for polytopes in 2-dimensional Euclidian space. Robustness is ensured through the use of exact arithmetic. The regularized operations are provided for two simple polygons with or without holes. The time complexity is $O(n^2)$ for simple polygons.

The algorithm by Liu, Wang et al. (2007) is for clipping arbitrary polygons with holes. The algorithm is based on segment manipulation and works by classification of intersection points into entry or exit points. Unlike solutions by Vatti and Greiner, this algorithm uses a single linked list data structure and performs better than Vatti's solution for smaller number of input points. The solution is limited to 2D polygons and modifications are needed for dealing multiple polygons and holes. The degenerate cases are specially treated and the methods are demanding having difficult implementation.

The algorithm by Martınez, Rueda et al. (2009) is based on classical plane sweep algorithm for computing intersections performing in time $O((n + k) \log (n))$. They claim the solution works for general polygons, although its working for polygons with holes is not demonstrated. Algorithmic details are given but the implementation issues are not discussed. The implementation seems difficult and edge overlaps are specially treated. Dimension independence and robustness issues are not discussed.

3 Preliminaries: Assumptions and Data Structure

Our approach for intersection computation between geometric objects is based on following assumptions and simplifications;

a) The objects are flat objects i.e. lines, polygons and polyhedra. For demonstration purposes we will confine ourselves to 2D polygons only.

b) The objects are simple polygons although the solution can handle self intersecting edges with few modifications (section 9).

c) We are focusing on the intersection computation problem. The intersection detection problem is not considered that itself is a challenging problem and algorithms exist in literature (e.g. (Chazelle and Dobkin 1987; Shamos and Hoey 1976)).

d) The objects may contain holes or nested holes e.g. a hole within a hole in which case the two regions have opposite orientation of edges. In Figure 2a, region 5 is within region 4 and both have opposite orientation.

e) The edge representations are based on "Left-handed" rule, the boundary edges are oriented anticlockwise while the hole edges are oriented clockwise.

f) A region is represented by a single or multiple polygons. We have demonstrated our solution for intersection computation on a pair of polygons. Our solution is extendible for intersection computation involving more than two polygons.

The data structure, CHT, is an arbitrary tree in which every node represents a convex hull. The AHD process decomposes any nonconvex polygon with or without holes into convex components which are arranged hierarchically in a CHT. Two basic functions *build* and *process* are used for populating the data structure and processing the data structure to retrieve the original object respectively. For implementation details of the AHD, CHT and, the associated functions the reader is referred to (Bulbul and Frank 2009). The example in Figure 2 shows the AHD process and the resulting CHT. Given a nonconvex polygon with nested holes (see Figure 2a), it is decomposed into its convex components using AHD process as shown in Figure 2b, and the resulting data structure, CHT, is shown in Figure 2c.

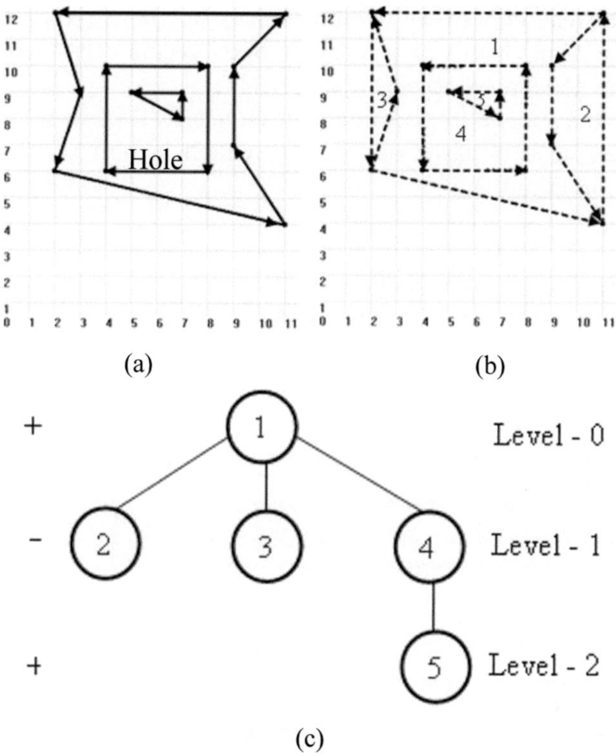

Fig. 2. Input polygon (a) decomposed input (b) convex hull tree data structure (c)

The nodes (convex hulls) in CHT at different levels are given positive and negative signs alternately as shown in Figure 2. The positively signed convex hulls represent the regions to be included while the negatively signed convex hulls represent regions to be excluded from their parent region represented by a positive convex hull. For example in Figure 2, the input nonconvex polygon is decomposed into five convex components, its convex hull node 1, two delta regions or concavities or notches (node 2, and node 3) and two nested holes (node 4, and node 5).

4 Our Approach

The closure of the basic intersection operation (Convex-convex intersection, CCINT) over convex sets is the most important property. Thus our approach exploits this property for computing convex-nonconvex Intersection, CNINT, by recursively applying CCINT between convex hull of the convex object and the component convex hulls of the nonconvex object.

Similarly, the intersection between two nonconvex polygons (Nonconvex-nonconvex intersection, NNINT) is computed by recursively traversing CHT of one of the nonconvex objects with CNINT.

The NNINT is the generic intersection operation, GINT, because it deals all the three cases as it is based on CNINT, which in turn is based on CCINT as shown in Figure 3.

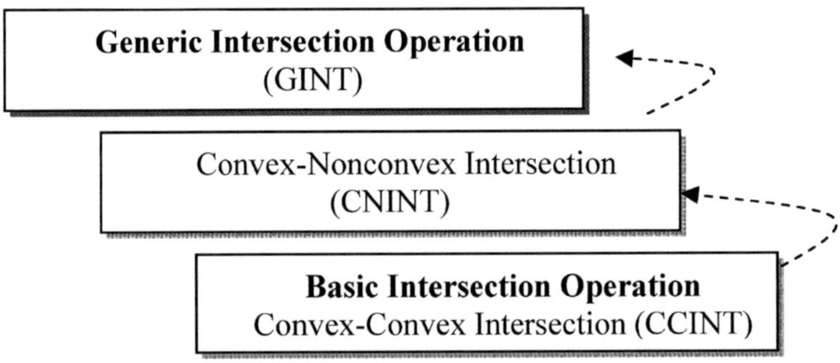

Fig. 3. Our approach for intersection computation

4.1 Basic Intersection Operation: Convex-Convex Intersection

The basic operation is closed under intersection and the result is always a convex region. The intersection computation between two convex polygons is the basic operation (for which known solutions exist). Any of the existing solutions can be used but we have provided an algorithm for intersection computation between two convex polygons using convex hulls.

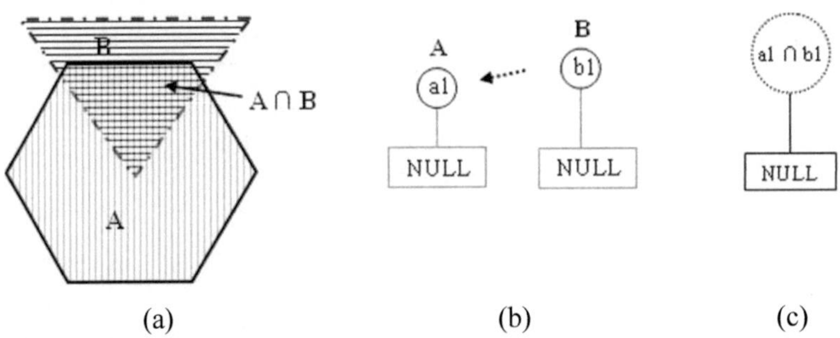

Fig. 4. Two convex polygons (a) CHTs of intersecting convex polygons (b) resultant CHT of A∩B

We have provided an algorithm for convex-convex intersection, the pseudocode of which is provided in section 5.1. Figure 4 shows two convex polygons and demonstrates their intersection at data structure level.

4.2 Convex- Nonconvex Intersection

The process of intersection computation between a convex polygon and a nonconvex polygon involves the repeated application of basic operation between the convex hull of the convex polygon and the convex hull components of the nonconvex polygon that are traversed recursively in CHT. The resulting convex hull of each basic operation is then placed in the resultant CHT at position same as the position of component convex hull of the nonconvex polygon involved in the basic operation. The further recursive processing of children trees of a component convex hull is stopped if the basic operation involving that component convex hull is null. This avoids unnecessary computations reducing the overall number of operations.

For example, Figure 5a shows a convex polygon *A* and a nonconvex polygon *B* and Figure 5b shows the decomposed inputs with their components numbered. Figure 5c shows the data structure representation of the input polygons *A* and *B*. The process starts by computation of basic operation between convex hull of convex polygon *A* (that is *a1*) and the convex hull of the nonconvex polygon *B* (that is the root *b1*). Since *b1*, the component convex hull of the nonconvex polygon is at root the result of basic operation between *a1* and *b1* will form the root of the resultant CHT as shown in Figure 5d. The process is then recursively applied to the children trees of *b1*. Since the intersection between convex hull *a1* and *b3* is null, the child tree of *b3* containing convex hull *b4* will not be processed further (Figure 5e). The resultant intersection is a nonconvex region that is represented by two convex hull components as shown in Figure 5f. The processing (merging) of the resultant CHT, containing two component convex hulls results in the resultant intersection region as shown in Figure 5g.

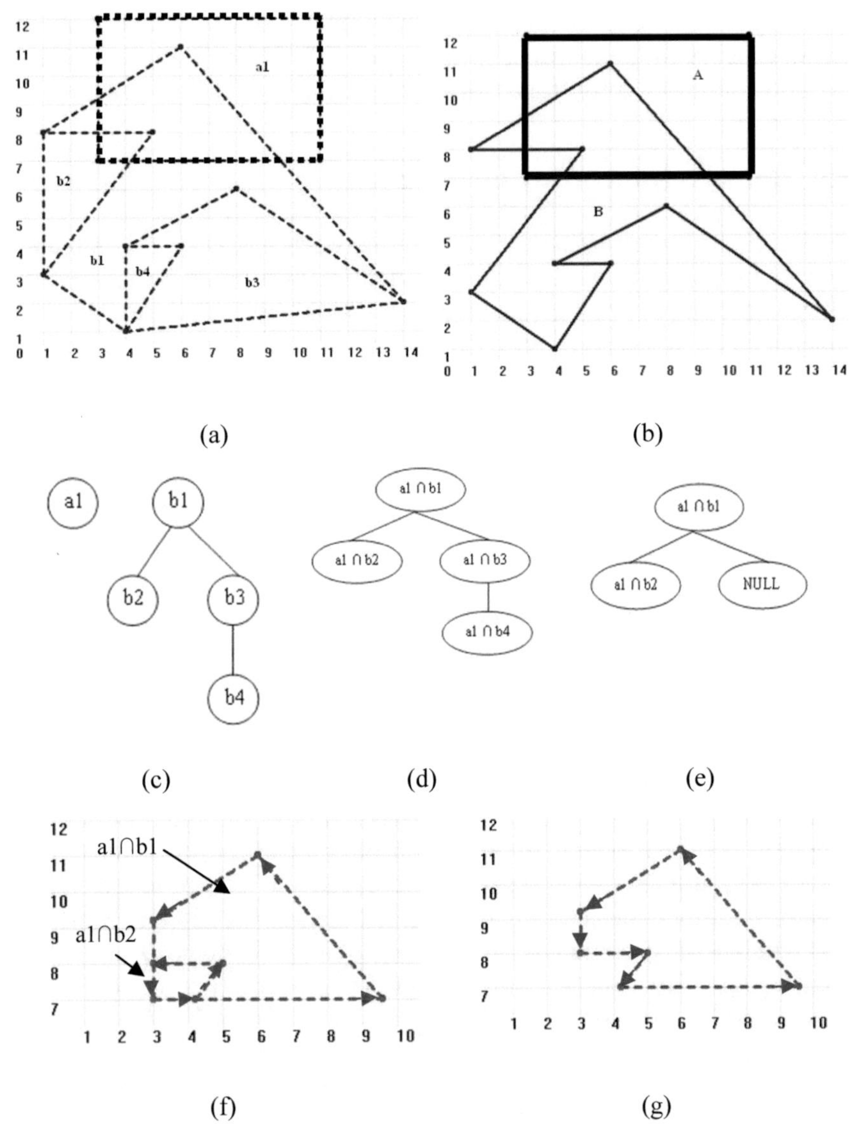

Fig. 5. Input polygons (a) decomposed inputs (b) CHTs of input polygons (c) intersection of convex hull of A, a1 and the CHT of B (d) CHT of intersection (e) intersection tree component hulls (f) intersection region A∩B (g)

Since the intersection of a convex and a nonconvex polygon may be a set of disjoint convex and/or nonconvex regions (e.g. see Figure 1d), the resultant CHT (if it does not represent a single convex region) is further processed for simplification, that results in multiple CHTs each representing the disjoint intersection region.

4.3 Generic Intersection Operation

The intersection operation for a convex polygon and a nonconvex polygon is used to compute the intersection between two nonconvex polygons represented by CHTs. Suppose two intersecting nonconvex polygons *A* and *B* as shown in Figure 6a. Figure 6b shows the decomposed convex hull components of both nonconvex polygons and their CHTs are shown in Figure 6c. The intersection process starts by taking the convex-nonconvex intersection between convex hull of nonconvex polygon *B* (*b1*) and the CHT of the nonconvex polygon *A* (Figure 6d).

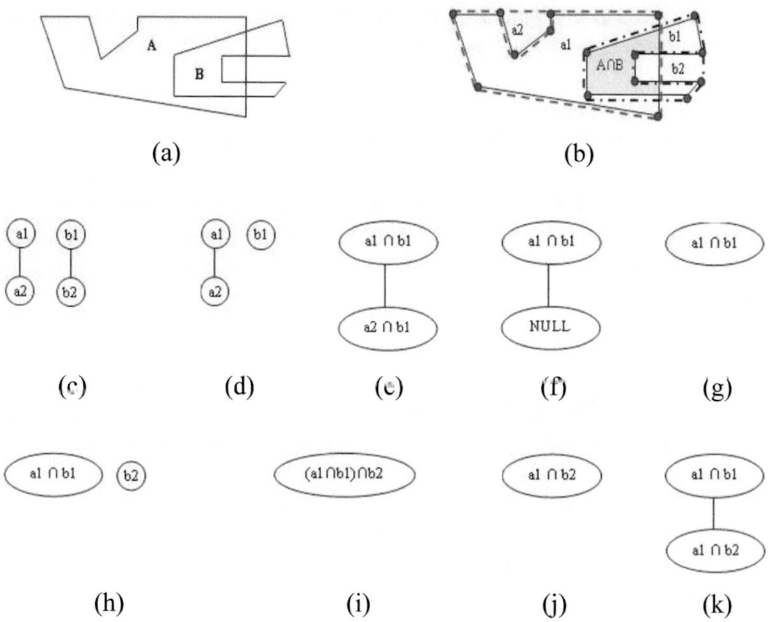

Fig. 6. Generic intersection operation

Since the intersection between b1 and a2 convex components is null (Figure 6e and Figure 6f), the result is a single node CHT (Figure 6g) having the convex hull ($a1 \cap b1$) which is then intersected recursively with all the children trees of convex hull *b1* of the nonconvex polygon *B* . Since, *b1* has only one child (which is also a single node representing a convex hull) the intersection of ($a1 \cap b1$) and child tree (Figure 6h) results in a convex hull ($a1 \cap b2$) represented in a single node CHT as shown in Figure 6i and Figure 6j. The resultant CHT containing ($a1 \cap b2$) is then added (grafted) back to the parent intersection CHT containing single node ($a1 \cap b1$). Thus, the result is a nonconvex region shown as shaded in Figure 6 b and the resultant CHT shown in Figure 6k.

If the result of convex-nonconvex intersection is a set of disjoint regions (a set of CHTs, each representing a region) then for each CHT of the disjoint region, the same process is repeated independently of the other CHTs representing the disjoint intersection regions.

5 Algorithms

In this section we provide the pseudocode of algorithms for each of the three operations discussed in previous section. An example is shown with each algorithm to describe algorithm.

5.1 Pseudocode: Convex- Convex Intersection

The pseudocode of the basic intersection operation between two convex polygons is shown in Figure 7. The algorithm uses the QuickHull algorithm (O'Rourke 1998) for convex hull computation and *compIntPoints* routine computes the intersection points by pairing the intersecting delta edges.

Input: Two convex polytopes (convex hulls)

1: Initialize set of intersection region points I as an empty set

2: $ch \leftarrow$ convexHull ($poly1 + poly2$)

3: $de \leftarrow$ (($poly1$ edges) + ($poly2$ edges)) − (ch edges)

4: $ip \leftarrow$ compIntPoints (de)

5: $irp \leftarrow (poly1 - poly2) + (poly2 - poly1) + ip$

6: $I \leftarrow$ convexHull (irp)

Output: Set of intersection region points I (always a convex hull)

Fig. 7. Pseudocode of convex-convex intersection

The algorithmic steps of basic intersection are shown in the example in Figure 8.

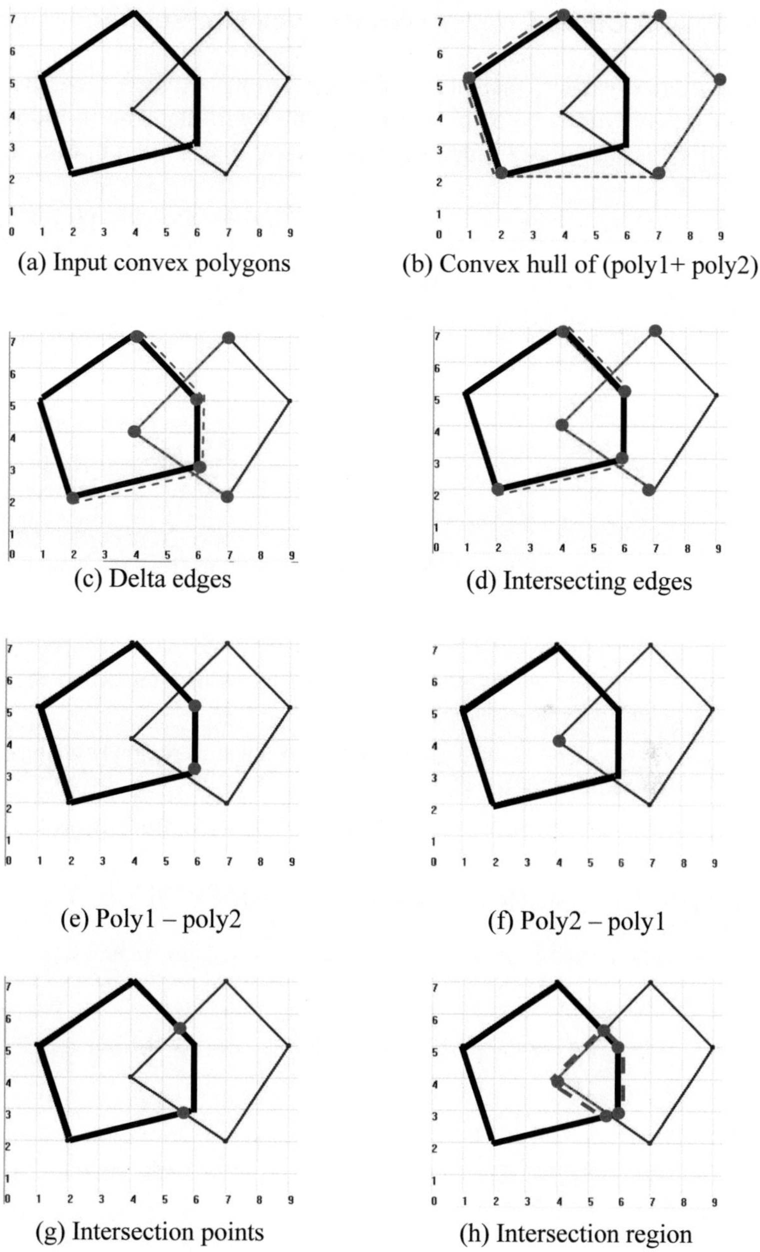

(a) Input convex polygons
(b) Convex hull of (poly1+ poly2)
(c) Delta edges
(d) Intersecting edges
(e) Poly1 – poly2
(f) Poly2 – poly1
(g) Intersection points
(h) Intersection region

Fig. 8. Convex-convex intersection example

5.2 Pseudocode: Convex –Nonconvex Intersection

As mentioned earlier, the intersection operation between a convex and a nonconvex polygon is not closed under intersection. The pseudocode of the CNINT is shown in Figure 9 and the example is shown in Figure 10.

Input: A convex polytope (convex hull) *poly1* and a nonconvex polytope *poly2* (CHT). poly1 = *ch and* poly2 = Node *x xts* where *x* is the convex hull of poly2 and *xts* is the set of children CHTs

Initialize *I* as an empty tree

if *xts* is empty

  then

    t ← Node (intersection of *ch* and *x*) [ ]

    I ← set of trees containing only *t*

else

    ct ← map (**CNINT** *ch*) *xts* // set of trees by recursively traversing *xts* with (**CNINT** *ch*)

    it ← *Node (***CCINT** of *ch* and *x) ct*

    eir ← process *it* // set of edges of intersection region obtained by processing *it*

    ir ← splitRegions eir // set of disjoint regions obtained by separating closed regions

    I ← build *ir* //set of trees obtained by building each region in *ir*

Output: Set of disjoint intersection regions *I* represented in *CHT*

Fig. 9. Pseudocode of convex-nonconvex intersection algorithm

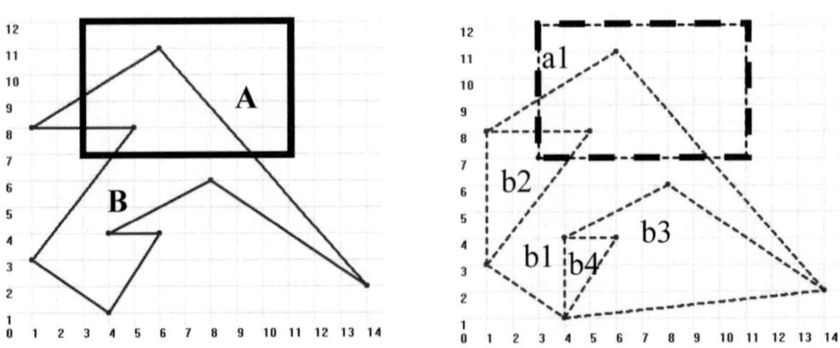

(a) Input polygons (b) Decomposed input

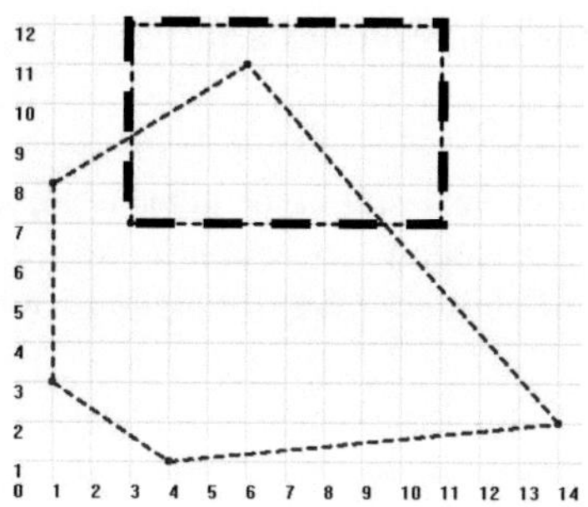

(c) a1 and b1

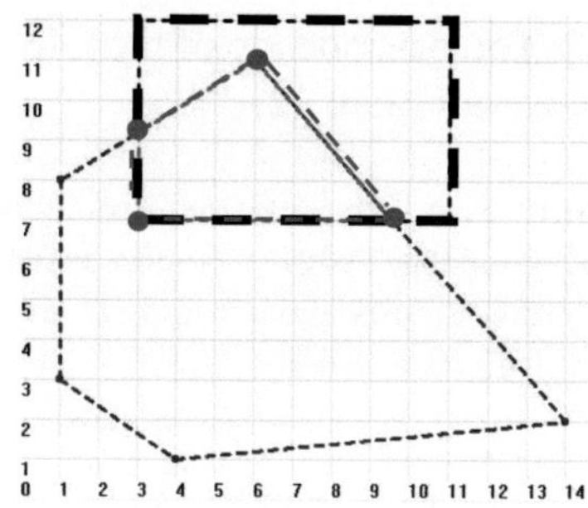

(d) a1∩ b1

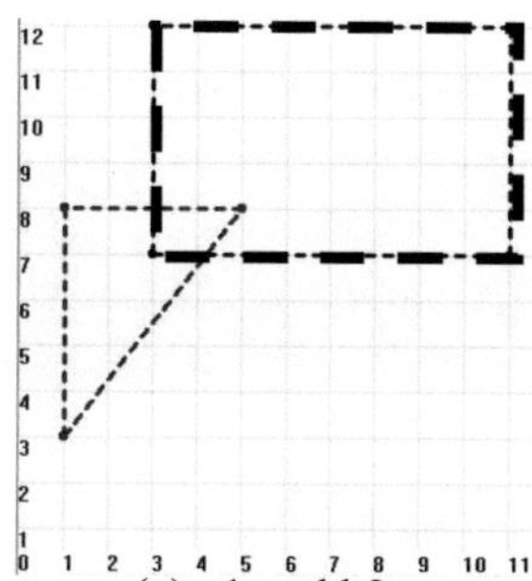

(e) a1 and b2

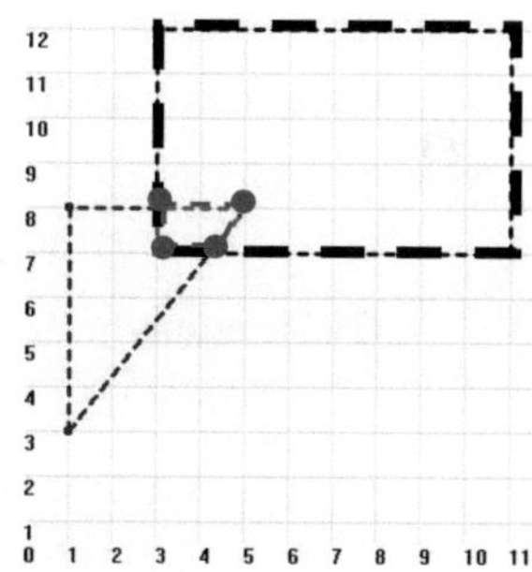

(f) a1 ∩b2

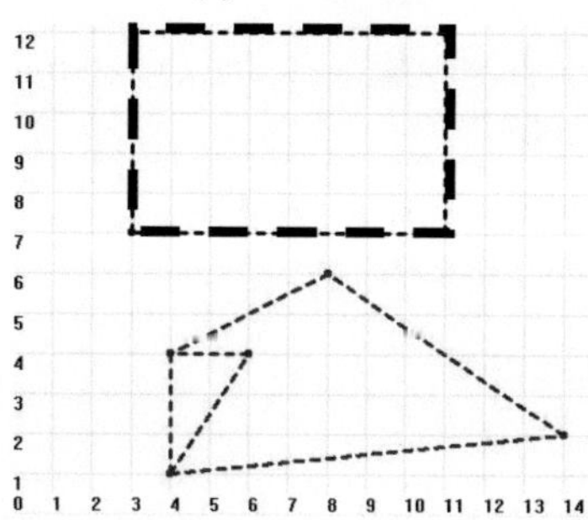

(g) a1∩b3 =NULL

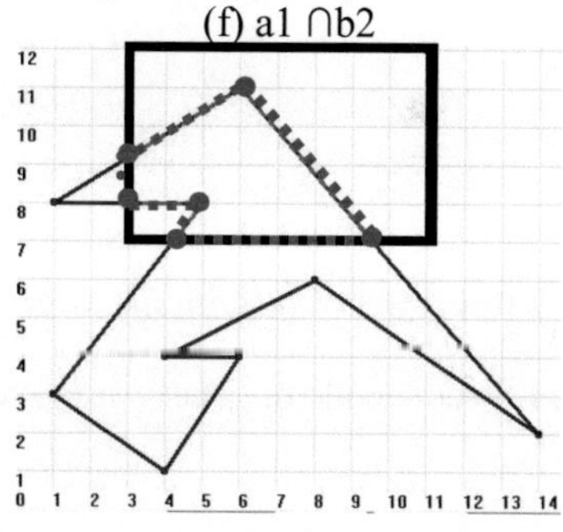

(h) A∩B

Fig. 10. Convex-nonconvex intersection example

5.3 Pseudocode: Generic Intersection Operation

The algorithm for generic intersection operation incorporates the previous two intersection operations. It is generic in the sense that it computes the intersection of two polygons (convex or nonconvex and with or without holes). The pseudocode of the GINT is shown in Figure 11 and an example is shown in Figure 12.

Input: Two polygonal regions (convex/nonconvex, simple or nonsimple) in *CHT* notation. Poly1 = Node *x xts* and poly2 = Node *y yts* where *x* and *y* are the convex hulls and *xts* and *yts* are the set of children trees of polygon 1 and polygon2 respectively

```
Initialize I as an empty tree
if xts is empty   //poly1 is convex
      then return CNINT x poly2
else if yts is empty   //poly2 is convex
      then return CNINT y poly1
else
      oi ← CNINT y poly1
      if oi is empty
            then return I
      else for every tree t_x in oi
                  for every tree t_y in yts
                        ct ← compute GINT t_x t_y
                        nt ← addtrees ct in tree t_x
                        I ← addtree nt into I
       return I
```

Output: Set of disjoint intersection regions *I* in *CHT*

Fig. 11. Pseudocode of generic intersection algorithm

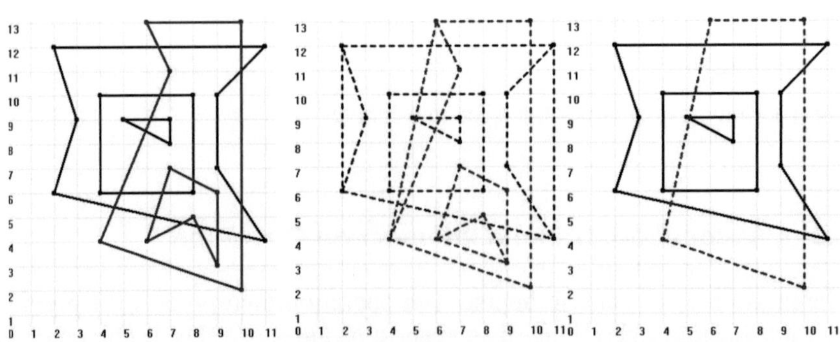

(a) Input polygons (b) Decomposed input (c) CNINT:CHT of A and convex hull of B

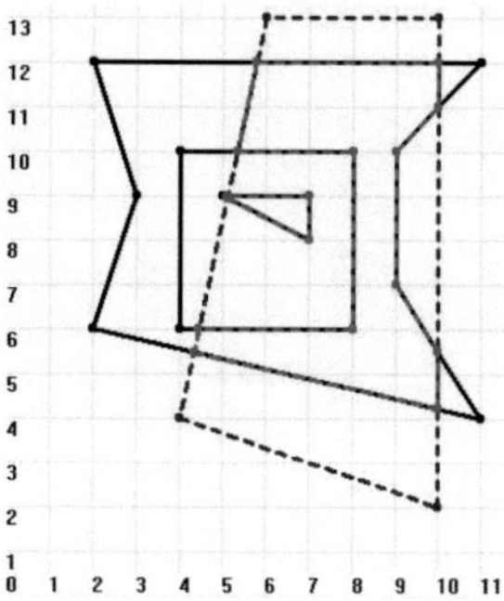

(d) Result of (c)

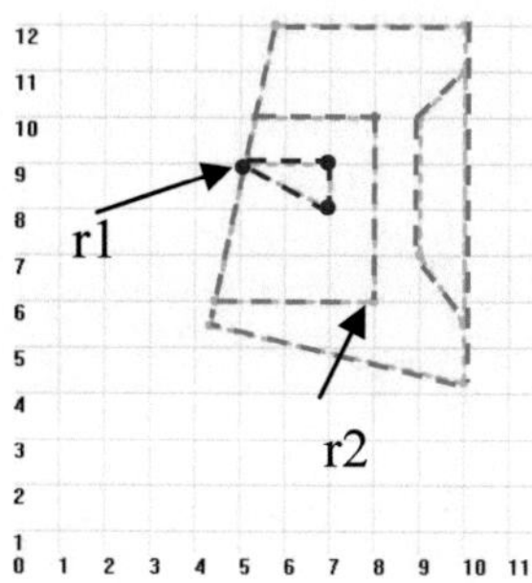

(e) result in (c) is simplified and two disjoint regions result

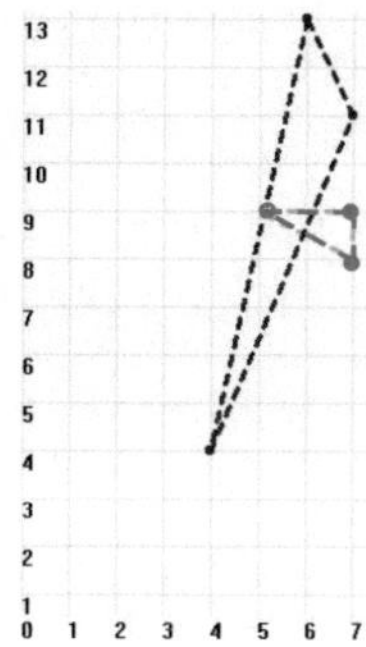

(f) r1 ∩ first child of B

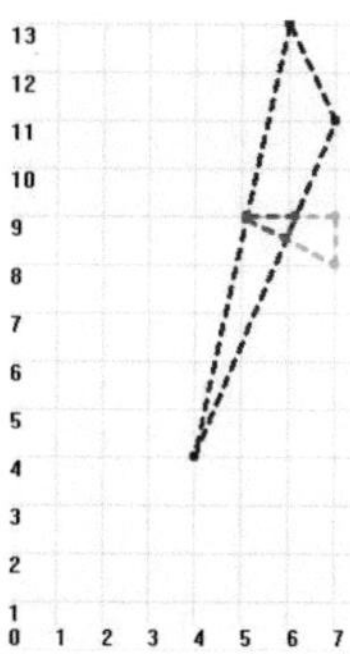

(g) Result of (f)

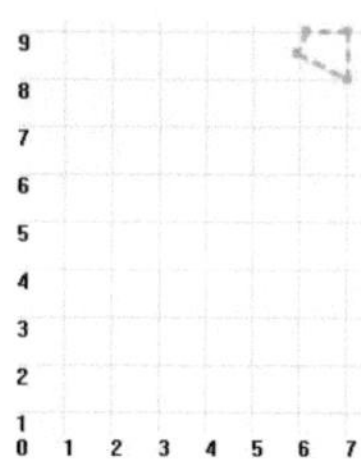

(h) Remove (g) from r1

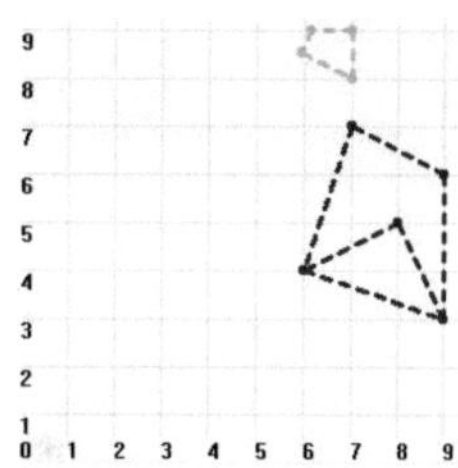

(i) Result in (h) ∩ second child tree of B

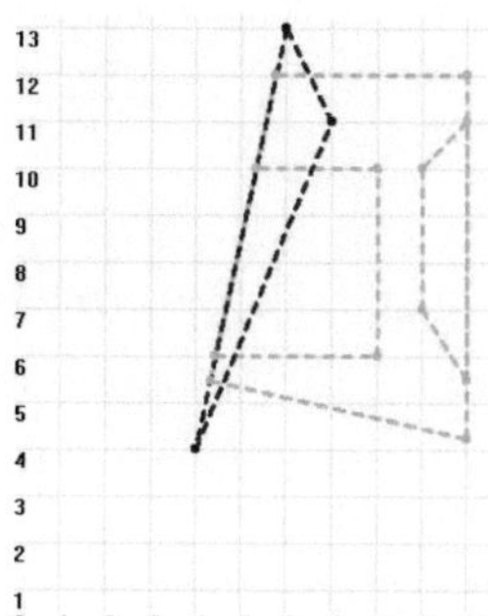

(j) r2 ∩ first child of B

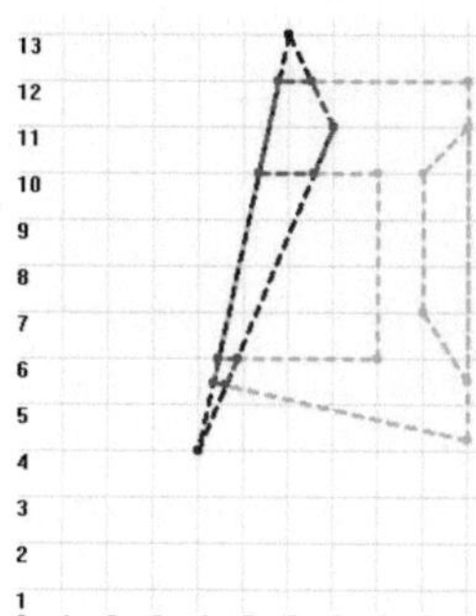

(k) Result of (j)

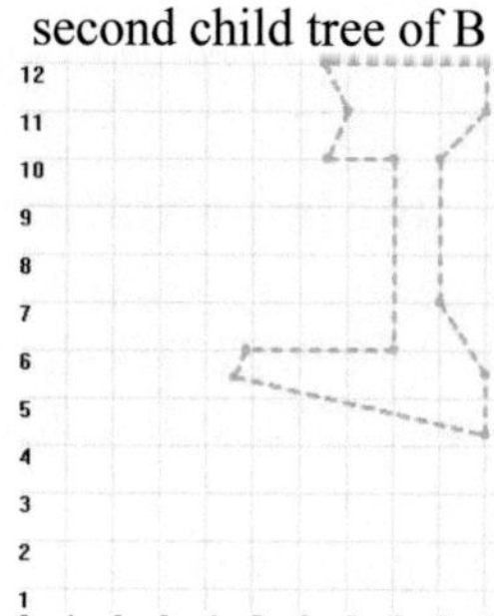

(l) Remove result in (k) from r2

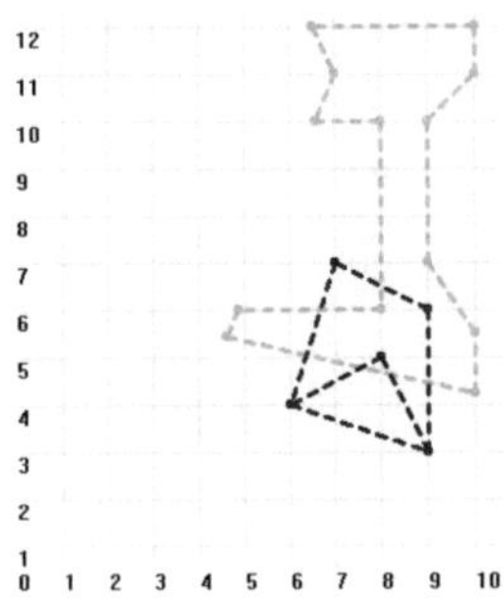

(m) Result in (l) ∩ second child tree of B

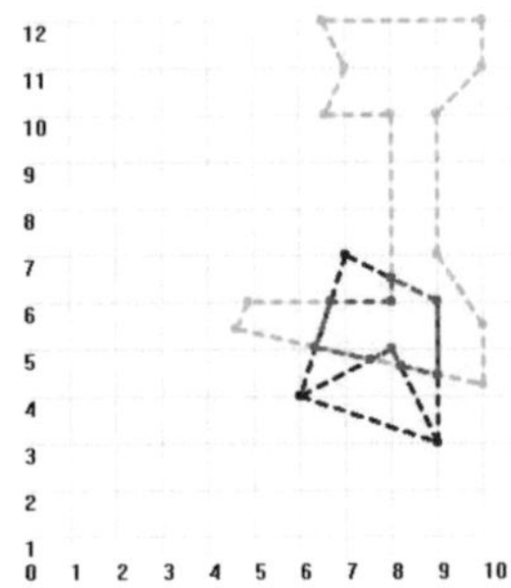

(n) Result of (m)

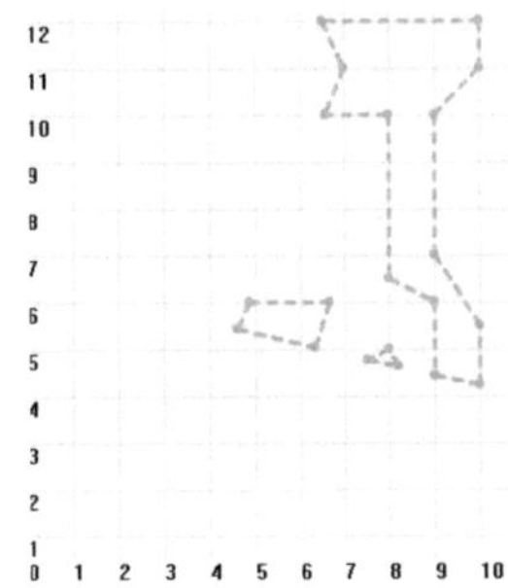

(o) Remove (n) from (l)

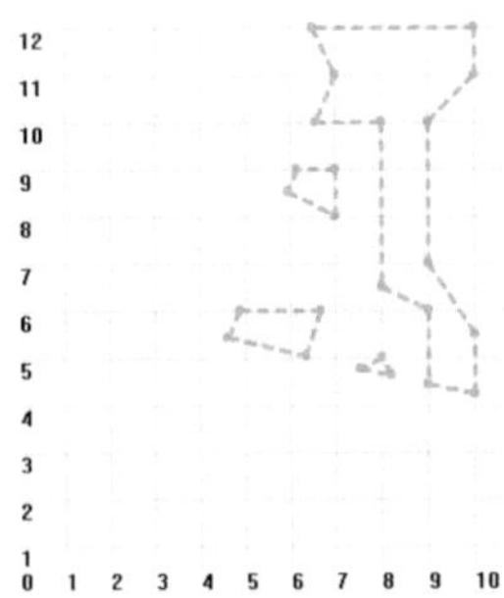

(p) Disjoint intersection regions

Fig. 12. Nonconvex-nonconvex intersection example

6 Implementation

The algorithms introduced in previous section are implemented in Haskell (Jones 2003). It is a functional programming language that supports lazy evaluation, higher order functions, and big numbers (big integers, big rational etc.). In Haskell, the CHT is defined recursively as;

data Tree = Node CHull [Tree]

The CCINT and CNINT operations are computed by recursively traversing the CHT of nonconvex polygon using *map* higher order function. Also, all the points are represented by their homogenous big integer coordinates. Big integers in Haskell are represented as "Integer" data type and they allow arbitrary precision arithmetic limited by the size of the main memory. Thus, the use of big integers allows robust intersection computations operations avoiding rounding errors.

7 Special Cases

The intersection of two polygons may be a point (Figure 13c), a line (Figure 13b), polygon or a combination consisting of the three (Figure 13f). In cases where the intersection region is not a polygon or a set of polygons, special treatment is needed. A variety of special cases for the basic intersection have been shown in (O'Rourke 1998) and few are shown in Figure 13.

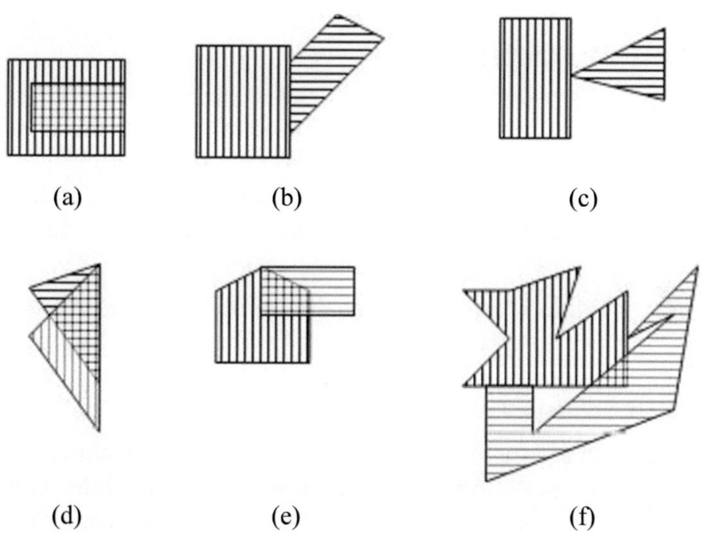

Fig. 13. Some special cases for intersection computation

Since our approach is based on basic intersection operation, we need not to tackle the special cases for convex-nonconvex or nonconvex-nonconvex intersection operations. If the special cases are tackled for basic operation, other operations based on it will also tackle the special cases.

For special cases like complete or partial overlap our approach do not need any special treatment. However, for cases when the intersection is a point or a line we need special treatment which is achieved by making changes not in the main intersection algorithms but in the *QuickHull* module and the *build* function of the AHD process. Table 2 lists special cases and whether these cases are specially treated in our algorithm or not.

Table 2. Special case treatment

Case	**Specially treated?**
Complete overlap	No
One is inside other	No
Edges overlap	No
Points overlap	No
Intersection is a line	Yes
Intersection is a point	Yes

8 Solution Characteristics

Our generic solution to perform Boolean intersection operations on general polygons has following properties;

1. Our approach is based on the basic intersection operation between two convex polygons. Thus it exploits the benefits associated with convexity.
2. It uses a single hierarchical tree data structure called *convex hull tree* which is an arbitrary tree of convex hulls. The data structure has two methods *build* and *process* for populating and accessing the stored data in the data structure.
3. The tree data structure allows reduced number of computations. We process the children trees of a node only if the intersection is not null. We proceed in a branch in CHT if the intersection is not null. Rule is "*an object A can not intersect with the component hulls of object B, if object A is not intersecting with the convex hull of object B*".
4. Our approach is robust as we are using homogenous big integer coordinates for representing points. Robustness is an important issue in geometric computations (David 1997). Most of the algorithms, as discussed in previous work section, do not cater the robustness issue. Only few address the issue (David 1997; Smith and Dodgson 2007).
5. In some applications, the result of Boolean operation may be needed to be decomposed e.g. Vatti (1992) mentions a case when

decomposed output is useful. So our approach is useful for applications where the decomposed output is needed in the form of convex components.

6. Our approach is easily implementable. The approach is implemented in Haskell and the code is compact having only 137 lines of code. Less code means less errors and ease of maintenance.
7. Most of the special cases need no special treatment like overlapping edges etc. Some require few modifications like when the result has dangling edges or points etc.
8. It is easily extendible for *n*-dimensions.

9 Suggestions for Improvement and Future Work

The approach can be improved by slight modifications. For example;

a) For cases where region is represented by multiple polygons we have to make the representation of a polygonal region as a list of convex hull trees rather than a tree.

b) For the self intersection cases, some preprocessing can be done involving the computation of intersection points of intersecting edges, updating the intersecting edges and then ultimately segregating the closed regions formed, so as to represent it with multiple simple polygons. The idea is to convert/decompose a non-simple polygon with multiple simple polygons or use any established method to deal nonsimplicity first and then continue with our solution.

The future goal is to devise a solution which has following properties;

1) Supports other Boolean operations union and symmetric difference etc. We have implemented the union and symmetric difference operations based on AHD and the work will be presented soon.

2) Dimension independent having single implementation. For implementing the dimension independent Boolean operations on polytopes, the dimension independent AHD (Bulbul et al. 2009) will be used.

References

Bentley, J. L. and T. A. Ottmann. 1979. "Algorithms for Reporting and Counting Geometric Intersections." IEEE Computer Society.

Bernard, Chazelle and Edelsbrunner Herbert. 1992. "An optimal algorithm for intersecting line segments in the plane." Journal of the ACM (JACM) 39(1):1-54.

Bulbul, Rizwan and Andrew U. Frank. 2009. "AHD: The Alternate Hierarchical Decomposition of Nonconvex Polytopes (Generalization of a Convex Polytope Based Spatial Data Model)." In 17th International Conference on Geoinformatics. Fairfax, USA.

Bulbul, Rizwan, Farid Karimipour and Andrew Frank. 2009. "A Simplex based Dimension Independent Approach for Convex Decomposition of Nonconvex polytopes." In 10th International Conference on GeoComputation (GeoComputation 2009). UNSW, Sydney, Australia.

Chazelle, B. and D. P. Dobkin. 1987. "Intersection of convex objects in two and three dimensions." Journal of the ACM (JACM) 34(1):1-27.

David, M. Mount. 1997. "Geometric intersection." In Handbook of discrete and computational geometry: CRC Press, Inc.

Fogel, Efi, Ron Wein, Baruch Zukerman and Dan Halperin. 2006. "2D Regularized Boolean Set-Operations." In In Cgal-3.2 User and Reference Manual, Cgal Editorial Board, Ed., http://www.cgal.org/Manual/3.2/doc_html/cgal_manual/Boolean_set_operations_2/Chapter_main.html.

FranciscoMartınez, Antonio Jesus Rueda and Francisco Ramo´n Feito. 2009. "A new algorithm for computing Boolean operations on polygons." Computers&Geosciences.

Greiner, Gunther and Kai Hormann. 1998. "Efficient Clipping of Arbitrary Polygons." ACM Transactions on Graphics (TOG) 17(2):71 - 83

Jones, Simon. 2003. Haskell 98 Language and Libraries: The Revised Report: {Cambridge University Press}.

Lauther, Ulrich. 1981. "An O (N log N) algorithm for Boolean mask operations." In Proceedings of the 18th conference on Design automation. Nashville, Tennessee, United States: IEEE Press.

Liu, Young Kui, Xiao Qiang Wang, Shu Zhe Bao, Matej Gambosi and Borut Zalik. 2007. "An algorithm for polygon clipping, and for determining polygon intersections and unions." Computers & Geosciences 33(5):589-598.

Margalit, Avraham and Gary D. Knott. 1989. "An Algorithm for Computing the Union, Intersection or Difference of two Polygons." Computers & Graphics 13:167-183.

Martınez, Francisco, Antonio Jesus Rueda and Francisco Ramo´n Feito. 2009. "A new algorithm for computing Boolean operations on polygons." Computers & Geosciences.

Nievergelt, J. and F. P. Preparata. 1982. "Plane-sweep algorithms for intersecting geometric figures." ACM.

O'Rourke, Joseph. 1998. Computational Geometry in C (Cambridge Tracts in Theoretical Computer Science): Cambridge University Press.

Peng, Yu, Jun-Hai Yong, Wei-Ming Dong, Hui Zhang and Jia-Guang Sun. 2005. "A new algorithm for Boolean operations on general polygons." Computers & Graphics 29(1):57-70.

Rappoport, Ari. 1991. "The n-dimensional extended convex differences tree (ECDT) for representing polyhedra." In Proceedings of the first ACM symposium on Solid modeling foundations and CAD/CAM applications. Austin, Texas, United States: ACM.

Rigaux, Philippe, Michel Scholl and Agnes voisard. 2001. Spatial Databases: With Applications to GIS: Morgan Kaufmann Publishers Inc. San Francisco, CA, USA.

Rivero, M. and F. R. Feito. 2000. "Boolean operations on general planar polygons." Computers & Graphics 24(6):881-896.

Shamos, Michael Ian and Dan Hoey. 1976. "Geometric intersection problems." In Proceedings of the 17th Annual Symposium on Foundations of Computer Science: IEEE Computer Society.

Smith, J. M. and N. A. Dodgson. 2007. "A topologically robust algorithm for Boolean operations on polyhedral shapes using approximate arithmetic." Butterworth-Heinemann.

Vatti, Bala R. 1992. "A Generic Solution to Polygon Clipping." Communications of the ACM 35(7):57-63.

Visual Analytics of Urban Environments using High-Resolution Geographic Data

Peter Bak[1], Itzhak Omer[2], Tobias Schreck[3]

[1] Visual Analytics Group, Dept. of Computer and Information Science, University of Konstanz, Germany
bak@dbvis.inf.uni-konstanz.de
[2] Urban Space Analysis Laboratory, Dept. of Geography and Human Environment, Tel Aviv University, Israel
omery@post.tau.ac.il
[3] Interactive Graphics Systems Group, Technische Universität Darmstadt, Germany
tobias.schreck@gris.informatik.tu-darmstadt.de

Abstract. High-resolution urban data at house level are essential for understanding the relationship between objects of the urban built environment (e.g. streets, housing types, public resources and open spaces). However, it is rather difficult to analyze such data due to the huge amount of urban objects, their multidimensional character and the complex spatial relation between them. In this paper we propose a methodology for assessing the spatial relation between geo-referenced urban environmental variables, in order to identify typical or significant spatial configurations as well as to characterize their geographical distribution. Configuration in this sense refers to the unique combination of different urban environmental variables. We structure the analytic process by defining spatial configurations, multidimensional clustering of the individual configurations, and identifying emerging patterns of interesting configurations. This process is based on the tight combination of interactive visualization methods with automatic analysis techniques. We demonstrate the usefulness of the proposed methods and methodology in an application example on the relation between street network topology and distribution of land uses in a city.

M. Painho et al., (eds.), *Geospatial Thinking*, Lecture Notes in Geoinformation and Cartography,
DOI 10.1007/978-3-642-12326-9_2,

1 Introduction

"All geographic data leaves its users, to some extent, uncertain about the nature of the real world" (Goodchild, 2005). This situation, which often stems from data resolution constraints and from their multidimensional character regarding space, time and objects, has affected socio-geographic research of the urban environment.

Until recently, urban environment research was limited to the use of large-scale aggregate data based on the level of administrative areas. The basic problem of using aggregated spatial data for geographical analysis stems from the fact that the distribution of objects within the areas is unknown. As a result, the aggregated data are not sufficient to capture the micro-scale situations, in which the main dimensions of urban environment – the built-up environment's properties, the individuals' socio-demographic properties and the individuals' behavior and perception – come together.

This situation has nonetheless changed recently due to improvements in GIS (Geographic Information System) technology and the construction of new geographic databases. Today it is possible to obtain geo-referenced high-resolution data (i.e. there is a link between the attribute of the object and is geographic location) on different types of urban objects in urban locations: (1) daily movement data, e.g., positioning data collected by global positing system (GPS) and location based services (LBS) technology to surrogate daily movement data from mobile usage patterns (e.g. Ratti et. el., 2006); (2) distribution of built-up environmental objects e.g., road networks, or housing, and (3) functional and socio-demographic objects, e.g., house-level socio-demographic and land-uses data.

The increasing availability of geographic data at high resolution and in good quality actually motivates the investigate the relation between these different types of urban objects at the same geographic scale, i.e. the house level scale, of diminishing gaps between data that represent different types of urban objects. This ability can be essential for identifying and understanding the variety of socio-demographic phenomena that incorporate different urban objects at different spatial dimensions. For example, it is possible to investigate how building types or the location of urban services are correlated with the spatial distribution of the populations' socio-demographic attributes. It is also possible to integrate high-resolution built-up and socio-demographic data with empirical data (i.e. data collected during interviews) to better understand the individuals' preference features – for instance, the choice of urban parks and commercial areas and their mode of transportation.

Nevertheless, use of high-resolution geographic data comes at a price. Despite their potential, it is rather difficult to use these data in research. Unlike the situation of aggregate data at the level of geographic areas, high resolution geo-referenced data at house level have no defined geographical boundaries and therefore, the main challenges of research are to identify patterns of interest based on huge amounts of urban objects with respect to their attributes and the complex spatial relation between them.

In this paper, we apply our methodology for assessing the spatial relation between high resolution geo-referenced urban land-uses (e.g. residence buildings, commercial open spaces as well as infrastructure objects such as street network) and the topological attributes of the urban street network, in order to identify typical or unique land-use spatial configurations and to characterize their geographical distribution. Configuration in this sense refers to the combination of the two geo-spatial attributes.

The proposed methodology uses a Visual Analytics framework, which combines interactive visual information displays with automatic data analysis techniques. The framework consists of first displaying geographic information, on which users can interactively segment the data into spatial configurations of interest. These are then clustered to reveal frequent or significant groups of configurations. In addition, appropriately designed cluster visualization is interactively linked with the original map display. Thus, the system is designed to support the user in understanding the joint properties of geospatial and multivariate data.

The structure of the paper is as follows: In Section 2, we provide a review of the related work from the geographic and visual analytics perspective. In Section 3, we introduce our methodology and an exemplary implementation. In Section 4, we apply our framework on a high resolution real world dataset of an urban environment, and demonstrate the flexibility it offers to analyze the data for various interesting correlations. In Section 5, we discuss our approach, the obtained results, and outline future work in the area.

2 Related Work

2.1 Geographic Data Analysis for Urban Environments

Visual and analytical comparison between spatial distributions of objects and attributes within GIS framework is an essential tool for understanding and explaining geographic phenomena in urban areas. To resolve the problems entailed by high-resolution data (i.e. neighborhoods definition and cartographic constraints resulting from mapping high resolution data over large area) previous studies suggested local geo-statistics methods. One of the main methods for analyzing and presenting high-resolution data is to use *local indices of spatial association* (LISA). These indices are based on the comparison of the characteristic of a given spatially located object and its neighbors (Anselin, 1995; Benenson & Omer, 2003). Applying these measures helps the observer identify spatial variance and small clusters in the spatial distribution of socio-demographic attributes in urban areas (Talen & Anselin, 1998). A production of thematic maps at different aggregation scales enables also to identify the natural geographic scale for the investigated phenomena. Previous studies suggested such local geo-statistics measures and aggregations for analyzing and overcoming the cartographic constraints of high-resolution data for assessing accessibility to urban services (e.g. Talen & Anselin, 1998; Omer, 2005), residential segregation (e.g. Omer & Benenson, 2002; Wong, 2003) etc.

2.2 Analysis the Effect of Street Network Topology

Much evidence has been collected indicating that the topological characteristics of a street network have the potential to affect the spatial distribution of activates and land-uses in the city. Studies have found that these topological properties of individual streets are significantly correlated to the spatial distribution of retail and services (Porta et. el., 2006) and human movement rates (e.g. Hiller et. el, 1993; Jiang, 2007). However, we have no sufficient knowledge on these relations i.e. why certain topological properties are more appropriate than others for predicting human activities in the city.

2.3 Visual Analytics of High-Dimensional Data

Owing to our data transformation applied (see Section 3), this work relates also to the wider area of visual analytics in high-dimensional data sets. Work in this area is concerned with finding appropriate visual representations for data sets, which by their dimensionality exceed the number of direct visual variables available (e.g., position, color, and shape (Ware, 2004)). *Mapping* approaches such as Parallel Coordinates (Inselberg and Dimsdale, 1990), Iconic Displays (Everitt and Nicholls, 1975), Dimensional Stacking (LeBlanc et al, 1990), or Scatter Plot Matrices (Wilkinson et al, 2005) define certain mappings from multiple dimensions to visual variables and geometric arrangements that represent the properties of all dimensions simultaneously. *Interaction techniques* address the problem of high-dimensional data analysis by allowing efficient navigation through the space of low-dimensional projections. Examples for these kinds of approaches can be found in (Wilkinson et al, 2005, and Elmqvist et al, 2008). *Dimension reduction* approaches operate by first reducing the number of dimensions before visualization takes place. Prominent dimensionality reduction techniques include Principal Components Analysis (Jolliffe, 2002) or Multidimensional Scaling (Cox and Cox, 2001). The general aim of dimensionality reduction is to capture as much information as possible in a limited number of dimensions. However, the reduced dimensions often are a linear or non-linear combination of all input dimensions, and therefore not straightforward to interpret by the user.

2.4 Self-Organizing Map Approach in Geographic Context

The SOM algorithm (Kohonen, 2001) is a technique which combines dimensionality and data reduction, and implicitly yields a mapping of data elements to position. The algorithm is especially suited for visualization of its output (Vesanto, 1999). It has been successfully applied to many data analysis problems including in geo-temporal data (Guo et al, 2006), textual data (Honkela et al, 1997), and financial data (Deboeck and Kohonen, 1998). The SOM approach has also been leveraged in Geospatial data analysis before. In (Spielmann, 2008) an interactive system linking a standard land-covering map with a SOM-based clustering of demographic records was introduced. It allowed the user to select demographic records and simultaneously display them on the SOM (allowing inspection of multivariate properties) and on the map (allowing inspection of spatial distribution). The system was shown to be useful for joint analysis of multivariate and geospatial data properties. In (Bacao, 2005), the basic SOM algorithm was extended by a learning constraint that considers for each data record also its geospatial coordinates. This algorithm produces clusters, which also partially reflect the geospatial position of data records in the SOM network, a property which SOM usually does not consider. Our approach relates to both works (Spielmann, 2008; Bacao, 2005) in that we apply the SOM algorithm in a joint geospatial and multivariate data analysis. In extension to (Spielmann, 2008), we provide an improved SOM display, which allows perception of the distribution of multivariate records already on the SOM display, reducing the need for additional detail views to understand the multivariate data properties.

3 Analytic Methodology

Analysis of high resolution urban environments requires methods to define units of investigation (for example, buildings, streets, etc), and allows analysts to interactively extract configurations and patterns of interest. Here we propose a methodology that enables analysts to conduct their investigations exactly for this purpose based on a modular pipeline, in which information visualization and guided analysis constitute the core part. Consequently, the proposed methodology can be viewed as a modular framework, in which techniques can be chosen by their appropriateness for the research tasks and data.

The pipeline of analysis is triggered by analytic questions and suitable data sources. Here the first step is to find a visual mapping that allows analysts to view and understand the distribution of the data, which is a highly complex task since high resolution data in the geographic domain is large and heterogeneous. The second step requires analysts to define and segment units of investigation, in which configurations and pattern of interest may be possible. However, a high level of refinement through iterations is required, once the results are obtained. These steps are followed by automatic analysis techniques that can extract frequent patterns and significant configuration that reflect the analysts' expectations. The results of the automatic techniques have to be made accessible in an interactive visual way, in order to allow reasoning and refinement of previous decisions that are required to sharpen and finalize the results. As such, the pipeline and methodological framework are highly iterative and combine visual and automatic analysis techniques. The following diagram (Figure 1) provides a high level overview of the described pipeline.

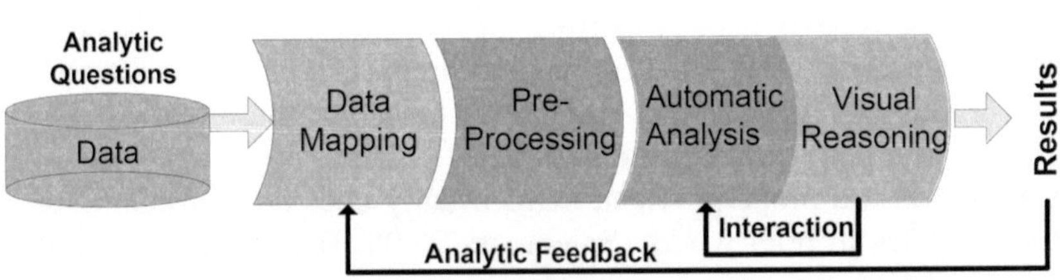

Fig.1. Visual Analytic Pipeline showing the interleaving stages of the analytic process

In the following, we describe a concrete instantiation of the presented methodology by describing the key analytic question of interest and the database schemes available in a selected domain. We then present the details of the applied analysis procedures, visual mappings, and user interactions supported by our system.

3.1 Analytic Questions

Within the developed methodological framework, we aim to investigate the relation between the topological structure of urban street network and the spatial distribution of urban land-use in the city of Raanana, Israel. We investigate whether the spatial distribution of topological properties in a given street network correlates with the spatial distribution of land-uses. Due to the multidimensional nature of the proposed approach it can be used to identify typical land-uses spatial configurations and to investigate how they are influenced by the topological properties of urban street network. This means to investigate the effect of topological structure, not only on the spatial distribution of one land-use, but also on the formation of land-use spatial configurations, with respect to their geographical location. Accordingly, the analytical questions in this research are:

- Are there typical and significant correlations between the topological properties of streets and land-uses' spatial configurations?
- What are the geographic patterns of identified typical land-use configurations?
- Which topological properties are more significant than others for formation of land-use spatial configurations?

3.2 Data Schema of Concern

To illustrate the potential of the proposed methodology, we conducted a detailed investigation of one city's land-use and street network. The data obtained for analysis was the spatial distribution of the land-use and the topology of the street network in the city of Raanana, Israel. We used two kinds of geographic data sets of Raanana. A street network data set (a total of 324 streets), and a land-use data set, at the level of individual buildings (a total of 8664 buildings). The source of the data is the 2002 Infrastructure Database of the Israeli Central Bureau of Statistics (www.cbs.gov.il), which are organized within a GIS framework.

On the land-use side, the data sets specify in real-world coordinates the presence of urban infrastructure elements. These include public-service installations such as education facilities, recreational areas, medical attention facilities, or recreational areas. They also include industrial area data, telecommunication installations and so on. By their nature, the infrastructure elements have a spatial extension and are therefore encoded by polygonal descriptions of the covered area.

The topology of urban streets takes individual streets as nodes (vertices) and street intersections as edges of a connectivity graph. The graph forms a basis for structural analysis using the centrality measures (Jiang & Claramunt, 2004; Jiang & Omer, 2007) initially developed for the description of social networks (Freeman, 1979). A graph G(V,E) is defined as a pair of a finite set of vertices $V = \{ v_1, v_2, \ldots, v_n\}$ and a finite set of edges $E = \{v_i , v_j\}$. Three centrality measures – degree, closeness, and betweenness – are used to describe the status of individual streets, in terms of which streets intersect with other streets. Degree indicates how many other streets are connected directly to a particular street, a characteristic that reflects the level of a street's integration with its neighboring streets. In a graph, the degree is the number of nodes that link a given node. Formally, the degree centrality for a given street (node) v_i is defined by:

$$C_D(V_i) = \sum_{k=1}^{n} r(v_i, v_k) \tag{3.1}$$

$$r_{ik} = \begin{cases} 1 : if\ Vi, Vj \in \mathrm{E} \\ 0 : otherwise \end{cases}$$

where n is the total number of streets (nodes) within a street network (vertices of the graph G).

Closeness indicates how close a street is to other streets by computing the shortest distances between every street node to every other street node, a feature that reflects how well a street is integrated within the network. Formally, the closeness measure is defined by:

$$C_C(v_i) = \frac{n-1}{\sum_{k=1}^{n} d(V_i, V_k)} \tag{3.2}$$

where d is the shortest (topological) distance between two given streets (v_i, v_k) in the street network (graph).

Betweenness centrality indicates the extent to which a street is located between pairs of streets; as such, it directly reflects the intermediate location of the specific street in the entire street network. Accordingly, we define the betweenness centrality as follows:

$$C_B(V_i) = \sum_j \sum_k \frac{P_{jik}}{P_{jk}} \tag{3.3}$$

where P_{jk} denotes the number of shortest paths from j to k and P_{jik} is the number of shortest paths from j to k that pass through street i, so C_B is the proportion of shortest paths from i to j that pass through k.

Figure 2 shows the topology and land-use of the urban environment. Topology is mapped to color using a heat map having red colors for high and yellow colors for low centrality values. The land-use is mapped to a diverging color schema for six categories (educational institutions, public services, culture, commerce, industrial buildings and parks).

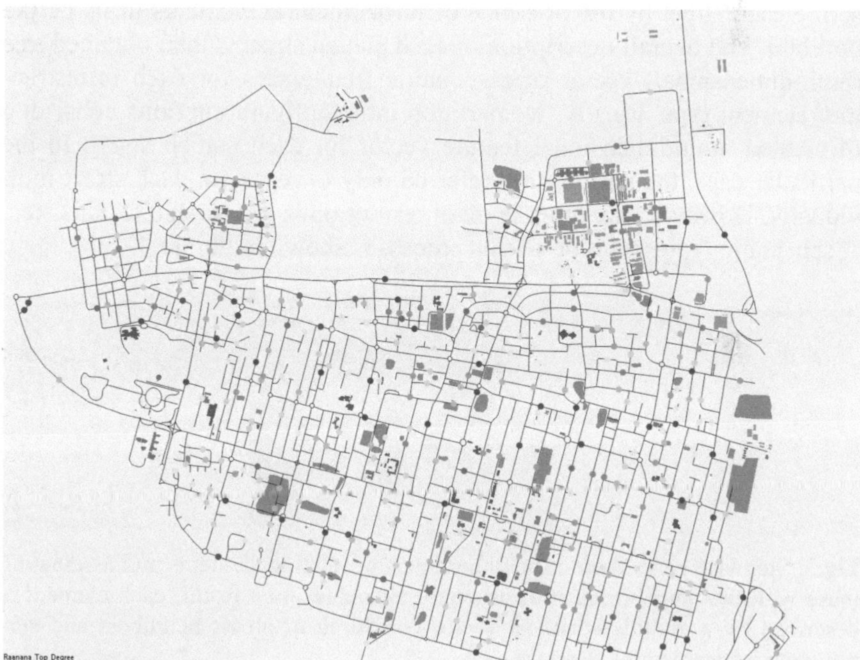

Fig. 2. Map representation of the considered urban environment of the city of Raanana: Topology (degree centrality) on the points and land-use on the polygons of the image. Values of topology are mapped to a diverging colomap going from yellow (low values) to red (high values). Land-use types are mapped to a discrete color schema for educational institutions, public services, culture, commerce, industrial buildings and parks

3.3 Segmentation and Definition

The data sets described are large and complex, requiring appropriate data segmentation to facilitate the analysis and visual representation. To facilitate the investigation, analysts have to define the combination of urban environmental variables of interest into spatial configurations. In order to find appropriate configuration units, we conduct a four stage process: Firstly, we structure the elements of our investigation by their spatial location and neighboring elements. In the particular case, the elements of our investigation are named streets described by the topological value. Secondly, we create a structure, in which each named street is described by its own topological value and those of the connected ones. Thirdly, we describe each street by the presence of infrastructural elements in its neighborhood. The overall description of each named street is then obtained as a (high-dimensional) vector of association frequencies for each infrastructure element type. Finally, we partition into configuration units consisting of created multidimensional feature vector for each named street. In the particular case, the partitioning referred only to represent each street individually. However, any other level of resolution is practicable for this step. A schematic representation of this process is shown in Figure 3.

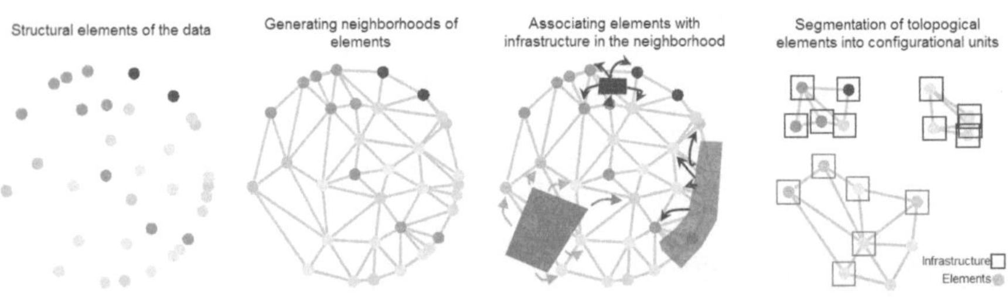

Fig.3. Stepwise generation of neighborhoods of relating elements and associating these with the neighboring infrastructural elements. As a result, each element is described by a multidimensional vector consisting of all its neighbors and surrounding infrastructural elements

3.4 Analysis of Configuration Units

Having applied the above mentioned preprocessing, we obtain a large number of street descriptors, which represent the local spatial pattern of land-uses and infrastructure of a given city. In order to perform a correlation analysis between these local patterns and a selected overall/global target variable, like the topological structure of street network, we first conduct a cluster analysis of all named street descriptions.

We chose to use the SOM algorithm (Kohonen, 2001) for cluster analysis. It is a combined vector quantization and projection algorithm. It produces a network of reference (prototype) vectors from a set of input data vectors by means of a competitive learning process. During the learning process, which takes place after an appropriate initialization of reference vectors has been performed, input data is sequentially presented to the network. Then, the currently best matching (in the nearest neighbor sense) cluster prototype is determined, and this prototype together with a neighborhood of prototypes is then adjusted toward the presented input (cf. Figure 4). As a function of time, during learning the degree of adjustment of prototype vectors is reduced, and stable results are obtained. The SOM reference vectors represent clusters in the input data set and are typically modeled on a 2D grid. Practically, one important property observed on the SOM analysis output is that the arrangement of prototype vectors approximates resembles the topology of data vectors in input space.

We perform SOM cluster analysis based on the vector descriptions of the infrastructure descriptions associated on average with each neighborhood (see Section 4). For setting of SOM parameters, we rely on rule-of-thumb settings typically recommended (Kohonen et al, 1996).

As a result, we obtain a network of clusters describing prototypical distributions of infrastructure elements over named streets in our data sets.

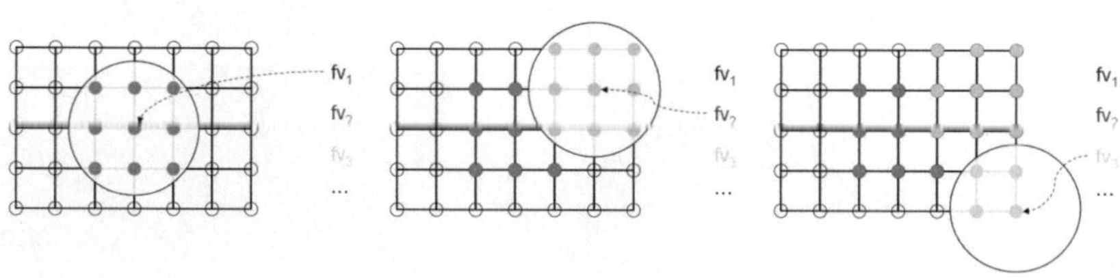

Fig.4. During the SOM learning process, sample input vectors are iteratively presented to the map, which is gradually adjusted to the presented input. The process yields set of cluster prototype vectors which are arranged on a regular grid that approximately represents the topology of the input data (Kohonen, 2001)

3.5 Visualization of Configuration Types

The SOM analysis yields an intermediate cluster result, which we visualize together with the quantitative information of the selected street network topology measure. We visualize land-use characteristics of each group (cluster) of named street patterns as a radial Parallel Coordinate Plot (Van Long, 2009). The basic idea is to map each land-use dimension to one axis emanating radially from the origin in an equally-spaced angular direction. Following the parallel coordinate approach, we connect the coordinate positions of each dimension in the SOM prototype vector by straight, bold lines. A high-dimensional glyph results in form of a radar-like chart. On this chart, we also overlay the set of street samples represented by the SOM prototype vector, by means of opacity bands (Fua, 1999). The street network topology value which is to be correlated with the land use prototypes is mapped to the background color of the group diagrams. To this end, we again use the yellow-red color-map introduced in Figure 2. The final analytic view is constructed by drawing radar charts for each group of named streets yielded by the SOM analysis, using the grid structure of the SOM. The display allows to visually assessing several data aspects. The distribution of land-use over the different groups can be assessed by comparing the shape of the diagrams. The land-use properties can be correlated with the network topology properties by means of the background coloring. The opacity bands allow to assess the crispness of the groups in terms of the spread of group member attributes around the prototypes. Figure 5 illustrates the construction of one group glyph.

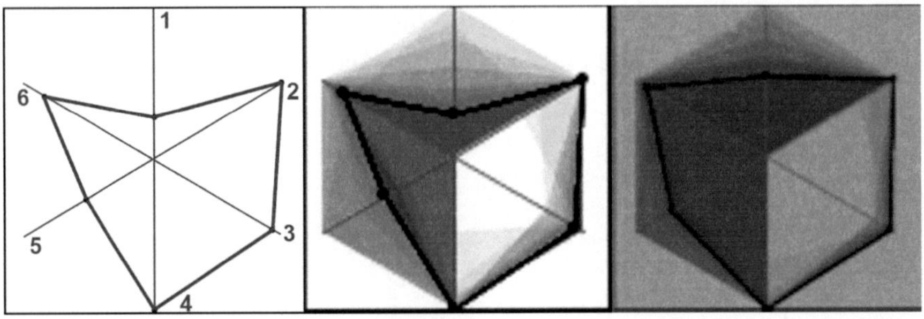

Fig.5. We show the properties of land-use types occurring in a group (cluster) of configuration types by means of a radar-plot (left image). The six radial axes refer the six land-use types. Samples of the cluster are overlaid by opacity bands, indicating the distribution of represented sample data points (second image from left). The street topology measure is mapped to the background color of the image (third image from left) using the same color-map as for the data themselves

3.6 Interaction Facilities

In order to facilitate user interaction, we implemented three major interaction techniques at different stages of the analytic pipeline (as shown the methodology section (Section 3) in Figure 1). In the data mapping stage users are able to select the relevant variables for their analysis. Users have to determine the "independent variable" of their analysis, which was the land-use in the current example. Users also have to select the "dependent variable" of the analysis, which was the topology of street networks in the presented example.

The main concern of the analysts is defining the neighborhoods of the elements, in the current example we used a Delaunay triangulation, which can obviously anytime be replaced by any other structural analysis technique or spatial clustering method. Choosing the right methods for this task is crucial and requires domain knowledge and optimal parameterization of the methods.

The number of automatic pattern extraction techniques is practically unlimited, if we take the combination of their parameters also into account. In the current example we successfully showed that Self-Organizing Maps are effective and useful for the current task and data. However, this method can be exchanged for different algorithms. Finding the best parameter settings is a highly iterative task, since only rule-of-thumb suggestions exist that require constant refinement. In order to facilitate this interaction, we provide a mouse-over function for the generated SOM-clusters which show the location of the cluster members in a geographic map. Consequently, the analysts can investigate the spatial features and distribution of the created clusters in addition to the distribution of the selected variables in the SOM-cluster itself. As a result, refinement of the properties of the clustering algorithm can be used to obtain smaller/larger units and higher/lower levels of distinction between cluster centers.

One of the required properties of information displays is that users can alter the color maps and their scaling in all visualizations. This feature is implemented at every stage of the analytic pipeline, in which visualization is involved. Currently we implemented a continuous heat-map color-map (from yellow to red) for the topological variable, and a discrete color map for the land-use representation. Users can also apply non-linear (square-root and logarithmic) scaling to the continuous color-mapping, in case this is required to compensate for skewed data distributions.

4 Results

In order to show the usefulness of the proposed methodology's instantiation, we present here its potential to identify typical and significant land-use spatial configurations, to locate land-use spatial configurations of interest and to characterize their geographical distribution. As described above, the investigation was conducted based on the relation between the spatial distribution of land-use and the topology of street network in the city of Raanana. The first action in implementing the methodology is a creation of clusters for different spatial land-use configurations using self-organizing maps. The resulting clusters are than colored with the three centrality measures of the street topology, as shown in Figure 6. Therefore, the resulting representations have the same cluster configurations, but different coloring for closeness, betweenness, and degree centrality measures. This possibility enables us to compare between these topological measures in term of their relation with each of the identified land-use configurations. In addition, the number of occurrences for each cluster is indicated in the upper left corner together with the average centrality measure in brackets.

Such presentation opens the possibility for identification of frequent and typical land-use spatial configurations with respect to the relation between land-use and street topology.

Fig.6. The correlation of Closeness (left), Betweenness (middle) and Degree (right) values with the spatial configurations of land-uses in Raanana

In general, the spatial distribution of education (1), culture & leisure (3) and parks and open spaces (6) is relatively high in all clusters and has limited variation between the clusters. Medium spatial distribution of the public services (2) is visible with high variability of the centrality measures. High variations in the spatial distribution of commerce (4) and industry (5) are visible, which is assumed to be influenced by the centrality measures of street network topology.

Closeness reveals a positive correlation with the availability of commerce, as shown in columns 3 and 4 of Figure 6 (left). Low closeness values show low availability, and high closeness values a high availability of commerce. This means that the difference between these configurations located on the 'edges of closeness centrality' – the most accessible places versus inaccessible places – is in the availability of commerce. Interestingly, the lowest closeness value (Column 3 – Row 1) has low availability of commerce and also of public services and industry. The highest closeness values (Columns 3-4 – Rows 3-4) show high availability of commerce and public spaces with low availability of industry.

When comparing these findings with the levels of betweenness and degree, three interesting configurations can be extracted as perceivable in Figure 6:

Configuration 1: High availability of commerce, with low availability of industry (Columns 3 and 4 – Rows 3 and 4) having high values of closeness, showing also high values for betweenness and degree. Such a configuration is expected to be located in the center of the city. In the case of Raanana, it is located along the main street (Weitzman Street) of the city. This configuration is shown in the left image in Figure 7.

Configuration 2: High availability of commerce and industry (Columns 1 and 2 – Rows 3 and 4) having high and medium closeness values and low or medium betweenness and degree values. This constellation describes the industrial area of Raanana in the north-east corner of the city. The described pattern is typical for industrial areas, which are accessible, but are located in a well separated district of the city. This configuration is shown in the middle image in Figure 7.

Configuration 3: High availability of educational institutions, culture and leisure and parks and open spaces (Columns 1 and 2 – Row 1) with tendencies to lower closeness values and higher betweenness values. Such configurations are mostly characterized residential areas with high socio-economic standards. This finding seems reasonable since such a combination of topological properties means to live in residence places which are close to other parts of the city but which are not served for movement or transit between other parts of the city. A geographical mapping of this configuration of interest (e.g., for spatial equity, socio spatial planning policy) shows clearly that it has an expected peripheral pattern. This configuration is shown in the right image in Figure 7.

Fig.7. Geographic location of different spatial configurations: High commerce area (left), industrial area (middle) and residential areas (right). The SOM-clusters showing the different configurations are shown in the left upper corner. The named streets included in the clusters are highlighted on the map in red color

These examples illustrate how locating the identified typical configurations on the geographic map are helpful in defining empiric findings. The explanation may concern previous knowledge on the development history and planning policy of Raanana as well as theoretical models that are suggested to elucidate on the spatial structure of land-uses in the city. Thus, the methodology enables us to identify typical and significant correlations between the topological properties of streets, or their combinations, and land-use spatial configurations, and further to explore their geographic patterns. This framework also helps to determine how the topological properties of a given street and its interrelation affect the functional content of its surroundings, i.e., which topological properties are significant for the formation of land-use spatial configurations?

5 Discussion and Conclusion

We described an analytic framework to assess urban spatial configurations. The methodology is applied on local land-use spatial configurations, and the local (i.e. degree centrality) and global (i.e. closeness centrality and betweenness centrality) of street networks' topologic properties in the city of Raanana. The methods used for analysis are based on SOM-clustering to group similar configurations, and on geographic views, which support analysts to iteratively extract interesting configuration patterns. These views, the more abstract cluster visualization, and the more concrete geographic map are highly interactive and strongly coordinated to each other. The contribution of the suggested methodological framework is clear: traditional local spatial analysis methods (i.e. local geo-statistics measures and aggregation at different scales) for analyzing and presenting high-resolution geographic data are typically limited to geographical presentation of one attribute only and blur the results by aggregation. Against this background, the proposed methodology has potential to shed light on the relation between multiple structural and geographical dimensions of an urban environment by keeping the individual objects as the level of investigation. Thus, the methodology enables investigation of local spatial relations between huge amounts of individual buildings with respect to their local and global attributes without loss of data as a result of aggregation. It is also possible to apply this methodology at different geographic scales and to explore the 'natural' geographic scale for the investigated phenomena. For now, we suppose that the main practical application of the proposed methodology is a an examination of a variety of urban spatial forms in order to reveal their unique spatial configuration with respect to functional and social composition of different city types. Such application has the potential to improve our knowledge on the relationship between urban forms and the formation of land use spatial distribution in cities and can be used to support urban spatial policy.

Future work will explore additional multivariate visualization options, e.g., other glyph based approaches and graph-based representations. Currently, we support the visual correlation analysis between multidimensional variables in an abstract manner (Section 3.5), and in a separate geographic visualization of selected clusters (Section 4). Joint representation of these views is a challenge, which we like to address in the future. The method proposed on (Bacao, 2005) seems an interesting starting point to this end. On the algorithmic part, many options are open to implement additional algorithms (multi dimensional scaling, principal component analysis, etc.), which might be appropriate for different analytic questions in this context. Applying our approach on other domains is certainly a long-term perspective.

References

Anselin, L. (1995) Local Indicators of Spatial Association – LISA, *Geographical Analysis 27*(2), 93–115.

Bacao, F., Lobo, V., Painho, M. (2005). The Self-Organizing Map, the Geo-SOM, and Relevant Variants for Geo-Sciences. *Computers & Geosciences,* 31, 155–163.

Benenson, I., & Omer, I. (2003). High-Resolution Census Data: A Simple Way to Make Them Useful. *Data Science Journal* 2 (26), 117-127.

Cox, M. & Cox, M. (2001). Multidimensional Scaling. *Chapman and Hall.*

Deboeck, G., Kohonen, T. (1998). Visual Explorations in Finance With Self-Organizing Maps. *Springer.*

Elmqvist, N., Dragicevic, P., Fekete, J.-D (2008). Rolling the Dice: Multidimensional Visual Exploration Using ScatterPlot Matrix Navigation. *IEEE Transactions on Visualization and Computer Graphics*, 14, 1141–1148.

Everitt, B. S. & Nicholls, P. (1975). Visual Techniques for Representing Multivariate Data. *The Statistician*, 24(1), 37-49.

Fua, Y.-H., Ward, M., Rundensteiner, E. (1999). Hierarchical Parallel Coordinates for Exploration of Large Datasets. *Proceedings of IEEE Conference on Visualization (VIS)*, 43-50.

Guo, D., Chen, J., MacEachren, A. M., Liao, K (2006). A Visualization System for Space-Time and Multivariate Patterns (VIS-STAMP). *IEEE Transactions on Visualization and Computer Graphics*, 12(6):1461–1474.

Hillier, B., Penn A., Hanson J., Grajewski T. and Xu J. (1993) Natural Movement: Configuration and Attraction in Urban Pedestrian Movement, *Environment and Planning B,* 20, pp. 29-66.

Honkela, T., Kaski, S., Lagus, K., Kohonen, T. (1997). WEBSOM—Selforganizing Maps of Document Collections. *Proceedings Workshop on Self-Organizing Maps*, 310–315.

Inselberg, A. & Dimsdale, B. (1990). Parallel Coordinates: A Tool for Visualizing Multi-Dimensional Geometry. *Proceedings Conference on Visualization*, 361–378.

Jiang, B. (2007). A Topological Pattern of Urban Street Networks: Universality and Peculiarity. *Physica A*, 384, 647-655

Jiang, B. & Harrie L. (2004) Selection of Streets from a Network Using Self-Organizing Maps, *Transactions in GIS*, 8(3): 335–350

Jolliffe, I. (2002). Principal Components Analysis. *Springer*, 3rd edition.

Kohonen, T. (2001). Self-Organizing Maps. *Springer*, 3rd edition.

Kohonen, T., Hynninen, J., Kangas, J., Laaksonen, J. (1996). SomPak: The Self-Organizing Map Program Package. *Helsinki University of Technology Technical Report.*

LeBlanc, J., Ward, M.O., Wittels, N (1990). Exploring N-dimensional Databases. *Proceedings Conference on Visualization*, 230 - 237.

Omer, I., & Benenson I., (2002). Investigating Fine-Scale Residential Segregation by Means of Local Spatial Statistics. *Geography Research Forum* 22, 41-60.

Porta, P. Crucitti, P. & Latora V. (2006). The network analysis of urban streets: A primal approach. *Environment and Planning B: Planning and Design* 2006, 33, 705–725

Ratti C, Pulselli RM, Williams S, Frenchman D (2006) Mobile Landscapes: using location data from cell phones for urban analysis. Environment and Planning B: Planning and Design 33: 727-748

Spielman, S. & Thill, J.-C. (2008). Social Area Analysis, Data Mining, and GIS. *Computers, Environment and Urban Systems,* 32, 110–122.

Talen, E. (2003). Neighborhoods as Service Providers: a Methodology for Evaluating Pedestrian Access. *Environment and Planning B: Planning and Design,* 30, 181- 200.

Talen, E., & Anselin, L. (1998). Assessing spatial equity: an evaluation of measures of accessibility to public playgrounds. *Environment and Planning A,* 30, 595–613.

Van Long, T. & Linsen, L. (2009). MultiClusterTree: Interactive Visual Exploration of Hierarchical Clusters in Multidimensional Multivariate Data. *Proceedings of Eurographics / IEEE Symposium on Visualization.*

Vesanto, J. (1999). SOM-based Data Visualization Methods. *Intelligent Data Analysis*, 3(2):111–126.

Ware, C. (2004). Information Visualization: Perception for Design. *Morgan Kaufmann*, 2nd Ed.

Wilkinson, L., Anand, A., Grossman, R. (2005). Graph-Theoretic Scagnostics. *Proceedings of the IEEE Symposium on Information Visualization.*

Wong D.W.S, (2003) Spatial Decomposition of Segregation Indices: A framework toward measuring segregation at multiple levels, *Geographical Analysis*, 35 (3), 179-184.

A Visual Query Language for Spatial Data Warehouses

Sandro Bimonte[1], Vincenzo Del Fatto[2], Luca Paolino[2], Monica Sebillo[2], Giuliana Vitiello[2]

[1] Cemagref, UR TSCF, 24 Avenue des Landais, 63172 Clermont-Ferrand, France
sandro.bimonte@cemagref.fr
[2] Dipartimento di Matematica e Informatica, Università di Salerno, Via Ponte don Melillo, 84084 Fisciano, Italy
{vdelfatt, lpaolino, msebillo, gvitiello}@unisa.it

Abstract. SOLAP systems are technologies intended to support geographic business intelligence. They allow decision-makers to explore and analyze a large amount of geo-referenced data usually through interactive browsing user interfaces. In this paper, we explore a new interactive methodology which supports users to query and aggregate this kind of structured data by means of specific visual notations. The widely recognized capability of visual query languages to bridge the gap between users' expertise and advanced query formulation adds effectiveness to the methodology, thus providing a useful support for posing correct SOLAP expressions by spatially arranging visual metaphors in a simple and intuitive way.

1 Introduction

A Data Warehouse (DW) is a centralized repository of data acquired from external data sources and organized following a multidimensional model [8] in order to be analyzed by On-Line Analytical Processing (OLAP) applications. The model is based on the concepts of dimension, fact and measure. A dimension corresponds to the analysis axis and it is described

M. Painho et al., (eds.), *Geospatial Thinking*, Lecture Notes in Geoinformation and Cartography, DOI 10.1007/978-3-642-12326-9_3, © Springer-Verlag Berlin Heidelberg 2010

through one or more hierarchies. A fact represents the analysis subject and it is described by several indicators (measures). The multidimensional model is then explored by using OLAP operators, namely Roll-Up, Drill-Down and Slice. Roll-Up and Drill-Down allow navigating into hierarchies and aggregate measures using SQL functions, while the Slice operator selects subparts of a DW. In the last decade, the model and the processing applications have been extended in order to effectively analyze the complex nature of spatial data. In particular, the Spatial OLAP (SOLAP) has reformulated the multidimensional model by defining spatial dimensions, spatial measures and spatio-multidimensional operators [17]. Moreover, from an architectural point of view, classical architectures based on three-tiers have been adapted to integrate Spatial DBMS, SOLAP Server and SOLAP client [17].

The research we are carrying out in this field is focused on the definition of new interaction methods meant to support expert users in their activities involving DWs and spatial data. In particular, our attention is focused on the SOLAP client tier that is responsible to display information by outputting results to users who may then visually analyze them. Specifically, current client applications allow for exploring a spatial DW through SOLAP operators by simply interacting with synchronized tabular and cartographic data [1]. However, setting parameters of SOLAP operators is a difficult task for non-computer science experts as it needs a deep knowledge of main SOLAP concepts [14]. Moreover, existing SOLAP clients do not allow for specifying complex spatial slice predicates through simple interactions with the user interface [17].

A promising means for allowing unskilled users to explore geographic databases, to interpret and possibly reuse recorded queries is represented by visual languages (such as [5, 10]. These systems are designed to bridge the gap between unskilled users and system interfaces. They are mainly based on the definition of graphical representations for the spatial properties associated with the geographic data manipulated, and for the involved spatial operators. Spatial properties are referred to the geometry of the geographic data and to their topology, which describes objects relative positions. When trying to add user-friendliness to Geographic Information Systems (GISs), the association of visual descriptions to such features seems to be quite a natural step.

Based on these observations, our opinion is that visual languages could improve analysis capabilities of SOLAP tools allowing for the definition of SOLAP queries in a very simple and intuitive way. Visual languages for spatial DWs should integrate SOLAP clients by providing useful instruments to set-up main SOLAP queries whose results can be consequently explored, refined and analyzed through tabular and cartographic displays.

In this paper, we present a new visual language meant to query spatial DWs. In particular, we introduce new visual metaphors to represent SOLAP concepts of spatial dimensions, spatial measures and aggregation functions. Moreover, visual metaphors are also adopted to support users to manipulate SOLAP operators (Drill and Slice), namely the *Condition Tree*, the *Nested Rectangles* and the *Grouping Metaphor*.

To best of our knowledge contrary to existing approaches, our new metaphors allow (i) a uniform representation of spatial DW elements (ii) and their interactive querying (spatial drill and spatial slice operators), by explicitly expressing spatial properties of data and hiding complexity of SOLAP data structures and queries. In such a way, spatial DW concepts are transparent to unskilled users who can focus exclusively on spatial problems and formulate SOLAP query by simply interacting with the interface elements. Interactivity allows decision-makers to easily and quickly create, modify and save different SOLAP queries according to SOLAP exploration analysis paradigm.

The paper is organized as follows. Section 2 presents the current status of the related work. Section 3 introduces the running example we exemplify in order to illustrate how the proposed visual query language works. Section 4 describes the visual language in terms of visual notations and metaphors. A global architecture is successively shown in Section 5. Conclusions are drawn in Section 6.

2 Related Work

In this Section we present the current status of the work concerning the management of visual approaches for querying and exploring (spatial) DWs. In [2] and [16] authors define OLAP queries by using a graphical representation of DWs. Users interact with graphical elements of the language by setting dimensions levels queries. Subsequently, the visual sentence is translated into a classical OLAP query. Moreover, in [16] authors combine the graphical representation of a DW with pivot table displays allowing for an interactive exploration of multidimensional data through tabular and graphical displays. Based on the same approach, in [19] authors use sequence diagrams to visually define sequence of OLAP queries. The authors associate one diagram for each dimension where a sequence is a set of Roll-Up/Drill-Down operations, represented by arrows, and a state which corresponds to a Slice operation visualized as a class object. Main drawbacks of these approaches are: (1) decision-makers have to be familiar with multidimensional model concepts to define their queries, (2) they do not take into account spatial data.

To best of our knowledge only [14] and [18] address SOLAP queries. [14] define a correspondence between geographic layers of the visual language and the spatial dimensions levels, allowing to define spatial predicates on spatial levels (spatial slice) and spatial Roll-Up and Drill-Down operations on the spatial dimension. However, the visual language does not provide a visual representation of all SOLAP concepts: aggregation functions, classical dimensions, (spatial) measures. Then, Drill and Slice operators on non spatial dimensions cannot be defined by means of the visual language, as well as the choice of measures and aggregation functions. [18] propose a visual query language for Spatial DWs where all the multidimensional elements and the spatial predicates are represented by Unified Modeling Language (UML) stereotypes. Then, they implement the spatial slice operator by means of Object Constraint Language (OCL) constraints on these stereotypes. However, this model does not allow users to visually define Drill operators. Moreover, using OCL to define topological and distance operators is a quite complex task for non-expert computer science users.

Hence, no visual language for spatial DWs allows defining spatial DW data structures and drilling and slicing SOLAP operators using simple visual icons arrangements or sketch.

On the other hand, the definition of spatial predicates using simple visual icons arrangements or sketch by means of visual spatial query languages is investigated in many works. More in general, many other works related to our research concerns with visual languages for querying GIS or spatial databases. An excellent although dated survey about visual query languages can be found in [3], where significant work has been analyzed and relevant issues have been outlined for the design of next generation visual query systems.

An important visual approach for GIS querying is represented by sketch based visual languages. Basically, they adopt the query-by–example approach where users draw particular configurations of the spatial elements that the system should be able to interpret. The depicted arrangement represents an example of the result that should be displayed. Sketch! is one of the first languages which adopted that approach for composing spatial queries [12]. In Spatial-Query-By-Sketch [4] users interact with a touch sensitive screen to sketch the example spatial configurations. They can augment or reduce the similarity ranking to modify the accuracy threshold for the resulting matches.

In general, sketch and drawing based approaches are suitable for expressing similarity-based queries. However, such methods become uncomfortable in the case of composite queries, because it may be difficult to sketch the sample query so that it includes all and only the characterizing elements the user is looking for. Besides, sketch and drawing based approaches rely on user's ability to express spatial relationships in a sketch. Indeed, even if some approaches offer support to the user during the drawing phase, exact queries can be generally ambiguous due to the several possible interpretations of the visual configurations [6].

A different approach, very close to our work, is followed by the Spatial Exploration Environment (SEE) [9]. SEE is an integrated framework that adopts the visual paradigm for spatial query specification and result visualization. It relies on a visual query interface for two-dimensional spatial data and an underlying visual query system, SVIQUEL, which allows the specification of topological and directional relationships between objects through direct manipulation. The issue of specifying any complex spatial query condition connected by Boolean operators, has been mainly faced with two basic approaches in literature. In the former, single conditions are combined just by the AND operator, and the final result indicates all conditions that must be satisfied to select objects. A further, merely visual, approach is proposed by Filter/Flow, where users use the pipe metaphor to describe Boolean logics [13]. Each condition is like a filter for the water flow: if two conditions must be satisfied at the same time (AND), then they are located as a sequence of cocks, while if at least one condition must be satisfied, then the flow is divided into two minor flows which may be interrupted by cocks, representing the conditions.

3 The Running Example

In order to simplify the comprehension of our proposal and illustrate how it works, in the following we exemplify a scenario based on the spatio-multidimensional schema depicted in Figure 1, which will be used throughout the paper. In Table 1 a subset of the extensive database is reported, whereas Figure 1 illustrates the schema of data, where facts, measures, dimensions and attributes composing the dimensional hierarchies are pointed. In particular, measures are attributes of a table which refers to detailed levels of dimensions (table Pollution), namely the dimensional attributes. The dimensions of the Spatial DW are three, namely location, time and pollutant. The first dimension consists of City, Dept and Region, and is represented through a normalized structure. The second dimension is represented by Time and is structured by a denormalized table as well as the pollutant dimension with the Pollutant table.

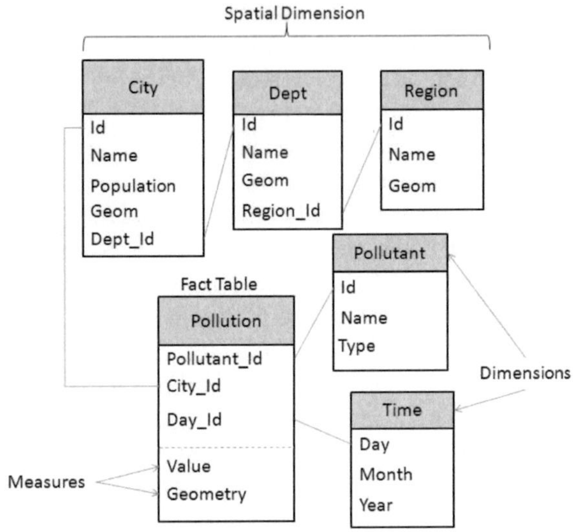

Fig. 1. The snowflake of a Pollution SOLAP application

Table 1. A subset of the extensive database

Pollutant Type	Pollutant Name	Pollution Value	Pollution Geom	City Name	Dept Name	Region Name	Time Day	Time Month	Time Year
Inorganic	Zinc	8	Id2	Napoli	NA	Campania	1-11-2001	11-2001	2001
Inorganic	Iron	2	Id3	Salerno	SA	Campania	10-3-2001	3-2001	2001
Inorganic	Iron	3	Id4	Salerno	SA	Campania	11-4-2001	4-2001	2001
Inorganic	Iron	5	Id5	Salerno	SA	Campania	11-4-2001	4-2001	2001

Based on the schema depicted in Figure 1, the following four queries are then formulated, built through an incremental approach.

Query 1 "*What is the average of pollution values and the whole zone to which this value is associated?*"

Query 2 "*For each Region, Year and Pollutant, what is the average of pollution values and the whole zone to which this value is associated?*"

Query 3 "*For each Region, Year and Pollutant and for each Dept, Month and Pollutant, what is the average of pollution values and the whole zone to which this value is associated?*"

Query 4: "*For each Region, Year and Pollutant and for each Dept, Month and Pollutant, what is the average of pollution values and the whole zone to which this value is associated by considering only pollution values measured within cities at most 25 kms far from the coordinates*

(40.197, 15. 058) or (41.311, 14.8806), within the Italian Campania region?"

In the next Section, the main concepts of the visual query language are described. The language is meant to allow the visual formulation of SOLAP queries and their subsequent translation into the underlying textual query language.

4 The Visual Notation

The main idea of our proposal is representing the SOLAP concepts of spatial dimensions, aggregate functions of measures, and SOLAP operators: Roll-Up, Drill Down and Slice, by visual notations. It is important to underline that in this work, we consider only spatial measures as set of spatial objects [11], and spatial dimensions characterized by spatial balanced hierarchies ("they have at the schema level, only one path where all levels are mandatory, and at instance level the members form a tree where all the branches have the same length" [11]). According to the traditional approach of many visual query languages, queries are expressed in terms of spatial arrangement of visual elements representing data, operators and functions. In particular, as for data, they are associated with a visual notation, whose structure takes into account the complex nature of geographic data by visually integrating the iconic and the property components, as shown in Figure 2 (a) as described in [10]. In order to adapt the interaction metaphors to our purposes, some visual notations have been modified and a new one has been defined. In particular, the management of the multifaceted structure of a SOLAP schema has required the extension of the Properties component to include the multidimensional model, as shown in Figure 2(b). This extension affects the source element, which now allows for retrieving information about measures and dimensions. According to such a structure, Figure 3 illustrates the visual representation of the measures, dimensions and dimensional attributes featuring the Pollution geometaphor.

4.1 Visual Spatial Drill Operator

Spatial Drill is the operator which allows for navigating into spatial dimensions and aggregating measures. Basically, it can be divided into two SQL subparts, namely the definition of one or more aggregate functions on some (spatial) attributes within the SELECT clause, and the definition of groups on which the aggregate functions have to be applied, using the

GROUP BY clause. In the following subsections, we describe the corresponding visual metaphors specified to visually compose the clauses. In particular, in 4.1.1, we introduce the concept of Nested Rectangle metaphor which allows to specify and aggregate alphanumeric measures as well as spatial measures. Subsection 4.1.2 describes the Grouping Metaphor through which users visually specify grouping and UNION operations.

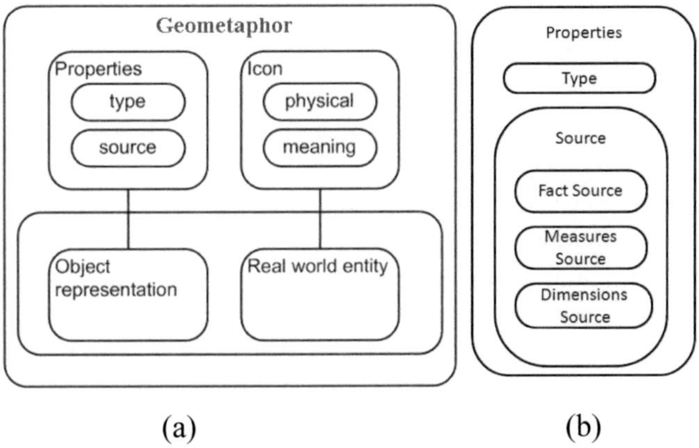

Fig. 2. (a) The original conceptual structure (b) The Properties component of the SOLAP structure

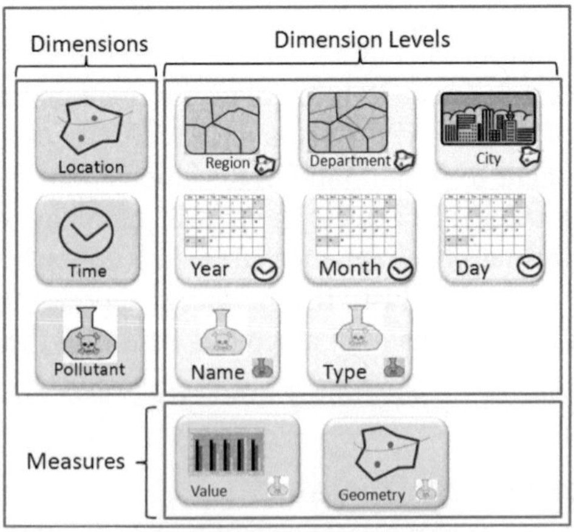

Fig. 3. A visual representation of the Pollution structure

4.1.1 Managing Measures

In order to visually support the user interaction targeted to specify an aggregate function, the Nested Rectangles metaphor has been adopted as interaction method.

Basically, every time an aggregate function has to be applied to a measure, a specific selector can be chosen, and a black rectangle can be drawn around the selected measure. It implies that the chosen function will be applied to the included measure. This operation can be easily repeated in order to use the result of an aggregate function as a measure for a new one. In terms of SQL syntax, a black rectangle represents a parenthesis within a SQL query, which can be recursively applied until the needed result is obtained.

Figure 4 shows the part of the visual query, composed according to the Nested Rectangle metaphor, which generates Query 1.

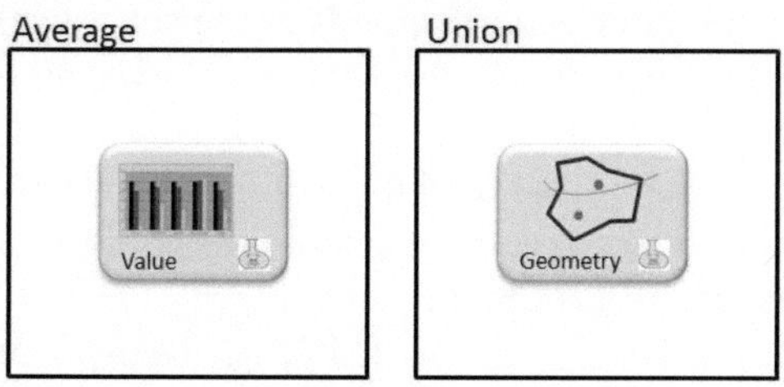

Fig. 4. The Nested Rectangle metaphor

The algorithm generates SQL code according to the DBMS implementation of the Oracle.

```
SELECT AVG(POLLUTION.Value), SDO_AGGR_UNION
(SDOAGGRTYPE(POLLUTION.Geometry,0.005)
FROM POLLUTION
```

4.1.2 Managing Dimensions

To the aim of managing grouping, a specific representation has been defined, named Grouping Metaphor, which allows for describing this operation in a visual and easy way disregarding the coding details and the spatial DW data and schema.

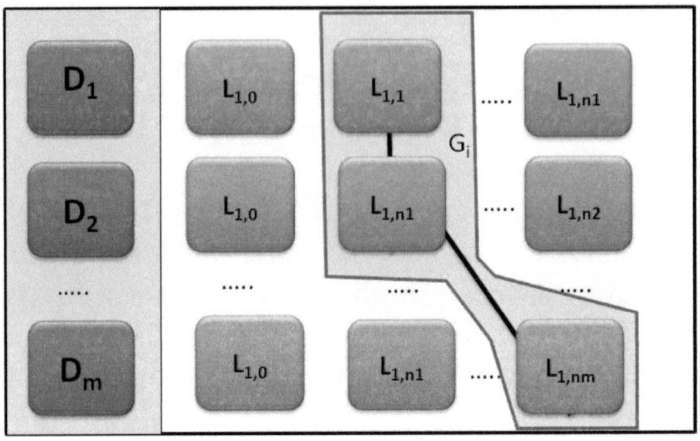

Fig. 5. A visual representation of grouping operation applied on dimensional attributes

Formally, we define two structures. The former D is the set of levels for each dimension. The latter G corresponds to the sets of selected levels such that each set contains *exactly one level for each dimension*. This is to say: D = { D1 {L1,0, L1,1, …, L1,n1}, D2 { L2,0, L2,1, …, L2,n2}, …, Dm { Lm,0, Lm,1, …, Lm,nm}}, where Di and Li,nk with $1 \leq i,k \leq m$ represent the dimensions with their levels, respectively, while Lj,0 with $1 \leq j \leq m$ represents the absence of grouping.

G={ G1 {L11, …, L1m}, …, Gk {Lk1, …, Lkm} }, where Lji belongs to Di ={Li,0, Li,1, …, Li,ni} and it is part of the jth group with $1 \leq j \leq k$. From a visual point of view, the dimension Di and its levels Li,j are represented by means of icons. Levels belonging to the same dimension are shown along the same line following a hierarchical order, while levels of different dimensions appear on different lines, as shown in Figure 5.

In case no grouping is required for a specific dimension, a precise icon is portrayed containing the empty symbol and a label "No Groups".

In order to compose the grouping operation it is sufficient to sketch a line which covers one or more levels of different dimensions. The levels covered by the line will participate to the grouping sequence in the GROUP BY clause. When required, the user may draw more lines in order to pose different groups which will be joint together through a UNION operation.

As for the running example, let us suppose that we have already composed the arrangement in Figure 4 by the Nested Rectangles metaphor. When the visual composition shown in Figure 6 is further made up, the translator produces the code corresponding to the Query 3, namely:

```
SELECT AVG(POLLUTION.Value), SDO_AGGR_UNION
(SDOAGGRTYPE(POLLUTION.Geometry,0.005), Re-
gion.Name, Time.Year, Pollutant.Name
FROM POLLUTION, CITY, DEPT, REGION,TIME
WHERE POLLUTION.city_id= CITY.id AND …
GROUP BY Region.Name, Time.Year, Pollutant.Name
UNION
SELECT …
FROM POLLUTION, CITY, DEPT, TIME
WHERE POLLUTION.city_id= CITY.id AND …
GROUP BY Dept.Name, Time.Month, Pollutant.Name
```

Just for sake of completeness, in case we consider only the left sided line, the algorithm generates Query 2.

When the grouping operations are effectively applied on data in Table 1, they produce the subsets shown in Table 2 and 3. In particular, Table 2 shows data which are grouped by Region name, Year and Pollutant name, whereas Table 3 presents data which are grouped by Dept Name, Month and Pollutant name.

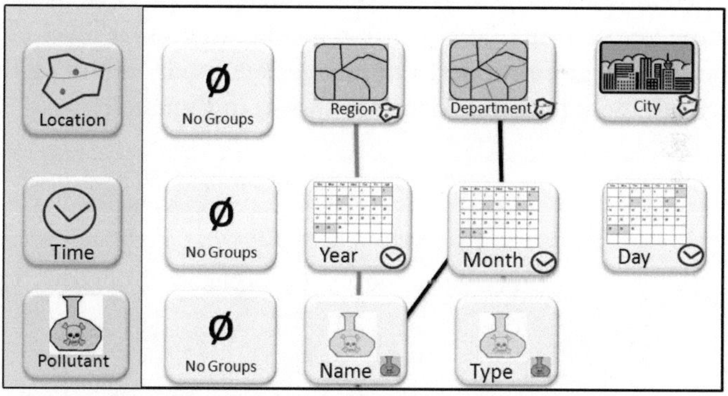

Fig. 6. The Grouping Metaphor applied on the example

Table 2. The result of the grouping operation by Region, Year and Pollutant name on data shown in Table 1

Pollutant Name	Pollution Value	Pollution Geom	Region Name	Time Year
Zinc	8	Id2	Campania	2001
Iron	3.3	Id3+Id4+Id5	Campania	2001

Table 3. The result of the grouping operation by Dept, Month and Pollutant name on data shown in Table 1

Pollutant Name	Pollution Value	Pollution Geom	Dept Name	Time Month
Zinc	8	Id2	NA	11-2001
Iron	2	Id3	SA	3-2001
Iron	4	Id4+Id5	SA	4-2001

In order to complete the description of the Roll-Up and Drill-Down operations through the Grouping Metaphor, we have to illustrate the interaction the user performs.

As for the Roll-Up operation, the user may change a connection by moving a vertex of a line towards a left sided icon, this is to say from any level to a more generalized one (i.e. from Department to Region). In case the user moves a line vertex from a level to the No Group level, s/he is going to perform an aggregation by removing a dimension attribute. On the other hand, the Drill-Down operation may be carried out by reversing the Roll-Up actions, namely the user may change a connection by moving a vertex of a line towards a right sided icon, this is to say, from any level to a more specialized one (i.e. from Department to City). In case the user moves a line vertex from the No Group level to any other level of the same dimension, s/he is going to perform an aggregation by adding a dimension attribute. Both the interaction tasks are shown in Figure 7.

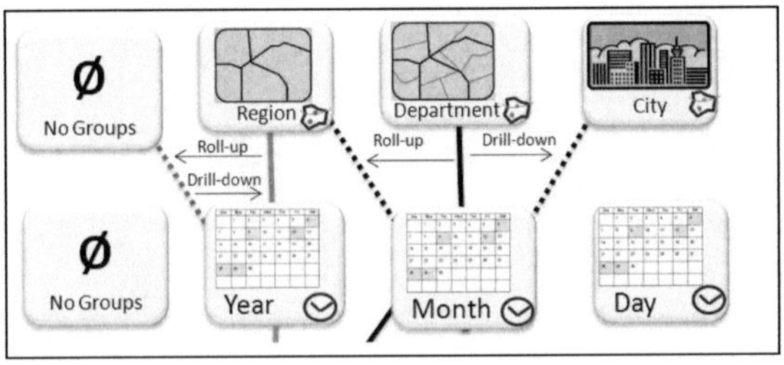

Fig. 7. A detail of the Grouping Metaphor to perform Roll-Up and Drill-Down operations

4.2 Visual Spatial Slice

One of the most significant problems of visual languages for (spatial) database querying is the low expressiveness of visual techniques to represent complex conditions involving Boolean operators, such as (P1 AND (P2 OR P3)) OR P4. Such a problem is also relevant for the SOLAP systems where the classical WHERE clause is used to perform Spatial Slice operations.

In this Subsection we present the visual technique, named Condition Tree, which faces this problem and helps users to compose complex logical expressions through a fully visual approach, no textual sentence is required.

As depicted in Figure 8, a Condition Tree C is defined as a Root Node R whose value is always set to the Boolean value True. Such a node may be connected with one or more Tree T. In turn, a Tree T is either an empty tree Ø or a Condition Node N, whose value is a predicate which may result either true or false. The Condition Node N may be also connected with zero or more Tree T. From a graphical point of view, the Condition Tree supports users in defining visual complex conditions through its structure where nodes represent simple conditions, edges represent AND connectors and edges starting from the same node are ORed connected to each other.

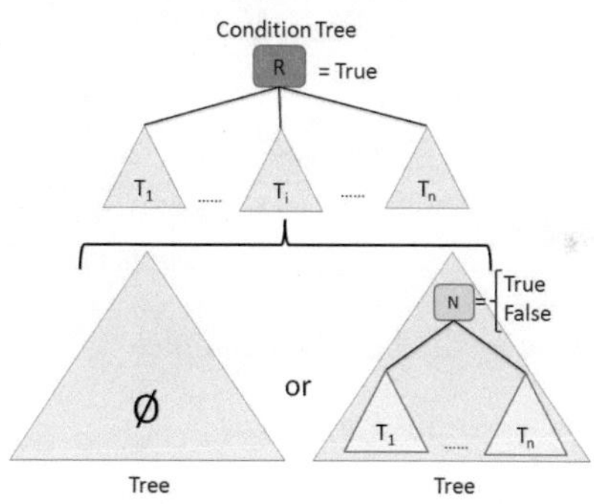

Fig. 8. Visual definition of a Condition Tree

As for the running example, the goal is to complete the query introduced in Section 3 (Query 4) by providing the visual arrangement which selects the Pollution values related to the Campania Italian region, which are in cities placed 25 km far from either the coordinates (40.197, 15. 058) or (41.311, 14.8806). The tree representing the visual query is shown in Figure 9. It is made up of three conditions connected to the root node. Starting from the left, the first geometaphor indicates the condition specifying the selection of the Campania region, it is then followed by two branches connected to a geometaphor selecting the cities which lie on a buffer of 25 km around the coordinate (40.197, 15. 058) and a geometaphor selecting the cities which lie on a buffer of 25 km around the coordinate (41.311, 14.8806). The SQL code for the WHERE clause produced by the visual representation of Figure 9 is the following:

```
WHERE Region.name like 'Campania' AND
(SDO_WITHIN_DISTANCE(City.geom,
  SDO_GEOMETRY(2001,NULL,
      SDO_POINT_TYPE(40.197,15.058, NULL),
      NULL,NULL)),'distance<25') = 'TRUE' OR
SDO_WITHIN_DISTANCE(City.geom,
  SDO_GEOMETRY(2001,NULL,
      SDO_POINT_TYPE(41.311, 14.8806, NULL),
      NULL,NULL)),'distance<25') = 'TRUE')
```

It is important to note that the Condition Tree metaphor fits the SOLAP spatial hierarchy schemas and instances thanks to the visual hierarchical structure of spatial predicates it provides.

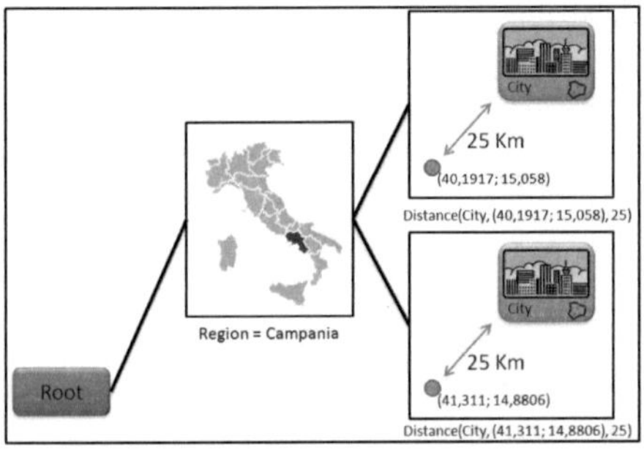

Fig. 9. The Condition Tree applied on the example

5 A Global Architecture

In this Section we propose a possible the architecture of a SOLAP system which integrates the new visual language. Our proposal allows user to define SOLAP queries in a simple and intuitive way. However, SOLAP analysis is an iterative process, the user explores data looking for unknown and/or unusual patterns through the interaction with interactive maps and tabular display which formulate and validate his/her hypothesis [1]. Then, we propose to couple the visual language to a SOLAP system as defined in Figure 10. The user can define his/her queries by using the visual language when spatial predicates are very complex and then exploring the spatial DW through a spatial analysis client (SOLAP client).

Moreover, since our visual queries can be translated into SQL statements it is possible to couple the visual language to any Relational SOLAP

Server (at most translating SQL queries into MXD ones) such as GeoMondrian [7]. The so formulated queries are then processed by the SOLAP tool that allows for exploring the SDW and visualize results through a SOLAP client.

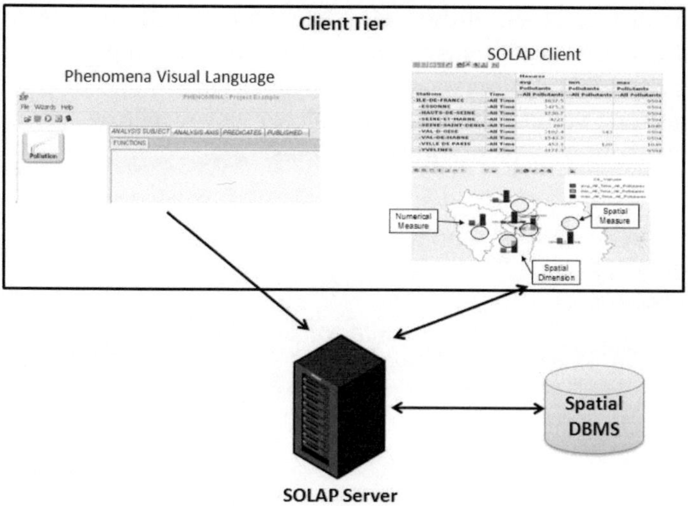

Fig. 10. The SOLAP Architecture

Currently, we are implementing the environment embedding the new visual language in Phenomena [10].

The interface of the original version of the visual environment will be extended with some visual notations available for both operators and operands can be spatially manipulated (Figure 11). In particular, it is divided into four parts, the Dictionary, containing the iconic representation of data on which visual queries can be posed, and four interactive working areas, named ANALYSIS SUBJECT, PREDICATES, ANALYSIS AXIS and PUBLISHED, where users may respectively select objects, elaborate filtered data and display the final visual query. In particular, the ANALYSIS SUBJECT working area allows users to build up a SELECT clause. The output generated through the function panels may be sent either to other function panels in order to be further computed, or to the PUBLISHED area, which contains the set of composed visual representations that will eventually appear within the final SELECT clause. Similarly, the PREDICATES area enables users to visually define the content of the SQL WHERE clause, which expresses a condition of a typical SELECT… FROM… WHERE statement. It contains both some basic panels allowing users to visually build representations of simple conditions, and a special panel, named COMBINE, where those representations may be merged in order to build more complex queries. Moreover, the ANALYSIS AXIS has been added for managing the grouping and union operations.

Finally, it is very important note that in our proposed architecture, performance of translation of the visual queries into SQL queries do not depend on performance of answering SOLAP queries. Indeed, they are resolved by the SOLAP Server and the Spatial DBMS and they depend on several factors: volume of data, materialized views, indexes, etc… This allows us to focus only on usability problems.

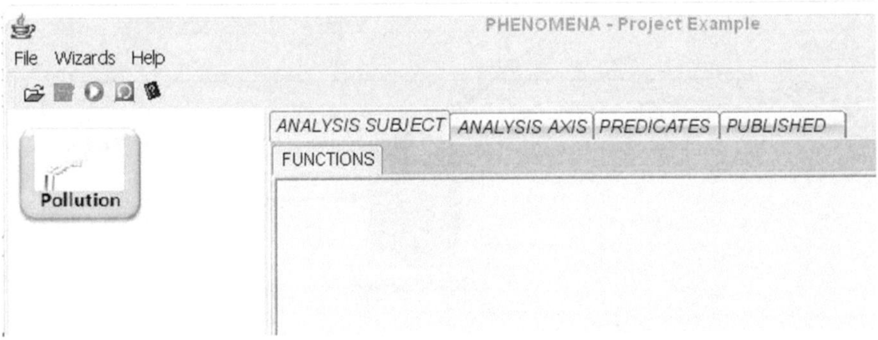

Fig. 11. The Visual Environment for SOLAP users

6 Conclusion

In this paper we have presented a new method for exploring data collection organized as a SOLAP multidimensional schema. Differently from other experiences which are based on browsing, we have described a method based on the query composition. Basically, we have: (1) described the conceptual structure for including the relationships with dimensions and measures, (2) defined the Nested Rectangle and the Condition Tree visual metaphors for aggregating measures and complex conditions, and (3) defined the Grouping Metaphor, a visual metaphor able to easily describe grouping operations. The example shown in the paper highlights the easiness and the adaptability of the system to be modeled on the specific SOLAP structure. Moreover, since our visual queries can be translated into SQL statements it is possible to couple the visual language to any Relational SOLAP Server (at most translating SQL queries into MXD ones).

Our future plan includes the introduction of visual notations for the management of more complex spatial dimensions and measures [11]. We also plan to introduce other SOLAP complex operators which are directly related (pivot, etc.) or not (push-pull, dice, etc.) to SOLAP client pivot tables [15]. Finally, at the moment we have implemented the metaphors for managing measures and dimensions, and we are connecting it to the visual language proposed in [10] for spatial slice operator. However, in order to demonstrate performance in real SOLAP application, we are planning to conduct usability studies comparing SOLAP tools with and without our extension implementation.

References

1. Bédard, Y., Proulx, M., Rivest, S., Badard, T. (2006) Merging Hypermedia GIS with Spatial On-Line Analytical Processing: Towards Hypermedia SOLAP. Geographic Hypermedia: Concepts and Systems, Springer, Berlin-Heidelberg, 167-185
2. Cabibbo, L. and Torlone, R. (1998) From a Procedural to a Visual Query Language for OLAP. In Procs of the Intl. Conference on Scientific and Statistical Database Management, Capri, Italy. IEEE Computer Society, 74-83.
3. Catarci, T., Costabile, M.F., Levialdi, S., Batini, C. (1997) Visual Query Systems for Databases: a Survey. Journal of Visual Languages and Computing 8(2) 215-260.
4. Egenhofer, M.(1997) Query Processing in Spatial Query by Sketch, Journal of Visual Languages and Computing 8(4) 403-424.
5. Ferri, F. Grifoni, P., Rafanelli, M. (2005) The Sketch Recognition and Query Interpretation by GSQL, aGeographical Sketch Query Language. In Proceedings of CIT 2005, 34-38
6. Ferri, F., Pourabbas, E., Rafanelli, M. (2002) The syntactic and semantic correctness of pictorial configuration query geographic databases by PQL, in: Proceedings of the 17th ACM Annual Symposium on Applied Computing (ACM SAC2002), Madrid, Spain, 432-437.
7. GeoMondrian, available at http://geosoa.scg.ulaval.ca/en/index.php?module=pagemaster&PAGE_user_op=view_page&PAGE_id=19
8. Inmon W.H. (1996) Building the Data Warehouse. Wiley, New York.
9. Kaushik, S., Rundensteiner, E. (2001) SEE: A Spatial Exploration Environment Based on a Direct-Manipulation Paradigm, IEEE Transactions on Knowledge and Data Engineering, 13 (4), 2001, pp: 654 – 670
10. Paolino L., Laurini, R., Sebillo, M., Tortora, G., Vitiello, G. (2009) Phenomena – A Visual Environment for Querying Heterogenous Spatial Data, Journal of Visual Languages and Computing, Vol. 20(6) December 2009, pp.420-436

11. Malinowski, E., Zimányi, E. (2008). Advanced Data Warehouse Design From Conventional to Spatial and Temporal Applications. Berlin, Springer
12. Meyer, B. (1992) Beyond Icons: Towards new metaphors for visual query languages for spatial information systems, in: Proceedings of International Workshop on Interfaces to Database Systems (IDS 92), Glasgow, UK, 113-135.
13. Morris, A.J., Abdelmoty, A.I., El-Geresy, B.A., Jones C.B. (2004) A Filter Flow Visual Querying Language and Interface for Spatial Databases, GeoInformatica 8(2) 107-141.
14. Pourabbas, E. and Rafanelli, M. (2002) A Pictorial Query Language for Querying Geographic Databases using Positional and OLAP Operators. SIGMOD Record,31,2,22-27
15. Rafanelli, M. (2003). Operators for Multidimensional Aggregate Data. In: Multidimensional databases: problems and solutions. Hershey, PA, USA: IGI Publishing.
16. Ravat, F., Teste, O., Tournier, R. and Zurfluh, G. (2007) Graphical Querying of Multidimensional Databases. In Proceedings of the Advances in Databases and Information Systems, September 29-October 3, 2007, Bulgaria. Berlin, Springer, 298-313.
17. Rivest, S., Bédard, Y., Proulx, M., Nadeaum, M., Hubert, F. and Pastor, J. (2005) Solap: Merging Business Intelligence With Geospatial Technology For Interactive Spatio-Temporal Exploration And Analysis Of Data. Journal Of International Society For Photogrammetry And Remote Sensing, 60, 1, 17-33.
18. Trujillo, J., Glorio, O. (2009) Designing Data Warehouses for Geographic OLAP Querying by Using MDA. ICCSA (1): 505-519
19. Trujillo, J., Palomar, M., Gomez, J., Song, Il-Y. (2001) Designing Data Warehouses With OO Conceptual Models. Computer, 34, 12, 6

The Impact of Data Quality in the Context of Pedestrian Movement Analysis

Adriano Moreira[1], Maribel Yasmina Santos[1], Monica Wachowicz[2], Daniel Orellana[2]

[1] Algoritmi Research Centre, University of Minho, Campus de Azurém, 4800-058 Guimarães, Portugal
{adriano, maribel}@dsi.uminho.pt
[2] Centre for Geo-Information, Wageningen University and Research, The Netherlands
{monica.wachowicz, daniel.orellana}@wur.nl

Abstract. Positioning data sets gathered from GPS recordings of moving people or vehicles and usage logs of telecommunications networks are being increasingly used as a proxy to capture the mobility of people in a variety of places. The purpose of use of these data sets is wide-ranging and requires the development of techniques for collaborative map construction, the analysis and modelling of human behaviour, and the provision of context-aware services and applications. However, the quality of these data sets is affected by several factors depending on the technology used to collect the position and on the particular scenario where it is collected. This paper aims at assessing the quality and suitability of GPS recordings used in analysing pedestrian movement in two different recreational applications. Therefore, we look at two positioning data sets collected by two distinct groups of pedestrians, and analyse their collective movement patterns in the applications of a mobile outdoor gaming and as well as a park recreational usage. Among other findings, we show that the different reading rates of the pedestrians' position lead to different levels of inaccuracy in the variables derived from it (e.g. velocity and bearing). This was significant in the case of bearing values that were calculated from GPS readings which, in turn, has shown a strong impact on the size of clusters of movement patterns.

M. Painho et al., (eds.), *Geospatial Thinking*, Lecture Notes in Geoinformation and Cartography,
DOI 10.1007/978-3-642-12326-9_4,

1 Introduction

Recreational places are extremely complex social systems that require the observation of the movement patterns and the development of behavioural models. In relation to the identification of movement patterns and behaviour, Giannotti and Pedreschi (2008) identify two key problems that usually arise, independently of the context where movement occurs. They are: i) how to collect mobility data, having in mind that this data are associated with complex, often chaotic, social or natural systems integrating large populations of moving entities, and ii) how to analyse this data in order to find useful patterns that are related to a collective movement behaviour. Several studies have been undertaken to analyse individual people trajectories or to group similar individual trajectories in paths that are relevant to some application domain (Alvares et al., 2007; Lee et al., 2007; Piciarelli and Foresti, 2006). In contrast, very few attempts can be found in evaluating the impact of pre-processing on the analysis of mobility data (Wachowicz et al., 2008a).

Mobility data can be obtained through several location technologies (based on GSM and UMTS), satellite based position technologies (GPS), and indoor positioning systems (Wi-Fi and Bluetooth). Therefore, a pre-processing of these large data sets of mobility is required before exploring collective movement patterns. As part of our research, we are interested in studying how groups of pedestrians move together in space, how they interact, and how the geographic context where they are moving affects the movement behaviour. Towards this end, we have focus on the pre-processing of two sets of GPS positioning data collected in two different applications. The first data set was collected in the city of Amsterdam (The Netherlands) during an outdoor mobile game. Data was collected during ten different days, along June of 2007, totalising 63,470 records for 419 players. Along this paper, this will be named as the mobile outdoor gaming application. The second one is named park recreational usage application, and focuses on the movement of visitors in the Dwingelderveld National Park in The Netherlands. This data set includes GPS recordings collected in seven different days from May to August 2006. More than 140,000 records are available forming the trajectories of more than 370 visitors. While pre-processing these data sets we have faced several data quality problems which will be further described in this paper. In the remaining of this paper we describe some of these problems and the impact of the data quality on spatial density-based clustering techniques.

This paper is organised as follows. In the next section, some of the ongoing work is described in the domain of human movement analysis. Section 3 describes the statistical properties of the positioning data sets and the main problems we have encountered in pre-processing them. In section 4 we present a few examples of the impact that data quality has in some of the methods used for movement analysis. Based on these results, some suggestions on how to overcome the limitations of the GPS positioning data are described in section 5. Finally, we draw some conclusions on this work and provide some insights for future research.

2 State-of-the-art on Movement Analysis

Specific studies have been undertaken to analyse individual trajectories or to group similar individual trajectories into paths that are relevant to a particular application domain (Alvares et al., 2007; Lee et al., 2007; Piciarelli and Foresti, 2006). These studies, although contributing to the understanding of how people move, are constrained to the analysis of trajectories and as a result, a general theory on movement analysis is not available yet. In (Andrienko et al., 2008), some of the general problems encountered in analysing movement data are described as being one of the following: i) lack of generalisation procedures; ii) recordings are usually associated to trajectories; and iii) the identification of a model that characterises the human movement in space is still missing.

Looking at the movement of an individual, some mathematical models have been already proposed for the analysis of movement behaviour of pedestrians. The main assumption is that the movement of pedestrians shows certain regularities and that their decisions are not completely random (Helbing, 1991; Helbing et al., 2001). These models represent an important step in the formalisation of a human movement model, which is restricted to individual pedestrian movement.

Giannotti et al. (2007) developed an extension of the sequential pattern mining paradigm to analyse trajectories of moving objects. Trajectory patterns are descriptions of frequent behaviours both in space (regions visited during the movements) and in time (the duration of the movements). This approach is based on the notion of regions of interest and the typical travel time between regions, leading to a sequence of spatial regions that are visited in a specified order.

On the other hand, Lee et al. (2007) proposed a partition and group framework that partitions a trajectory into a set of line segments followed by grouping similar line segments into clusters. One of the advantages of this approach is the possibility to discover common sub-trajectories from a trajectory database, since similar portions of the trajectories can be identified.

Alvares et al. (2007) argue that meaningful patterns can only be extracted once the geographic semantics of where the trajectories are located is considered. They proposed a reverse engineering framework for mining and modelling semantic trajectory patterns. This approach only identifies well known patterns, since background geographic information is considered through a process of association rule mining.

Another work found in the literature develops a spatial knowledge representation of the movement of mobile players by developing an ontological formalism based on concepts and a set of axioms that must be true in every possible location of a player (Wachowicz et al., 2008b). In this approach, the authors characterise the spatial patterns of the trajectories followed by players.

All of the above mentioned approaches have used GPS positioning data, or positioning data collected through mobile cellular networks. However, little systematic attention was given to the quality of the data set being used to derive patterns or models. Although limited assessment and validation of the results were provided in some cases, the impact of the underlying data quality is not reflected on the methods that were used to produce the final results. One exception is the work described in (Dias et al., 2008), where an analysis of the errors in positioning recordings obtained by two GPS receivers is carried out by computing the distribution of distances between the measured positions and their projections into a given track. In this case, where the position recordings are aggregated along discrete segments of the track, the GPS errors have not shown a significant impact on the pedestrian movement analysis. However, this is not always the case, as shown in section 4.

3 Characteristics of positioning data sets

In this section we describe the main statistical proprieties of GPS positioning by using as examples the data sets gathered for two applications. They are: mobile outdoor gaming and the park recreational usage.

3.1 Sampling rate and accuracy of readings

A typical GPS positioning data set is a collection of positioning readings obtained through a GPS receiver. Each of those recordings includes a position in the geographic space, expressed as a pair of coordinates (usually latitude and longitude in the WGS84 datum), and a timestamp representing the time instant when the position signal has been acquired by the GPS receiver.

While the timestamp can be assumed to be very precise due to the characteristics of the GPS system, the same cannot be assumed for the spatial position. The accuracy of each position recording is dependent on several factors, including the number of satellites within line-of-sight of the receiver, the weather conditions, and the type of GPS receiver. In general, we can assume that a GPS recording is always affected by a positioning error, with non stationary statistics (both in time and in space).

In addition, the ability of a sequence of GPS recordings to capture the movement of an object also depends on the frequency or sampling rate. Most commercial GPS receivers support three modes for data collection: constant time sampling rate, where each record is stored every *n* seconds, constant distance sampling rate, where each record is stored each *n* meters, and "automatic" mode, where a particular algorithm determines when it is the right moment to store a record in order to turn the sequence of records into a good representation of the trajectory of a receiver.

Next we present the characteristics of two GPS data sets, first in terms of the sampling rate, and then by discussing the type of the positioning errors encountered within each data set. These data sets were obtained through the use of several GPS receivers simultaneously. Each of these receivers was carried by a pedestrian for some amount of time, and eventually carried by a different pedestrian later on. Therefore, our data sets are organised as a collection of timely ordered sets of records, each one of these sets representing the positions visited by a pedestrian during one session. For the analysis of the sampling rate, we took each of these data sets and computed the time difference between consecutive records. The histogram of these time differences is shown in Fig. 1.

In the mobile outdoor gaming application, the majority of the recordings were taken at an almost constant rate of every 10 seconds. Actually, 91% of the cases lie between 10 and 12 seconds. In the park recreational usage application, the situation is completely different. Although the most frequent recording period is between 3 to 4 seconds (29% of the records), there is a much broader interval of recording periods with high probability. This suggests that the "automatic" mode was used in this case, while a constant sampling rate mode was used in the former scenario. However, even having a constant sampling rate, 9% of the recordings have been taken at different sampling rates.

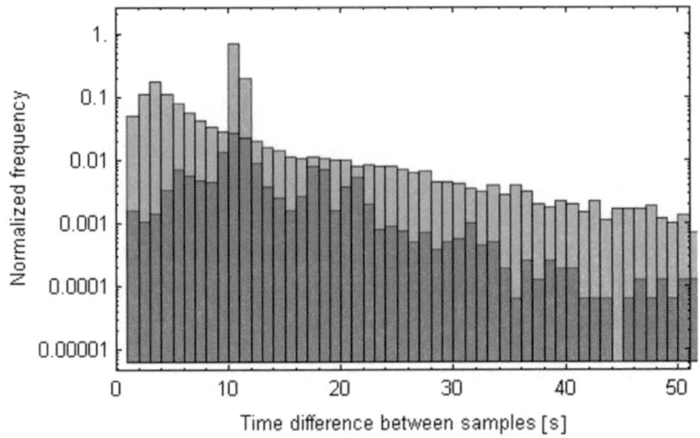

Fig. 1. Histogram of the time difference between consecutive records for a subset of the positioning data sets for the mobile outdoor gaming application with 16,104 records (in dark grey), and for the park recreational usage application with 15,903 records (in light grey). Note the log scale on the Y axis.

Moreover, although some extreme sampling periods have been observed in both data sets (up to 459 seconds for the first scenario and up to 3,399 seconds in the second scenario), suggesting the existence of outliers, the range of sampling periods extends to much larger values in the case of the park recreational usage application. This can be explained by the differences in receiving the GPS signals where the movement has taken place. For example, whenever the distance between two consecutive points is larger than the one a pedestrian can travel, the computed velocity between these points is always higher than a certain value (around 5 to 6 m/s for pedestrians). As a result, we can assume that one, or both, of the points are affected by a large positioning error.

We have carried out the pre-processing of these two data sets, searching for points where positioning errors have occurred, and the results are depicted in Fig. 2. The grey triangle represents the area of invalid readings, that is, where the distance between readings is larger than the distance a pedestrian can travel in the time interval between two readings. The limit was set to 5 meters per second which, for a pedestrian, can be considered as a high speed. These graphs show that 1.7% and 1.1% of the readings (Fig.1a and Fig. 1b respectively), are invalid readings that have occurred within a wide range of sampling periods.

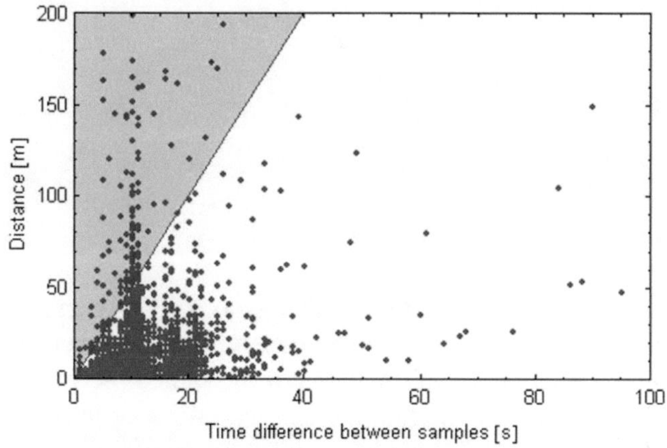

a. A subset of the mobile outdoor gaming application

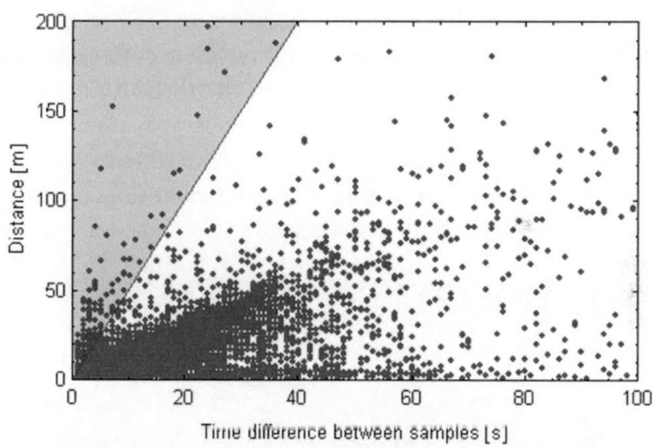

b. A subset of the park recreational usage application

Fig. 2. Distance between readings vs. sampling period

3.2 Representing movement by a set of vectors

In this paper, a multidimensional vector representation was used for modelling the collective movement of pedestrians. Each vector represents a point in space and time, the speed of movement, and the direction of movement. One can derive such a representation of movement from GPS data by considering consecutive pairs of GPS readings to compute the speed and the direction of movement (bearing). This representation is depicted in Fig. 3.

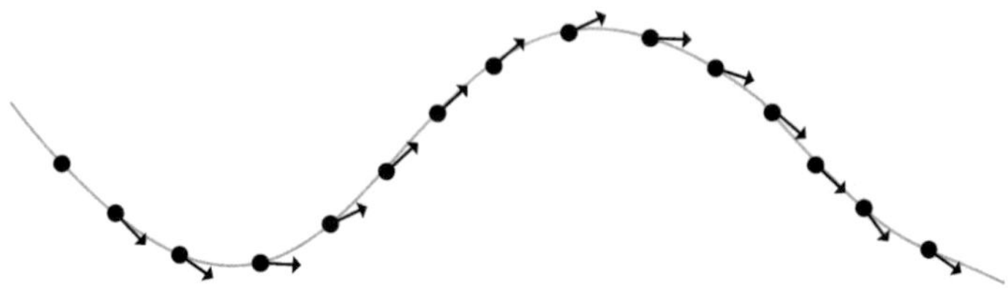

Fig. 3. Representing movement by a set of vectors

By converting sequences of GPS readings to a set of vectors, eventually from multiple objects moving simultaneously, the movement is no longer seen as a timely ordered set of positions of the same object but, instead, as a set of independent and instantaneous observations of the movement behaviour. This is particularly interesting for the analysis of aggregates, such as flocks of moving objects.

During the process of converting sequences of GPS readings into vectors, a few other aspects of the data quality have emerged. One of these problems is related with the computation of the direction of movement (or bearing values). While computing these values we have noticed the appearance of a quantisation noise, especially in the case of slow moving objects. This effect is illustrated in Fig. 4, where each dot represents the pair of speed and bearing values of each vector.

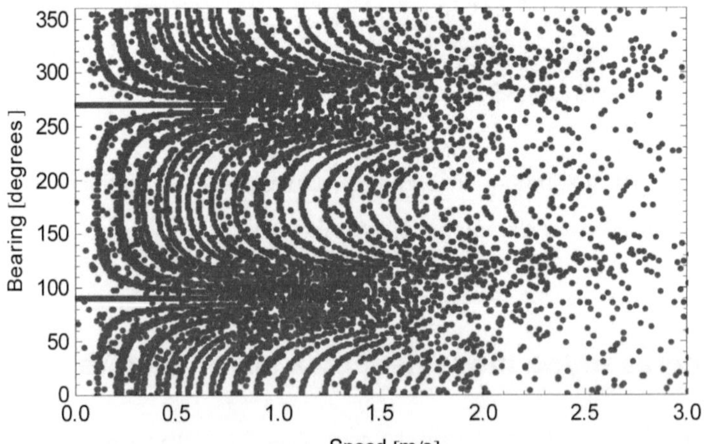

a. A subset of the GPS recordings of the mobile outdoor gaming application

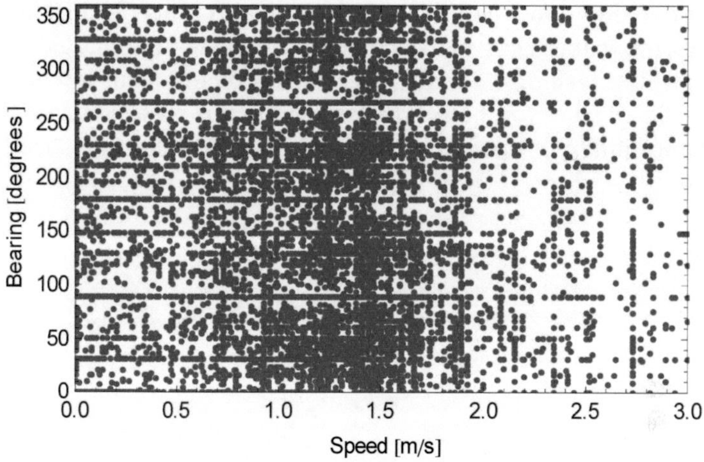

b. A subset of the GPS recordings of the park recreational usage application

Fig. 4. The quantisation effect on the computed bearing values

In Fig. 4a it is clear that, for low speed values, the bearing values are arranged on a regular pattern. In contrast, a different pattern shape is shown in Fig. 4b. One possible explanation for the occurrence of these patterns is due to the limited precision of the original position readings. Actually, this is clearly observed by zooming into a region with a high concentration of position records as illustrated in Fig. 5.

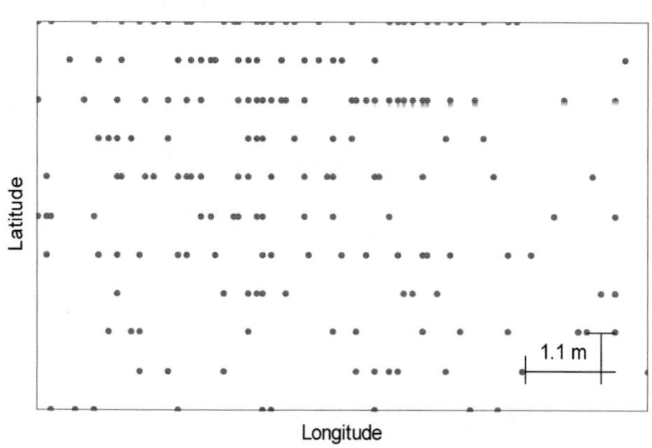

Fig. 5. The quantisation on the Latitude values from the recordings of the mobile outdoor gaming application

When the two consecutive points that are used to compute the bearing value are too close (in spatial terms), the moving object is actually not moving or moving very slowly, and as a result, the possible values for the bearing are reduced to 8 cases (0, 45, 90, …, 315). This quantisation effect in the bearing values (Fig. 6) is actually the result of the limited precision on the position readings. For this reason, for vectors with low speed, the most probable values for the bearing are within this set of 8 cases. Note the strong concentration of points at 90 and 270 degrees for low speed values in Fig. 4a.

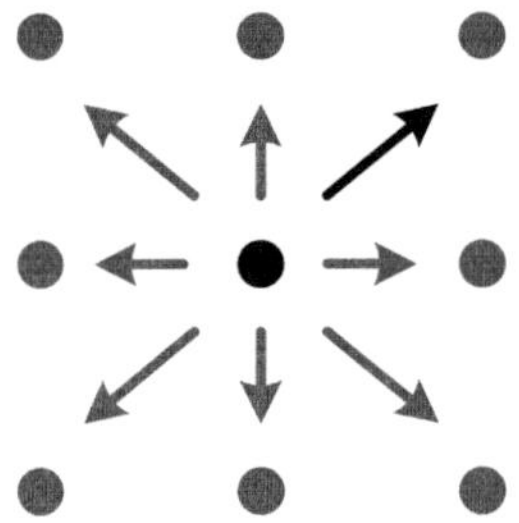

Fig. 6. Quantisation noise resulting from the limited precision of the position readings

Therefore, whenever the distance between the two readings used to compute the bearing values is shorter than 1.5 meters (i.e. less than the expected GPS error), the computed bearing values have very low confidence levels. For example, for the mobile outdoor application, 13% of the vectors were computed from consecutive points that are within 1.5 meters or less from each other (black points in Fig. 7). Consequently, the corresponding bearing values should be considered as having a low confidence level.

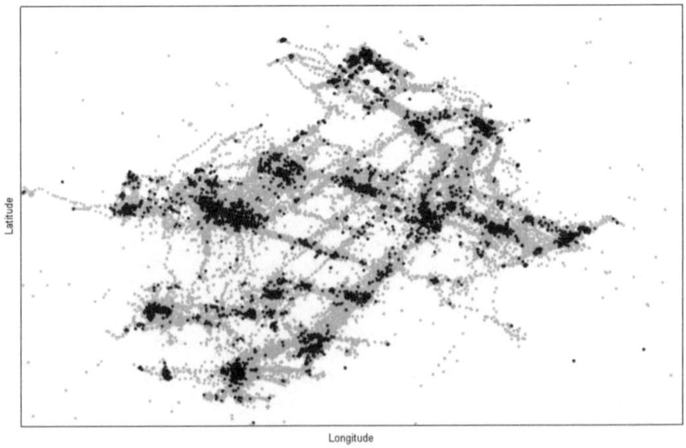

Fig. 7. In the mobile outdoor gaming application, the vectors with low confidence bearing values are depicted in black (maximum distance between consecutive points = 1.5 meters)

Another aspect that must be taken into account while converting sequences of GPS readings into vectors is the sampling rate. If the sampling rate is too low, the resulting vectors may exhibit characteristics that do not represent the characteristics of the real movement, as shown in Fig.8.

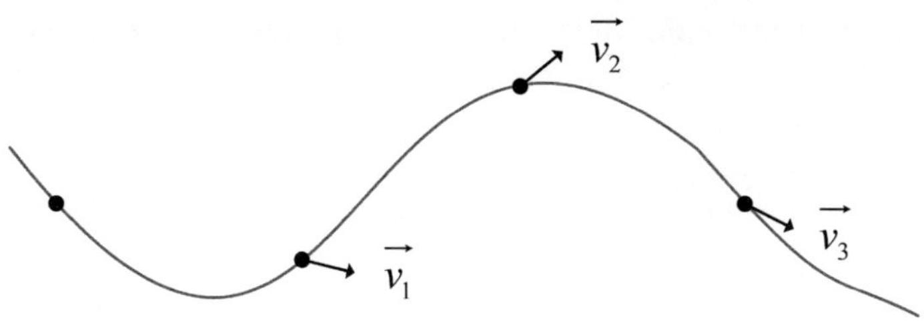

Fig. 8. An example of a low sampling rate

In this example, vectors v_1 and v_2 are characterised by bearing values that are too different from the real direction of movement and, therefore, they do not appropriately represent the movement at those specific positions.

4. Detection of movement patterns

In this section, the analysis of clustering patterns is presented in order to show the impact of the bearing quantisation noise in the final results. We limited the clustering process to the use of two attributes, position and bearing, in order to clearly demonstrate that the obtained results were not influenced by other parameters. Moreover, the clustering process was firstly performed with the original data set, and then with its respective pre-processed data set in which a small amount of Gaussian noise was added to the X and Y values before the bearing was calculated. We refer to this pre-processed data set as the "shaken" one. Consequently, the original data set is referred as the "unshaken" one.

4.1 Uncovering patterns through spatial density-based clustering

Clustering is the process of grouping a set of objects into clusters in such as way that objects within a cluster have high similarity with each other, but are as dissimilar as possible to objects in other clusters (Zaït and Messatfa, 1997; Grabmeier, 2002). In our research, the Shared Nearest Neighbour (SNN) algorithm was used (Ertoz et al., 2002). This algorithm is adequate for this particular task due to its capabilities of identifing clusters with convex and non-convex shapes, having different sizes and densities, as well as due to its ability to deal with noise points. The similarity between points is obtained by looking at the number of nearest neighbours that two points share. Using this similarity measure, density is defined as the sum of the similarities of the nearest neighbours of a point. Points with high density become the core points, while points with low density represent noise points. All groups of points that are strongly similar to core points will be included in the clusters.

The SNN algorithm uses 3 inputs parameters: *k*, *EPS* and *MinPts*. The number of neighbours that need to be analysed in each step of the clustering process is defined by *k*; *EPS* defines the value for the threshold density and *MinPts* defines the threshold that allows the classification of a point as a core point. After defining the input parameters, the SNN algorithm first finds the *k* nearest neighbours of each point of the data set. Then the similarity between pairs of points is calculated in terms of how many nearest neighbours the two points share. Using this similarity measure, the density of each point is calculated as being the number of neighbours of the current point with which the number of shared neighbours is equal or greater than *EPS* (density threshold). Next, the points are classified as being core points if their density is equal or greater than *MinPts* (core point threshold). At this stage, the algorithm has all the information needed to build the clusters. The clusters start to be built around the core points. Points that do not integrate any cluster are classified as noise points. Since no input parameter is used to determine the number of clusters, the number of clusters emerges directly from the data and not from a number previously defined based on the domain knowledge of a user.

For the identification of the *k* nearest neighbours of a point, a distance function must be defined. In the original algorithm, this function is based on the Euclidean distance among points. In our case, the distance function was redefined in order to accommodate the several dimensions of a vector: position, speed, bearing, and time. Besides, the definition of weights for each one of these variables was also implemented in the SNN algorithm mainly because different weights might lead to different types of clusters, which in turn, represent different types of movement.

Given the points $p_1 (x_1, y_1, s_1, b_1, t_1)$ and $p2(x_2, y_2, s_2, b_2, t_2)$, the distance between them is calculated using Equation 1.

$$DistFunction(p_1, p_2) = w_1 * (\sqrt{(x_1 - y_1)^2 + (x_2 - y_2)^2} / mDist)$$
$$+ w_2 * (|s_1 - s_2| / mSpeed) + w_3 * (\Phi(|b_1 - b_2|) / mBearing)$$
$$+ w_4 * (|t_1 - t_2| / mSeconds) \quad (1)$$

with

$$\Phi(\alpha) = \begin{cases} \alpha & ,\alpha \leq \pi \\ 2\pi - \alpha & ,\alpha > \pi \end{cases} \quad (2)$$

where x_i and y_i represent the position, s_i represents the speed, b_i represents the direction of movement (bearing), and t_i is the timestamp associated with each point.

In this function, w_1, w_2, w_3 and w_4 are the weights assigned to the position, speed, angle and time, respectively. To normalise the obtained values, *mDist*, *mSpeed* and *mSeconds* are computed as the maximum difference between the values of any two points in the data set according to the Euclidean distance, speed and time, respectively.

4.2 The park recreational usage application

This section demonstrates the impact of the bearing quantisation noise on finding the clusters in the data set gathered for the park recreational usage application. The clustering process was carried out considering two attributes: position and bearing. Referring to Equation 1, the weights were w_1=95%, w_2=0%, w_3=5% and w_4=0. The SNN input parameters were k=10, *Eps*=3 and *MinPts*=5. These values were chosen in order to cluster vectors that, being close to each other in terms of position, are also pointing into similar directions. Fig. 9a, shows the clusters obtained from the original data, meanwhile the clusters in Fig. 9b are for the pre-processed data set (i.e. the shaken data set).

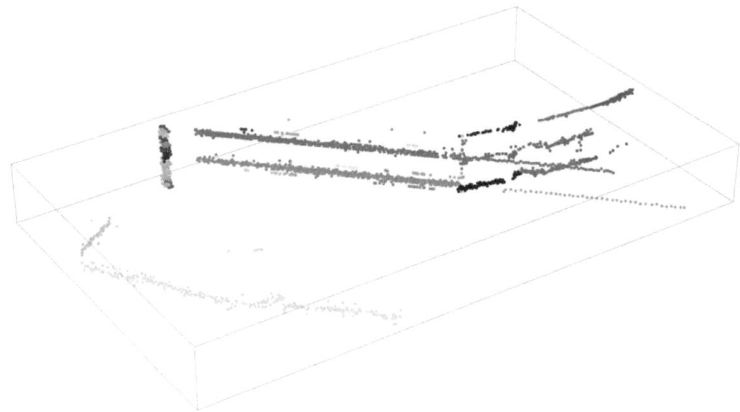

a. Before shaking – 34 clusters

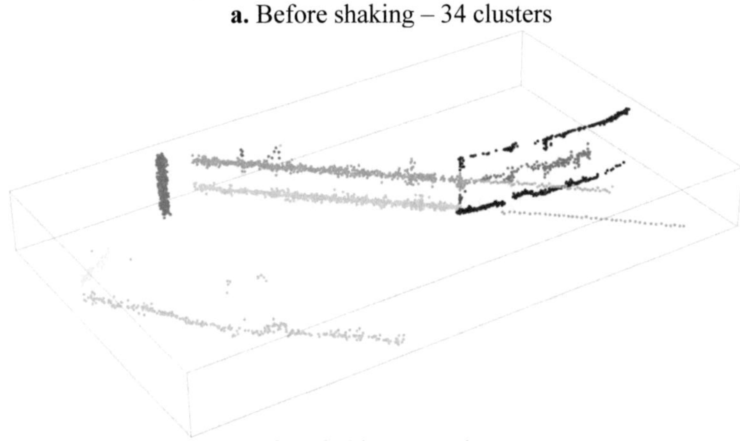

b. After shaking – 12 clusters

Fig. 9. Clusters before (a) and after shaking (b): Colours represent different clusters.

In Fig. 9, the Z coordinate is used for the bearing, separating vectors pointing in different directions. These examples show that, for the same input parameters, two different sets of clusters have been obtained. Note that for the original data set (unshaken), a large number of small clusters were identified by the SNN algorithm, mainly located around the largest clusters depicted in red and orange in Fig. 9a. After eliminating the bias produced by the bearing quantisation noise, most of the vectors previously belonging to small clusters have been included to the larger clusters. The same results were observed using other combinations of the weights.

4.3 The mobile outdoor gaming application

In the mobile outdoor gaming application, we have also analysed the bearing quantisation effect with a subset of 3,875 points. The clusters identification process also used the weights w_1=95%, w_2=0%, w_3=5% and w_4=0, and the SNN input parameters of *k*=10, *Eps*=3 and *MinPts*=5. The clusters obtained from the original data are shown in Fig. 10a, meanwhile the clusters obtained after the shaking process are presented in Fig. 10b.

It is interesting to point out the difference between this data set and the one from the park recreational usage application. In this case, we have data associated with movement between areas of interest that are part of a game. The players visited those areas and stayed there for a while. It also seems that between the areas of interest they stopped many times. This is why we have a data set with a huge concentration of different bearing values. The result of the clustering process reflects this reality. In both clustering processes, the obtained clusters integrate vectors with different bearing values, generating what seems to represent zones of suspension of movement and zones of transition between them. After the clustering process, once again we have fewer clusters in the shaken data subset.

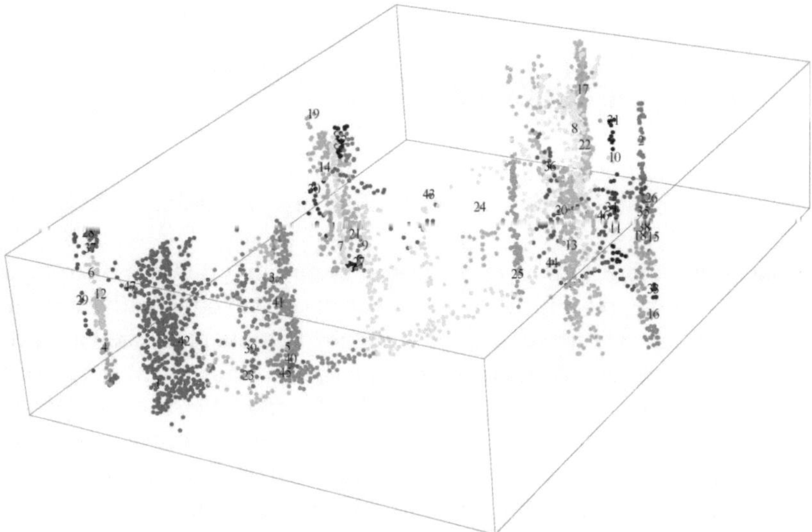

a. Before shaking – 47 clusters

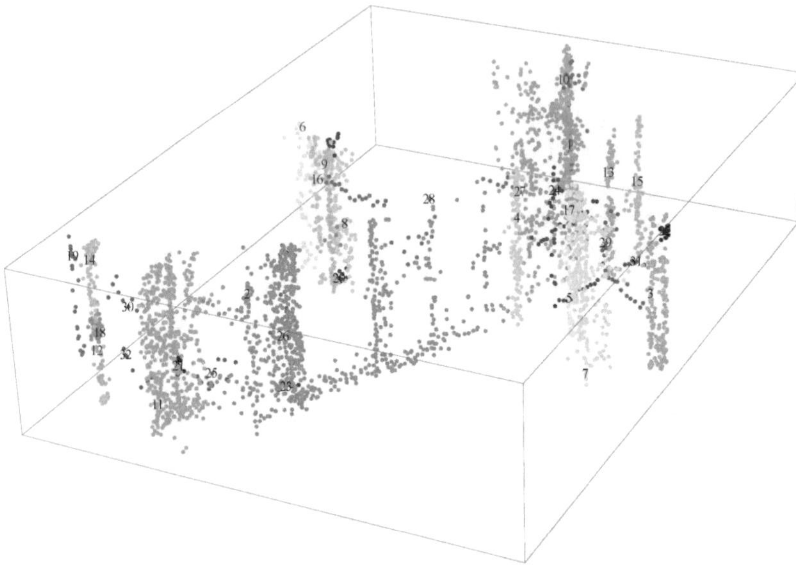

b. After shaking – 32 clusters

Fig. 10. Clusters before (a) and after shaking (b)

5. Discussion and Conclusions

The outcomes of this research work have allowed us to identify a set of optimal characteristics for the data collection, pre-processing and clustering tasks.

Concerning the data collection task, the sampling rate has a strong impact on the calculation of the bearings based on two consecutive readings. As the sampling rate decreases, the tendency is to increase the bearing inaccuracy. The vector representation of pedestrian movement requires that the sampling rate to be as high as possible (for example, one recording per second) and be taken within a regular period (for example, every day). This is important to guarantee the computation of accurate vectors.

Concerning pre-processing, the inherent bearing quantisation noise has a strong impact on the computation of clusters. There is always a need for a shaken procedure in order to avoid the overestimation of clusters. Moreover, spatial outliers have also a strong impact on computed speed-values. Therefore, the definition of a threshold could help to remove unreal high-speed vectors. After this pre-processing process, a re-computation of movement parameters is recommended to ensure the coherence of the data.

Finally, for any clustering technique based on bearing values, it is important to be aware that any bearing computed from two very near consecutive points (vectors) will present a low confidence level. An adequate threshold could be defined by using a "circular error probability" for a GPS recording. This circle is defined by its radius, inside of which the true horizontal coordinates of a position have a 50-percent probability of being located. This approach is frequently used as a description of the overall quality of a GPS data collection. Since the accuracy depends on diverse factors such as the geometry of satellites, the structure of the surrounding space, and the use of augmentation systems, the adequate values for this threshold should be based on the characteristics of each data collection.

Further research work will be focused on analysing movement data of vehicles in transportation management applications.

References

Alvares, L. O., Bogorny, V., Macedo, J. and Spaccapietra, S. (2007) Dynamic Modeling of Trajectory Patterns using Data Mining and Reverse Engineering, Proceedings of the 26 International Conference on Conceptual Modeling (ER'2007), Auckland, pp. 149-154.

Andrienko, N., Andrienko, G., Pelekis, N. and Spaccapietra, S. (2008) Basic Concepts on Movement Data, in Giannotti, F. and Pedreschi, D. (Eds.): Mobility, Data Mining and Privacy, Springer-Verlag, pp. 15-38, 2008.

Dias, E., Edwards, A. J. and Purves, R. S. (2008) Analysing and aggregating visitor tracks in a protected area, in A. Stein, J. Shi and W. Bijker (Eds): Quality Aspects in Spatial Data Mining, CRC Press, Taylor & Francis Group.

Ertoz, L., Steinbach, M. and Kumar, V. (2002) Finding Clusters of Different Sizes, Shapes, and Densities in Noisy, High Dimensional Data, Proceedings of the Second SIAM International Conference on Data Mining, San Francisco.

Giannotti, F., Nanni, M., Pedreschi, D. and Pinelli, F. (2007) Trajectory Pattern Mining, Proceedings of the Knowledge Discovery in Databases (KDD'07) Conference, San Jose, pp. 330-339.

Giannotti, F. and Pedreschi, D. (2008) Mobility, Data Mining and Privacy: A Vision of Convergence, in Giannotti, F. and Pedreschi, D. (Eds.): Mobility, Data Mining and Privacy, Springer-Verlag, pp. 1-11.

Grabmeier, J. (2002) Techniques of Cluster Algorithms in Data Mining, Data Mining and Knowledge Discovery, 6(4), pp. 303-360.

Helbing, D. (1991) A Mathematical Model for the Behaviour of Pedestrians, Behavioral Science, 36, pp. 298-310.

Helbing, D., Molnár, P., Farkas, I. and Bolay, K. (2001) Self-Organising Pedestrian Movement, Environment and Planning B: Planning and Design, 28, pp. 361-383.

Lee, J.-G., Han, J. and Whang, K.-Y. (2007) Trajectory Clustering: A Partition-and-Group Framework, Proceedings of SIGMOD Conference (SIGMOD'07), Beijing, pp. 593-604.

Piciarelli, C. and Foresti, G. L. (2006) On-line trajectory clustering for anomalous events detection, Pattern Recognition Letters, 27, pp. 1835-1842.

Wachowicz, M., Ligtenberg, A., Renso, C. and Gürses, S. (2008a) Characterising the Next Generation of Mobile Applications Through a Privacy-Aware Geographic Knowledge Discovery Process, in Giannotti, F. and Pedreschi, D. (Eds.): Mobility, Data Mining and Privacy, Springer-Verlag, pp. 39-72.

Wachowicz, M., Orellana, D., Renso, C., Moraga, E. and Parada, J. (2008b) The spatial knowledge representation of players movement in mobile outdoor gaming, Proceedings of the 4th International Conference on Monitoring and Management of Visitors Flows in Recreational and Protected Areas, Italy, pp. 456-460.

Zaït, M. and Messatfa, H. (1997) A comparative study of clustering methods, Future Generation Computer Systems, 13(2), pp. 149-159.

Visit Potential: A Common Vocabulary for the Analysis of Entity-Location Interactions in Mobility Applications

Christine Körner, Dirk Hecker, Michael May, Stefan Wrobel

Fraunhofer IAIS, Schloss Birlinghoven, 53754 Sankt Augustin, Germany, {firstname.lastname}@iais.fraunhofer.de

Abstract. A growing number of companies and public institutions use mobility data in their day-to-day business. One type of usage is the analysis of spatio-temporal interactions between mobile entities and geographic locations. In practice the employed measures depend on application demands and use context-specific terminology. Thus, a patchwork of measures has evolved which is not suitable for methodological research and interdisciplinary exchange of ideas. The measures lack a systematic formalization and a uniform terminology. In this paper we therefore systematically define measures for entity-location interactions which we name *visit potential*. We provide a common vocabulary that can be applied for an entire class of mobility applications. We present two real-world scenarios which apply entity-location interaction measures and demonstrate how the employed measures can be precisely defined in terms of visit potential.

1 Introduction

Every day people interact with the environment by passing or visiting geographic locations. Over the past years, technologies which trace personal movement have found their way into everyday life, as for example GPS, GSM or RFID. From the designated usage of such technologies in navigation systems or mobile communications it is only a small step to build a comprehensive collection of movement information and to apply the data in further domains. Some business companies and governmental institu-

M. Painho et al., (eds.), *Geospatial Thinking*, Lecture Notes in Geoinformation and Cartography, DOI 10.1007/978-3-642-12326-9_5, © Springer-Verlag Berlin Heidelberg 2010

tions have already recognized the value of wide-ranging mobility data and commissioned nationwide mobility surveys (ag.ma 2010; SPR+ 2010; Marchal et al. 2008). In consequence, it is indispensable to provide adequate analysis tools for the interpretation of such data.

One possible analysis is the study of spatio-temporal interactions between a set of mobile entities and a set of locations. By spatio-temporal interactions we simply mean the passage of an area or the visit of a location. Measures for this kind of interactions are, for example, applied in outdoor advertising to determine the price of poster locations. Outdoor advertisers are thereby not only interested in the performance of a single location but also of a set of locations, i.e. the effect of a poster campaign. One measure of interest hereby is the percentage of population which sees at least one poster of a campaign and is thus "reached" by a given advertisement. However, interaction measures can be applied in a broad class of applications. In comparison to classic frequency measures which simply state the number of visiting or passing entities at some location, measures that use mobility data have the advantage that they can exploit movement histories of mobile entities. It is thus possible to analyze visits with respect to the origin of an entity. For example, with different parameterizations we can use interaction measures to determine which part of traffic in a city is caused by locals and which part by commuters. Further, the measures can identify regular visitors and quantify their number of repeated visits. Such information is interesting for the owners of restaurants or of retail chains, which can analyze the portion of customers that return on a regular basis. Due to the availability of movement histories, such an analysis is not restricted to a single location but can be applied to a set of subsidiaries as well. Finally, interaction measures can model the dependency within a group of locations. Imagine a drug store chain which plans to open a new subsidiary. Of course, the chain prefers highly frequented locations. However, it is not interested to provide alternative shopping facilities for existent customers. Instead it aims at reaching people which rarely pass any of their present subsidiaries. Such a measure transports a similar meaning as the above mentioned measure in outdoor advertising. The few examples already show the usefulness of mobility-based interaction measures. However, they also show the variety of domains in which they are applied and let assume the patchwork of measures which has emerged.

In this paper we provide a systematic definition and common vocabulary of measures that express entity-location interactions. We formulate the common underlying concepts and develop a consistent set of definitions which we name *visit potential*. We have selected two real-world applications which demonstrate the present use of entity-location interaction measures and which show the implementation of measures based on our general vocabulary.

Our paper is organized as follows. We begin with the introduction of two application domains that employ entity-location interaction measures. We provide a common vocabulary and precise definition of the measures in Section 3 and show their application to the presented domains in Section 4. Section 5 discusses related work in the area of trajectory data mining. We conclude our paper with a summary and an outlook on challenging research questions.

2 Application Domains

In this section we present two real-world domains which use entity-location interaction measures. We have selected the domains to show the variety of employed measures and applications.

2.1 Outdoor Advertising

The pricing of poster sites is a critical business task in outdoor advertising and must be justified by objective performance measures. Clearly, the more people pass a poster location, the higher the price a vendor may ask. In order to rate campaigns and to optimize their placement, predominately three indicators are used: gross rating points (GRP), opportunities to see (OTS) and reach. Gross rating points specify the total number of poster contacts that 100 persons of a respective population produce with a given campaign in a given period of time. Opportunities to see measure the intensity by which a person encounters a campaign. They state the average number of poster contacts once a person beholds a poster of the campaign. Reach finally states the percentage of a given population which sees at least one poster of the campaign (Sissors and Baron 2002). Reach plays an important role in outdoor advertising because it measures the spread of information within a population. Fig. 1 displays two examples of poster campaigns in Cologne, Germany, and Table 1 contains the respective performance measures. Although both campaigns contain the same number of posters, the values for OTS and reach differ substantially. This difference is due to the different distribution of posters in space. As human movement shows regularities in temporal and geographic space, the clustered campaign (Fig. 1 right) reaches less people, however, with a higher intensity than the dispersed campaign (Fig. 1 left).

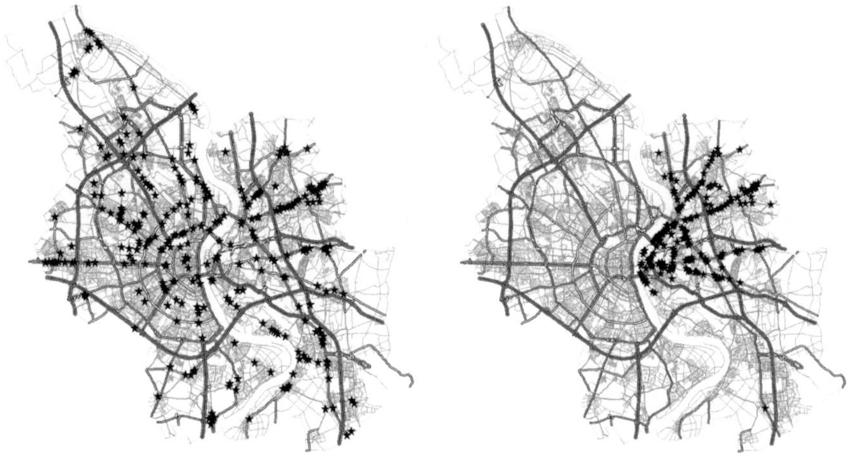

Fig. 1. Poster campaigns in Cologne; left: dispersed campaign; right: clustered campaign; source: press conference “ma 2007 Plakat”, Jan. 16th 2008 (FAW 2010)

Table 1. Performance measures of dispersed and clustered poster campaign in Cologne; source: press conference “ma 2007 Plakat”, Jan. 16th 2008 (FAW 2010)

campaign	# posters	GRP	OTS	reach
dispersed	321	953	10.2	92%
clustered	321	1115	20.6	54%

For Switzerland the Swiss Poster Research Plus AG (SPR+ 2010) commissioned a GPS-based mobility study with more than 10,000 participants in order to provide reliable performance measurements. The measurements took place in 12 Swiss conurbations for a period between 7 and 10 days and form a representative sample within the measured areas. Fig. 2 left shows the total GPS traces that were recorded. In addition, empirical data of about 52,000 poster sites are available. The data contain geographic coordinates and a visibility area for each panel from which the poster can be seen (Fig. 2 right). By geographic intersection of mobility data and visibility areas, all poster passages can be calculated. As the passage of a poster does not imply that a person actually looks at the poster, passages are qualified by a weight which accounts e.g. for the angle and speed of passage. A thus qualified passage constitutes a poster contact and serves as basis to evaluate the GRP, OTS and reach of a campaign. Further details on the project can be found in Pasquier et al. (2008).

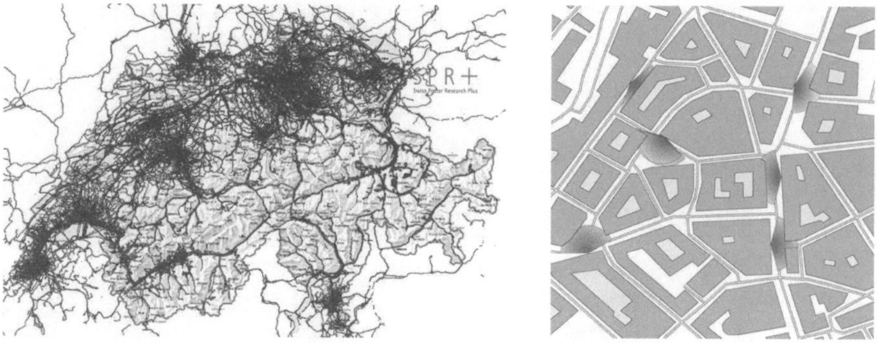

Fig. 2. Left: GPS traces in Swiss mobility study; right: visibility areas of poster panels after intersection with building layer; source: SPR+ (2010)

2.2 Evaluation of Bird Recordings

BirdTrack (2010) is a joint project of the British Trust for Ornithology (BTO), the Royal Society for the Protection of Birds (RSPB) and BirdWatch Ireland to record migration movements and the distribution of birds in Great Britain and Ireland. The project relies on volunteers to watch and record birds because it aims at a nationwide observation.

In order to participate in BirdTrack, a volunteer has to register one or more sites that he / she visits regularly to watch birds. The sites are internally mapped to a 10 by 10 km grid on which all evaluations take place. Records about bird sightings can be entered over an online form and are analyzed on a daily basis. As the origin and number of reports cannot be controlled by BirdTrack due to voluntary participation, BirdTrack analyzes not only the reported species but also the distribution and frequency of submissions. It provides maps about the covered areas and time series diagrams about the number of records. Examples of both statistics are shown in Fig. 3. The left figure is a map which shows all sites that have been visited within the year 2009. This map is called a coverage map. Fig. 3 right shows the number of submitted records by week and per day for the region of Wales for the years 2007, 2008 and 2009. These graphs are called coverage graphs and are available for different regions as well as for the whole survey area.

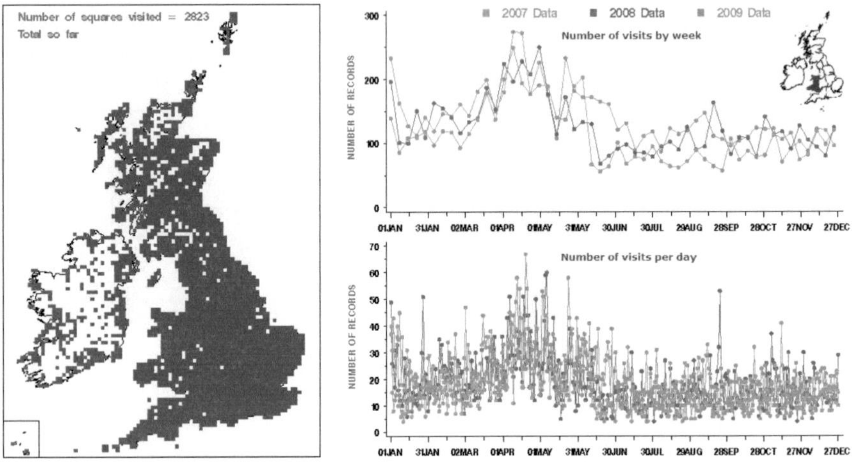

Fig. 3. Left: BirdTrack coverage map for Great Britain and Ireland in 2009; right: BirdTrack coverage graphs for Wales by week and per day for 2007, 2008 and 2009; source: BirdTrack (2010)

3 Formalization of Visit Potential

We now systematically define entity-location interaction measures. Subsection 3.1 develops the common underlying concepts of the measures and introduces the location and entity perspective of analysis. In Subsection 3.2 we give a systematic, application independent definition of the measures, and Subsection 3.3 extends the measures to the concept of visit classes.

3.1 The Visit Concept

Visit potential measures the degree of spatial interaction between geographic locations and mobile entities. Let $\mathcal{L}$ denote a given universe of geographic locations. We then call the finite subset $L = \{l_1, l_2, \ldots, l_m\} \subseteq \mathcal{L}$ of geographic locations of interest the location set. Similarly, let $\mathcal{E}$ denote some population of mobile entities and let $E \subseteq \mathcal{E}$ with $E = \{e_1, e_2, \ldots, e_n\}$ denote the subset of entities under consideration. We will denote the cardinality of both sets by $|L|$ and $|E|$, respectively. Entity movement can be expressed as trajectory function $tr_e\colon \mathbb{R} \rightarrow \mathbb{R}^d$, which maps each moment in time to the geographic location of an entity in d-dimensional geographic space. We can now define a visit of an entity e to a location l.

Definition (Visit) Given a location l, a mobile entity e and the entity's continuous trajectory function tr_e: $\mathbb{R} \rightarrow \mathbb{R}^d$, a visit is a tuple (l, e, t_l, t_u) for which holds that

1. the spatial intersection of l and $tr_e(t)$ is non-empty for all $t \in [t_l, t_u]$, i.e. $tr_e(t) \cap l \neq \emptyset \;\; \forall t \in [t_l, t_u]$,
2. the time span $[t_l, t_u]$ is maximal, i.e. there exist no time points $t_l^* < t_l$ or $t_u^* > t_u$ so that $tr_e(t) \cap l \neq \emptyset \;\; \forall t \in [t_l^*, t_u^*]$,
3. the duration of passage is greater or equal to some given minimum time span ε, i.e. $t_u - t_l \geq \varepsilon$.

In the above definition t_l and t_u define the lower and upper temporal bound of the visit. The maximality criterion ensures that an uninterrupted stay at some location cannot be split into an infinite number of visits of shorter duration. Finally, depending upon application demands, a visit may be required to last a minimum period of time. In this way it is possible to consider application requirements and to distinguish, for example, the actual visit to a theater from a simple passage or the picking up of tickets. Note that our definition relies on a functional description of movement. It is independent of the actual representation and quality of trajectory data.

The visit of an entity to a location forms the fundamental event of interest when analyzing visit potential. Considered over time, the number of visits can be modeled as a counting process.

Definition (Counting process) Given a random variable $N(t)$ that denotes the number of occurrences of a specified event within time span $(0, t]$, the set $\{N(t)\}$ of random variables with $t \in [0, \infty)$ forms a counting process. Without loss of generality we define $N(0) = 0$.

In our case $N(t)$ denotes the number of visits that an entity $e \in E$ realizes with a single location $l \in L$ within t days. In order to distinguish the counting processes of different entity-location pairs, we extend the notation of $N(t)$ to $N(t,l,e)$. Clearly, $N(t,l,e)$ increases monotonically over time, i.e. if $t_1 < t_2$ then $N(t_1,l,e) \leq N(t_2,l,e)$.

Given the counting processes of all entity-location pairs, we can determine the number of visits for sets of locations or sets of entities. We will use the variable

$$Vs(t,L,e) = \sum_{l \in L} N(t,l,e) \qquad \forall e \in E$$

to refer to the number of visits that an entity $e \in E$ produces with a location set L until time moment t. Similarly, we use the variable

$$Vt(t,l,E) = \sum_{e \in E} N(t,l,e) \qquad \forall l \in L$$

to denote the number of visits that a location $l \in L$ receives from E until time moment t. We will also refer to this number as the visitors of L. Note that an entity may be repeatedly counted as visitor.

Given the number of visits for each entity $e \in E$, we obtain the visit distribution $D_{Vs}(t,L,E)$. The visit distribution states for all $v \in \mathbb{N}_0$ the visit frequency, i.e. the number of entities with exactly v visits:

$$h_{Vs}^{v}(t,L,E) = \left|\left\{e \in E \mid Vs(t,L,e) = v\right\}\right|.$$

Similarly, we can determine the visitor distribution $D_{Vt}(t,L,E)$ by observing all visitor frequencies

$$h_{Vt}^{v}(t,L,E) = \left|\left\{l \in L \mid Vt(t,l,E) = v\right\}\right|.$$

Basically, the definitions of visits and visitors rely on the same concept, but implement it from different perspectives. Visits employ the entity point of view while visitors are location-oriented. In consequence, visit potential measures exist for both points of view as we will see in the next section.

3.2 Visit Potential Measures

From the visit and visitor distribution as defined in the previous section we can systematically derive measures for the spatio-temporal interaction of a set of locations with a set of mobile entities. The most basic measure is *gross visits*, which denotes the total number of visits within some time span t:

$$\begin{aligned} grVs(t,L,E) &= \sum_{e \in E} Vs(t,L,e) = \sum_{v \geq 0} v \cdot h_{Vs}^{v}(t,L,E) \\ &= \sum_{l \in L} Vt(t,l,E) = \sum_{v \geq 0} v \cdot h_{Vt}^{v}(t,L,E). \end{aligned}$$

Gross visits always reflect a contact volume and depend on the number of locations and entities involved. In order to compare the number of interactions of entity or location sets with different cardinality, we define the *average visits* of an entity set:

$$avgVs(t,L,E) = \frac{\sum_{v\geq 0} v \cdot h_{Vs}^{v}(t,L,E)}{|E|} = \frac{grVs(t,L,E)}{|E|}.$$

Similarly, the *average* number of *visitors* per location can be calculated:

$$avgVt(t,L,E) = \frac{\sum_{v\geq 0} v \cdot h_{Vt}^{v}(t,L,E)}{|L|} = \frac{grVs(t,L,E)}{|L|}.$$

So far, we have considered visits only with regard to the complete entity and location set involved. However, one may also ask how many entities produce visits with the location set or, respectively, how many locations are visited at all. This is quantified by the following two measures *entity coverage*

$$eCov(t,L,E) = \frac{\left|\left\{e \in E \,\middle|\, Vs(t,L,e) \geq 1\right\}\right|}{|E|} = \frac{\sum_{v\geq 1} h_{Vs}^{v}(t,L,E)}{|E|}$$

and *location coverage*

$$lCov(t,L,E) = \frac{\left|\left\{l \in L \,\middle|\, Vt(t,l,E) \geq 1\right\}\right|}{|L|} = \frac{\sum_{v\geq 1} h_{Vt}^{v}(t,L,E)}{|L|}.$$

Coverage states the proportion of entities (locations) with at least one visit and thus quantifies the influence of one set of the respective other set.

3.3 Visit Classes

The definitions given in the previous section allow only for a very general description of the visit (visitor) distribution. Easily, two different visit (visitor) distributions may result in the same number of average visits (visitors) or the same coverage. In order to disclose more characteristics of the distribution, the measures can be defined with respect to different *visit classes*.

Visit classes restrict the entity (location) set by introducing a lower bound for the number of visits an entity (location) must show in order to be included in some measure. For example, the number of average visits according to visit class $vc = 2$ results from averaging visits of all entities with two or more visits. In the following we extend the definitions of gross visits, average visits and visitors, entity and location coverage with respect to visit classes.

In the case of gross visits the extension to visit classes has two interpretations. On the one side, the visit volume of all entities with at least $vc \geq 0$ visits can be considered:

$$grVs(t,L,E,vc) = \sum_{v \geq vc} v \cdot h_{Vs}^{v}(t,L,E).$$

On the other side, the visit volume of all locations with at least $vc \geq 0$ visitors can be determined:

$$grVt(t,L,E,vc) = \sum_{v \geq vc} v \cdot h_{Vt}^{v}(t,L,E).$$

Similarly, the definitions of average visits and visitors can be extended:

$$avgVs(t,L,E,vc) = \frac{\sum_{v \geq vc} v \cdot h_{Vs}^{v}(t,L,E)}{\left|\left\{e \in E \,\middle|\, Vs(t,L,e) \geq vc\right\}\right|} = \frac{grVs(t,L,E,vc)}{\sum_{v \geq vc} h_{Vs}^{v}(t,L,E)},$$

$$avgVt(t,L,E,vc) = \frac{\sum_{v \geq vc} v \cdot h_{Vt}^{v}(t,L,E)}{\left|\left\{l \in L \,\middle|\, Vt(t,l,E) \geq vc\right\}\right|} = \frac{grVt(t,L,E,vc)}{\sum_{v \geq vc} h_{Vt}^{v}(t,L,E)}.$$

If no entity (location) exists which reaches the given visit class vc, i.e. $\sum_{v \geq vc} h_{Vs}^{v}(t,L,E) = 0$ or respectively $\sum_{v \geq vc} h_{Vt}^{v}(t,L,E) = 0$, then the average visits (visitors) are undefined. The extension of average visits and visitors to visit classes is especially useful for $vc = 1$. The measures then exclude all entities (locations) without visits, and the averages express the intensity with which a visiting entity frequents a location set (or respectively the intensity by which a visited location is frequented by entities of some entity set).

Finally, we extend the coverage measures. The definitions in Section 3.2 already required at least one visit for the considered entities or locations, therefore the adaptation to higher visit classes is straightforward:

$$eCov(t,L,E,vc) = \frac{\left|\left\{e \in E \,\middle|\, Vs(t,L,e) \geq vc\right\}\right|}{|E|} = \frac{\sum_{v \geq vc} h_{Vs}^{v}(t,L,E)}{|E|},$$

$$lCov(t,L,E,vc) = \frac{\left|\left\{l \in L \,\middle|\, Vt(t,l,E) \geq vc\right\}\right|}{|L|} = \frac{\sum_{v \geq vc} h_{Vt}^{v}(t,L,E)}{|L|}.$$

Comparing coverage across several visit classes provides information about the structure of repeated visits. For example, we can determine the proportion of infrequent or regular customers of a shop. If the location set consists of all shops of a retail chain, we can even identify regular customers across the whole chain.

The concept of visit classes generalizes the definition of visit potential measures as stated in Subsection 3.2. We obtain gross and average visits / visitors by setting $vc = 0$ and entity / location coverage using $vc = 1$. In the remaining sections we will use the shorter notations of Subsection 3.2 whenever appropriate.

4 Visit Potential in Practice

Visit potential measures generalize from specific application contexts and can thus be applied to a wide range of domains. In this section we show the generalization capability by application of visit potential in the two real-world domains presented in Section 2.

4.1 Visit Potential for Outdoor Advertising

In order to specify poster performance in terms of visit potential, we first need to instantiate the objects of interest and their type of interaction, i.e. locations, entities, trajectories and visits. The entire set of available poster locations in Switzerland forms the universal set $\mathcal{L}$ of geographic locations. Hereby each location consists of the poster's visibility area. The selection of a specific poster campaign instantiates a location set L. The population of the surveyed conurbations in Switzerland forms the populations $\mathcal{E}$ of mobile entities. All persons included within the mobility survey constitute the entity set E. The trajectories are given in form of (x,y,t)-tuples. We obtain a visit by the spatial intersection of a trajectory and a visibility area applying a minimum passage time span $\varepsilon > 0$. Depending on whether weights are introduced to account for passage speed, angle of passage with respect to the poster board etc., a visit represents a simple passage or a poster contact. Usually, performance measures are evaluated for a time period of $t = 7$ days. However, longer periods of 10, 14 or even 21 days are also possible. We can now precisely define the performance measures GRP, OTS and reach. GRP corresponds to the visit potential measure average visits (multiplied by 100) whereas OTS corresponds to average visits for visit class $vc = 1$. The reach of a poster campaign equals the measure entity coverage:

$$GRP = avgVs(t,L,E)\cdot 100 = \frac{\sum_{v\geq 0} v\cdot h_{Vs}^{v}(t,L,E)}{|E|}\cdot 100,$$

$$OTS = avgVs(t,L,E,vc=1) = \frac{grVt(t,L,E,vc=1)}{\sum_{v\geq vc=1} h_{Vt}^{v}(t,L,E)},$$

$$reach = eCov(t,L,E) = \frac{\sum_{v\geq 1} h_{Vs}^{v}(t,L,E)}{|E|}.$$

Note that the application relies on a representative sample of test persons in the measured conurbations. Therefore, the measures in the data sample correspond to the point estimators of the measures in the population.

4.2 Visit Potential for the Evaluation of Bird Recordings

Although provided as graphics, coverage maps and coverage graphs are closely related to visit potential and can be defined more formally. The universal set $\mathcal{L}$ of discrete geographic locations consists of all squares in a 10 by 10 km grid of Great Britain and Ireland. The location set L is either equal to the universal set $\mathcal{L}$ or contains regional subsets of the grid which represent, for example, North Scotland, South Scotland or Wales. The population $\mathcal{E}$ consists of all people that live or travel through Great Britain and Ireland, and the entity set E contains all registered users of BirdTrack. As users join the project on a voluntary basis they are not necessarily representative for the population. However, representativity of volunteers is not the primary aim of BirdTrack. For the project it is more important that a sufficient number of reports are regularly available for all locations of the grid. The trajectories are given by the sequence of submitted reports of each volunteer. They take the form of snapshots in time and space and are collected over the whole year. Having defined the basic units of interest, we now translate the statistics displayed in coverage maps and graphs.

The coverage map statistic is available for $t = 1$ month or for $t = 1$ year and corresponds to the measure location coverage:

$$coverage\ map\ statistic = lCov(t,L,E) = \frac{\sum_{v\geq 1} h_{Vt}^{v}(t,L,E)}{|L|}.$$

Coverage graph statistics are displayed as time series and show aggregated visits for $t = 1$ day or t = 1 week. They correspond to the visit potential measure gross visits:

$$\textit{coverage graph statistic} = grVs(t, L, E) = \sum_{v \geq 0} v \cdot h^{v}_{Vt}(t, L, E).$$

5 Related Work

Measures for entity-location interactions always depend on the available data sources. Without mobility data one may consider, for example, population density to estimate the frequency of visitors. As people naturally spend large parts of their time in their living neighborhood, locations in densely populated areas are likely to be highly frequented. However, such approximations cannot account for trips farther away from the place of living as may be necessary for shopping or work. For example, factory outlet malls are usually located outside of cities in sparsely populated areas, yet they are highly frequented.

Frequency measurements allow to quantify the number of visitors for a given location directly. For example, supermarkets or train stations regularly determine the number of daily visitors. Frequencies can also be determined for the street network. May et al. (2008a, 2008b) developed a nearest neighbor-based spatial data mining algorithm to predict nationwide vehicle and pedestrian frequencies for Germany. The Fachverband Außenwerbung e.V. (FAW 2010), a special interest group for outdoor advertising in Germany, uses these frequencies to quantify the performance of poster boards. However, the informational value of frequencies is limited. Frequency counts cannot state the origin of visitors or provide information about their movement history. It is thus impossible to determine the number of repeated visits or to state the number of different entities that visit a location in a given time span.

Mobility data in form of trajectories as provided e.g. by GPS, RFID or GSM have drawn the attention of the data mining community recently. However, current developments in trajectory data mining concentrate on the analysis of mobility patterns and not on measures for interactions between mobile entities and locations. Algorithms are predominately presented for clustering of (parts of) trajectories (Rinzivillo et al. 2008; Pelekis et al. 2007; Nanni and Pedreschi 2006), detection of relative motion patterns (Gudmundsson et al. 2007; Hwang et al. 2005; Laube and Imfeld 2002) or sequential analysis of movement (Zheng et al. 2009; Giannotti et al. 2007; Yang and Hu 2006). The latter is in part related to visit potential. Frequent spatio-temporal sequential patterns are sequences of geographic locations that occur at least a given number of times in the trajectories of some mobile entities. Spatio-temporal sequential pattern mining requires in the beginning a set of disjoint geographic locations which represent relevant places for some application. These locations are used to transform the trajectories into sequences of visited locations. Afterwards, frequent transitions between the locations are detected. The locations may be provided externally (e.g. a collection of points of interest (POI)), by an analysis of trajectories for regularly visited regions or by a combination of both approaches (Giannotti et al. 2007). Giannotti et al. (2007) and Palma et al. (2008) provide algorithms to extract such a set of locations directly from a set of trajectories. Alvares et al. (2007) provide an algorithm to extract all stops from the trajectories of a mobile entity for a given set of locations. Algorithms for the sequential analysis of movements are related to visit potential insofar, as they also consider movement information only with respect to a set of relevant locations. Our definition of visits as given in Section 3.1 is similar to the definition of stops by Alvares et al. (2007), however, it is made on a conceptual level and is not restricted to data in the form of space-time points. The work of Zheng et al. (2009) differs from frequent sequential pattern mining as they consider not the frequency of patterns but the interest in some location or movement sequence. The authors adapt the concept of hubs and authorities by Kleinberg (1999) for the rating of geographic locations. Their measure of interest may be interpreted as a semantically enriched entity-location interaction as the measure estimates and includes the local expertise of each person. However, the measure has been designed for recommender systems and is thus suitable only to measure entity-location interactions for established locations. In addition, when comparing entity-location interactions of arbitrary locations in a large area (e.g. a country), the concept of local experience weakens and the interpretation of the measure is not clear.

6 Summary and Future Research Challenges

A number of companies and research institutions apply entity-location interaction measures in their day-to-day business. However, the measures are tailored to specific applications, use context-dependent terminology and are often only informally defined. As a result, a number of measures have evolved which are not suitable for methodological research and interdisciplinary exchange as their common background is hard to identify. In this paper we therefore present a systematic definition of entity-location interaction measures and provide a common vocabulary under the name visit potential. We describe two real-world applications that use entity-location interaction measures and show how these measures can be precisely defined in terms of visit potential.

We intend this paper to provide a common basis for future research concerning the extraction of entity-location interactions from mobility data. Although visit potential defines a set of fairly simple count statistics, their usage in real-world applications is not necessarily easy. May et al. (2009) encountered the problem of missing measurement days and analyzed a method from survival analysis for its applicability to estimate a measure corresponding to entity coverage. A second challenge is the handling of non-representative mobility data. Naturally, practitioners are not interested in characteristics of the data sample, but want to infer knowledge about the underlying population. However, as mobility data are often provided from secondary sources and are not necessarily representative, methods must be developed that detect and compensate possible biases in the results. Finally, the optimization of some entity or location set according to one or more of the presented measures is of interest. For example, the reader may remember the outdoor advertising application, where the positioning of posters is of vital interest when planning a campaign.

Acknowledgement

The authors would like to thank the Swiss Poster Research Plus AG (SPR+), the German Arbeitsgemeinschaft Media-Analyse e.V. (ag.ma) and the German Fachverband Außenwerbung e.V. (FAW) as well as the British Trust for Ornithology (BTO), the Royal Society for the Protection of Birds (RSPB) and BirdWatch Ireland for providing comprehensive materials about the application domains.

References

ag.ma (2010) Arbeitsgemeinschaft Media-Analyse e.V. (German working group for media analysis) http://www.agma-mmc.de, last date accessed Jan 2010.

Alvares LO, Bogorny V, Kuijpers B, de Macedo JA, Moelans B, Vaisman A (2007) A model for enriching trajectories with semantic geographical information. In Proc. of the 15th Annual ACM international Symposium on Advances in Geographic information Systems (GIS'07). ACM, pp 1-8.

BirdTrack (2010) http://www.birdtrack.net, last date accessed Jan 2010.

FAW (2010) Fachverband Außenwerbung e.V. (German special interest group for outdoor advertising) http://www.faw-ev.de, last date accessed Jan 2010.

Giannotti F, Nanni M, Pedreschi D, Pinelli F (2007) Trajectory Pattern Mining. In: Proc. of the 13th ACM SIGKDD International Conference on Knowledge Discovery and Data Mining (KDD'07). ACM, pp 330-339.

Gudmundsson J, Kreveld M, Speckmann B (2007) Efficient detection of patterns in 2D trajectories of moving points. In: Geoinformatica 11(2):195-215.

Hwang SY, Liu YH, Chiu JK, Lim EP (2005) Mining mobile group patters: a trajectory-based approach. In: Proc. of the 9th Pacific-Asia Conference on Knowledge Discovery and Data Mining (PAKDD'05). Springer, pp 713-718.

Kleinberg JM (1999) Hubs, authorities, and communities. ACM Computing Surveys 31(5). ACM.

Laube P, Imfeld S (2002) Analyzing relative Motion within groups of trackable moving point objects. In: Proc. of the 2nd International Conference on Geographic Information Science (GIScience'02). Springer, pp 132–144.

Marchal P, Yuan S, Flavigny PO (2008) Person-based GPS surveys in France: „Lille Experiment“ by ISL, and GPS Subset in the French National Travel Survey (ENTD 2007-2008). COST 355 project meeting. http://cost355.inrets.fr/IMG/ppt/WG3-Torino-051007-Marchal-Yuan-Flavigny-GPS-v2du05100700.ppt, last date accessed Jan 2010.

May M, Körner C, Hecker D, Pasquier M, Hofmann U, Mende F (2009) Handling Missing Values in GPS Surveys Using Survival Analysis: A GPS Case Study of Outdoor Advertising. In: Proc. of the 3rd ACM SIGKDD Workshop on Data Mining and Audience Intelligence for Advertising (ADKDD'09). ACM, pp 78-84.

May M, Hecker D, Körner C, Scheider S, Schulz D (2008a) A vector-geometry based spatial kNN-algorithm for traffic frequency predictions. In: Proc. of the 2008 IEEE International Conference on Data Mining Workshops (ICDMW '08). IEEE Computer Society, pp 442-447.

May M, Scheider S, Rösler R, Schulz D, Hecker D (2008b) Pedestrian flow prediction in extensive road networks using biased observational data. In: Proc. of the 16th ACM SIGSPATIAL International Conference on Advances in Geographic Information Systems (ACM GIS '08). ACM, pp 1-4.

Nanni M, Pedreschi D (2006) Time-focused density-based clustering of trajectories of moving objects. In: Journal of Intelligent Information Systems (JIIS), 27(3):267-289, Special Issue on Mining Spatio-Temporal Data.

Palma AT, Bogorny V, Kuijpers B, Alvares LO (2008) A clustering-based approach for discovering interesting places in trajectories. In: Proc. of the 2008 ACM Symposium on Applied Computing (SAC'08). ACM, pp 863-868.

Pasquier M, Hofmann U, Mende FH, May M, Hecker D, Körner C (2008) Modelling and prospects of the audience measurement for outdoor advertising based on data collection using GPS devices (electronic passive measurement system). In: Proc. of the 8th International Conference on Survey Methods in Transport.

Pelekis N, Kopanakis I, Ntoutsi I, Marketos G, Andrienko G, Theodoridis Y (2007) Similarity search in trajectory databases, In: Proc. of the 14th IEEE International Symposium on Temporal Representation and Reasoning (TIME 2007). IEEE Computer Society Press, pp 129-140.

Rinzivillo S, Pedreschi D, Nanni M, Giannotti F, Andrienko N, Andrienko G (2008) Visually driven analysis of movement data by progressive clustering. In: Information Visualization 7(3):225-239.

Sissors JZ, Baron RB (2002) Advertising Media Planning. McGraw-Hill, chp 4-5

SPR+ (2010) Swiss Poster Research Plus AG. http://www.spr-plus.ch, last date accessed Jan 2010.

Yang Y, Hu M (2006) TrajPattern: mining sequential patterns from imprecise trajectories of mobile objects. In: Proc. of 10th International Conference on Extending Database Technology. Springer, pp 664-681.

Zheng Y, Zhang L, Xie X, Ma WY (2009) Mining Interesting locations and travel sequences from GPS Trajectories. In: Proc. of the 18th International World Wide Web Conference (WWW'09). ACM, pp 791-800.

Pre-School Facilities and Catchment Area Profiling: a Planning Support Method

Cláudia Costa, Paula Santana, Rita Santos, Adriana Loureiro

Geographical Studies Institute and Researcher in the Geography and Planning Studies Centre – University of Coimbra; Departamento de Arquitectura, Praça D. Dinis, Colégio das Artes; 3000-043 Coimbra, Portugal
paulasantana.coimbra@gmail.com

Abstract. For the effective implementation of social support facilities or infrastructures, knowledge of the physical and sociodemographic characteristics of the catchment area is required. However, it is precisely the definition and profiling of this catchment area that causes the main problems.

This paper raises the question of the Modifiable Areal Unit Problem (MAUP) and presents a method for catchment area profiling that takes account of socioeconomic and contextual attributes. Facility catchment areas are defined by a function that takes account of context indicators, enabling a broader yet more precise (and therefore more thorough) profiling of the area.

As a case study in the application of this method, this paper describes the existing supply of pre-school facilities in the municipality of Amadora and analyses the area's needs with regards to further nursery schools or kindergarten facilities.

Keywords: Group facilities, education and childcare establishments, MAUP, catchment areas, Geographic Information Systems.

M. Painho et al., (eds.), *Geospatial Thinking*, Lecture Notes in Geoinformation and Cartography,
DOI 10.1007/978-3-642-12326-9_6, © Springer-Verlag Berlin Heidelberg 2010

1 Group Facility Planning Norms: Educational and Childcare Facilities

Cities, as living organisms, produce dynamics, generate environments and create societies. Today, urban space as we know it is not only dense, impoverished and polluted, but also unsafe and crime-ridden (Santana, 2009). Consequently, town planning, quality of life and the well-being of the population have become pressing concerns, given the right of all people and communities to live in a safe healthy place with access to public facilities (Corburn, 2004:545).

A community has multiple dimensions (physical, social, economic, cultural, etc.) hence, new town planning methods are required that involve multilevel analysis, in order to identify the various factors contributing to territorial iniquity (Corburn, 2004:542; Santana, 2005). The identification and monitoring of environmental features essential for well-being and the quality of life should become features of the town planner's remit, enabling the creation of healthier places able to promote quality of life on all dimensions (Santana, et al., 2008). According to Weinstein (1980), "we must fit the city to man's needs rather than fit man to the city's needs".

The urban space must be organized in such a way as to respond to the real and actual needs of the population. As such, the cost/benefit ratio of urban planning practices needs to be made clear, transparent and rational (Teixeira & Antunes, 2008), with its main objectives being the improvement of the quality of life of the population, a reinforcement of the role of the community, and the cultivation of feelings of proximity or belonging to a neighbourhood (Santana & Nogueira, 2008). It is also necessary to raise awareness amongst the citizens about the importance of their contribution, at the behaviour level, to individual and community well-being (Santana *et al.*, 2007).

Such an approach involves re-thinking the attributes of the place (public and private infrastructures, leisure spaces, etc.) to ensure that they respond to the needs of potential users, and re-designing the identity of the place in order to improve feelings of self-esteem, trust and security. These aims are essential right now, i.e. planning for specific groups with specific needs, in the present and in the future (Barton & Tsourou, 2000).

In fact, one of the main aims of urban planning is to ensure a balanced distribution of the functions of accommodation, work, culture and leisure. The planning, creation and maintenance of group facilities[1] falls within this

---

[1] Some authors define group facilities as essential infrastructures that provide public services and satisfy basic needs, thereby providing the foundations of the urban and social fabric (Costa Lobo, Pardal, Correia, & Lobo, 1995; Partidário,

remit, and for this, account must necessarily be taken of the specific needs and characteristics of user populations, the accessibility of the infrastructure, and its capacity and suitability for the purpose for which it was designed[2] (DGOTDU, 2002. Consequently, the planning of public group facilities is based upon profiling studies that take account not only of their function but also of the user population.

These studies are normally carried out by the local authority or, at its request, are inserted into the various planning levels in which it is involved. Account must therefore be taken of the conditions that influence the decision, based upon the balance between supply and demand, while aiming to satisfy the needs of the population and offer a quality service. The decision about where to locate such facilities is therefore a complex one, which aims, on the one hand, to maximise their profitability, yield and use, while at the same time minimising effort and optimising the cost-benefit ratio (Simões Lopes, 2001: 139; Teixeira & Antunes, 2008).

According to the DGOTDU (2002), nursery schools should be located in a central area that is easy to access and safe for pedestrians and vehicles, so as to avoid the children being subject to extensive daily trips. They should also, if possible, be located along the usual daily route, as close as possible to points where parents can leave for work. It is also advisable to locate these facilities in parishes with a large female workforce, or with high birth or infant death rates (DGOTDU, 2002). No irradiation criteria are indicated for this type of facility[3]. Kindergartens, for their part, should be correctly inserted into the urban fabric of the town, near the residential areas that they are to serve, and in the vicinity of parks, car parks, sports/culture/leisure facilities, and the public transport system. In terms of irradiation, the DGOTDU stipulates that a journey on foot should not take more than 15 minutes, while a journey made using public transport should be no longer than 20 minutes. Taking account of the age group in question, the distance between the facility and the place of residence or parents' workplace should be subordinated to the general principle of great proximity (Table 1).

1999; DGOTDU - Direcção-Geral do Ordenamento do Território e Desenvolvimento Urbano, 2002).

[2] Article 6 of Law 48/98 of 11th August and Article 18 of Decree-Law 380/99, of 22nd September.

[3] Maximum time taken or distance travelled by users between the site of origin (normally their residence) and the facility (destination), on foot or using public transport, measured in minutes or kilometres.

Table 1. Norms for the Planning and Profiling of Educational and Childcare Facilities

Facility	Age Band covered	Residence-School Irradiation (minutes)		Base population (inhabitants)	
Nursery Schools	0 – 2 years			Min.	5000 (5 children)
				Max.	5000 (35 children)
Kindergarten	3 - 5 years	On foot	15	Min.	900 (20 children)
		Public transport	20	Max.	3600 (150 children)

Source: DGOTDU, 2002.

A number of authors have stressed the importance of this type of infrastructure, and the need for proper planning and assessment. Valle (1993-94), for example, believes that the aim of planning is to improve quality of life and increase the wellbeing of these populations, and that this fact will only be achieved when there is a balanced relationship between the populations and the supply of facilities and infrastructures required. As regards conditions of access and their consequences on equity, Simões Lopes (2001:152) presents the aims of the "Constitution of the Portuguese Republic" (1976): "reduce inequality and create conditions for the access to essential goods and services, implying an effective distribution of service facilities and the creation of conditions to enable populations to access goods and social services." However, as George (1989) points out, it is only possible to improve urban living if the decisions to implement facilities, infrastructures and services are synchronized and adequate.

1.1 The importance of scale in territorial profiling

Despite the norms established by the DGOTDU, the scale usually used for territorial profiling in group facility planning is the parish, and information tends to be used only at that level (Matias, 2005).

However, new planning paradigms emphasise the use and dissemination of methods within organizations as a central aspect of the implementation of good plans and public policies. Recent advances in the so-called Geographic Information Systems (GIS) can contribute to this goal, given their applicability as a tool for territorial profiling. That is to say, they are able to represent and operate data with associated spatial information, thereby developing an innovative culture for the processing and dissemination of information in different spheres of action. However, despite increased use of these technologies as a strategy for the storage and recovery of data, the technology is as yet rarely used in decision-making processes (Ramos, 2004: 1).

There is thus both a need and an opportunity to change the bureaucratic paradigm of traditional public management to make way for a model that gives priority to the quality of information used to support decision processes, over and above rigid institutional norms. That is to say, within this new paradigm, the importance of information is emphasised at different levels and moments of territorial management. In addition to identifying norms and establishing operative processes, it is necessary to define methods and strategies based on precision indicators that are sensitive to what is to be represented (Ramos, 2004: 2).

However, one of the most difficult problems facing urban planners is the accessibility, availability and reliability of conventional data (such as census information, statistics, maps, reports, etc), which provide the main source of information for spatial studies. This leads to a certain incoherence in the analyses undertaken prior to the planning of certain facilities (Júnior & Costa, 2007: 5435). In figure 1 it is possible to see that the scale used in a planning project can affect the profiling of the population that will use it: it is different to plan a facility for 17,144 children (municipality) than for 1.232 children (foot travel time).

This is the case, particularly in socially-complex or large-scale contexts, where the significance levels of the information provided are difficult to establish. These limitations often arise because the information is processed in a homogeneous way, with no account taken of the diversity of situations that inevitably arise (Ramos, 2004: 2). Thus, while general data of a socioeconomic nature may be acquired from the National Institute of Statistics (NIS) at the level of the statistical subsection (the most precise level of information), the same does not occur for specific or contextual socioeconomic indicators, such as the location coefficient, urban fabric and land use. Despite the fact that such data can induce, influence or explain socioeconomic information, it is not always possible to collect it at parish level. Thus, the results of any quantitative analysis based on these data will be spatially conditioned by the definition of the boundaries of the territorial subdivisions to which they are associated, and will in turn affect the results of the statistical indicators. This issue has come to be known as the *Modifiable Areal Unit Problem* – MAUP – (Openshaw, 1978); i.e. the drawbacks that arise from the different ways of delimiting spatial units derived from different unit designs, implying different analyses and therefore policies (Ávila & Monasterio: 2). This problem arises from the need to combine individual spatial data or from the scale of analysis itself.

Scale: Municipality

Profiling: Population 0 – 9 years (2001) covered: 17,144 children

Scale: Parish

Profiling: Population 0 – 9 years (2001) covered: 2542 children

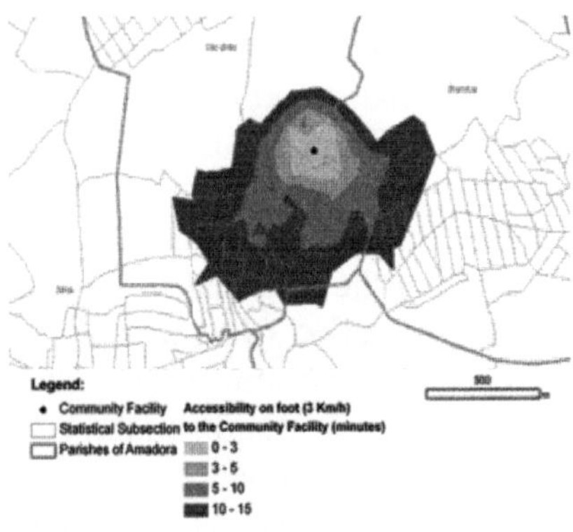

Scale: Travel Time

Profiling: Population 0 – 9 years (2001) covered: 1232 children

Note: Value resulting from the analysis using method explained below.

Fig. 1. Demographic profiling of a new group facility in the Municipality of Amadora, taking account of different scales. Source: Prepared using data from the Amadora Municipal Council (AMC) and 2001 Census (NIS)

As Haining (2003) points out, a single statistic that has been calculated using two spatial units will generate differences in accordance with the territorial limit used, even when the original data are the same. This author indicates two types of MAUP in the statistical properties: one arises from the different levels at which spatial units may be combined (the problem of *scale*[4]); the other concerns the limit of the units (the problem of *partition*[5]) (*Idem*, 150).

On the level of area-associated census data, the significance of the MAUP should be considered when planning a collective facility. However, it is necessary that the aim of inferences is established, i.e. whether these concern the areas or the individuals that live in them. If the aim is to infer the characteristics of the individuals that live there, there are no theoretical guarantees that the results obtained will be good estimations (Openshaw, 1984). This effect is called the "ecological fallacy"[6].

To solve this problem, two approaches are found in the literature: one recognises the problem as intrinsically related to the zonal support and aims at the formulation of optimal criteria for territorial zoning projects (Openshaw, 1978; 1996; Martin, 1995; 2000); the other seeks to develop algebraic operators to estimate and reduce the instabilities that the MAUP imposes on the cluster of variables within the original zonal support (Openshaw, 1984; Bailey & Gattrel, 1995; Wrigley *et al.*, 1996; Green & Flowerdew, 1996).

As Martin (1995) has pointed out, excessive territorial fragmentation in the polygon model imposes limitations upon the perception of the whole socioeconomic and contextual phenomenon. Thus, territorial profiling should not be restricted to this type of representation, but should be complemented by spatial representations through images and surfaces, based upon minimal units that combine similar occurrences: the Neighbourhood. This represents the spatial distribution of the population in a continuous form. With this reading of the situation, it is more difficult to establish boundaries between different territories, because transitions between them

[4] The scale effect refers to the tendency, within a system of modifiable territorial units, where different statistical results are obtained from the same variables, when the information is grouped into different levels of spatial resolution, such as census divisions, neighbourhoods, parishes, counties, etc. (Wrigley *et al.*, 1996).

[5] Or the effect of zoning. This is the variability of statistical results obtained in a system of modifiable territorial units as a result of the different ways that these units may be grouped at a particular scale, rather than due to variations in their size (Wrigley *et al.*, 1996).

[6] For Wrigley et al. (1996), "the ecological fallacy involves inappropriate relation inferences on the level of aggregated results in geographical units. This usually occurs when the aggregate data is the only available source and when individual characteristics and relationships form the object of study. Due to the effects of scale and zoning of the MAUP, relationships measured at the level of area units by means of correlation coefficients generally tend to present absolute values that are greater than the unknown correlations on the level of individuals".

are gradual and contain relative degrees of uncertainty, and it allows a more adequate reading of the global tendency as regards the spatial distribution of variables. In a way, it is appropriate to think of the population as being continuously distributed, while recognising that roads, urban zoning and geographic accidents often constitute abrupt boundaries between one territory and another (Ramos, 2004 7).

The fundamental question related to the application of this representation is due to the impossibility of recognising or collecting individualized values of the variables at all points of the territory and for all its inhabitants. Thus, these surfaces should be constructed from data associated to area units through processes of interpolation, estimating the spatial distribution of indicators (Ramos, 2004: 7).

This analytic process breaks with the traditional programmatic perspective of spatial representation and advances towards an analytic perspective that is dynamic and interactive. By making use of the possibilities for the computational representation of space, indications are sought for the existence of patterns of spatial distribution of particular features such as concentration, spatial dependence, persisting aspects and pattern transitions in geographic space. Such patterns, revealed through computational techniques, may constitute an innovative way of processing information, as space is now incorporated into the analysis, treated as a variable rather than as a mere neutral physical support (Ramos, 2004: 8; Santana *et al.*, 2008).

2 Data and Methods

The information used includes reports published by the Amadora City Council (AMC), showing the location of facilities, maps, orthophotomaps, reports, etc; and demographic, economic and other data from the National Institute of Statistics (NIS) on the level of the statistical subsection. This is the largest scale of analysis provided by the NIS, and comes closest to defining the Neighbourhood in an urban area such as Amadora (Geirinhas, 2001; Santana, *et al.*, 2007).

Within the Municipality of Amadora, the supply and accessibility of educational and childcare facilities for a target public of children aged up to 6 years were gauged, taking as references the legal standards laid out in"Regulations for the Planning and Characterization of Group Facilities"[7] published by the DGOTDU (2002), previously mentioned.

[7] Free translation of the title.

According to the methodology followed in this study (Figure 2), the facilities [Facility] were identified by adapting and processing databases of the Amadora City Council, available on line at the Council's website[8], namely a map of the educational and childcare facilities prepared by them in 2007; streets and roads of the municipality [Network] and by the use of the *Geocoding* extension of the *ArcGis 9.2* (produced by ESRI) computer platform. From this spatialization, areas accessible on foot were constructed using the extension *Network Analyst* of the platform *ArcGis 9.2*, described below.

This tool, through its function *Service Area*, allows levels of accessibility to be identified, taking into account distance on foot at an average speed of 3 km/h[9], using the road network. This produced the following analysis intervals: 0-3 minutes, 3-5 minutes, 5-10 minutes and 10-15 minutes. These accessibility areas were then profiled, using information on the level of the statistical subsection relating to population, accommodation[10] [Censos_Stat_Subsection], sociomaterial deprivation [Deprivation_Index] and municipal property tax [Coefficient_location].

The statistical subsections doesn't always correspond to the accessibility areas. However, at the basis of the definition of the statistical subsections is the consideration of the homogeneity of the area (Dias, 2002). This way, population and accommodation distribution were considered to be homogeneous within the subsection and those indicators where extrapolated. Through the analysis of these two dimensions (population and housing), it was possible to arrive at an approximate number of individuals and residences existing in the accessibility areas.

---

8 www.cm-amadora.pt

9 The average speed of 3km/h was obtained by considering the minimum and average speeds suggested by Austroads (1995), in order to have a single parameter for the distances travelled on foot by children.

10 Data about population and accommodation were taken from the 2001 census (NIS). The first was analysed in terms of age band (0-4 years) and family groups with children under 6 years old. As for the second, two indicators of inadequate housing were used: substandard accommodation and main residence without indoor toilet.

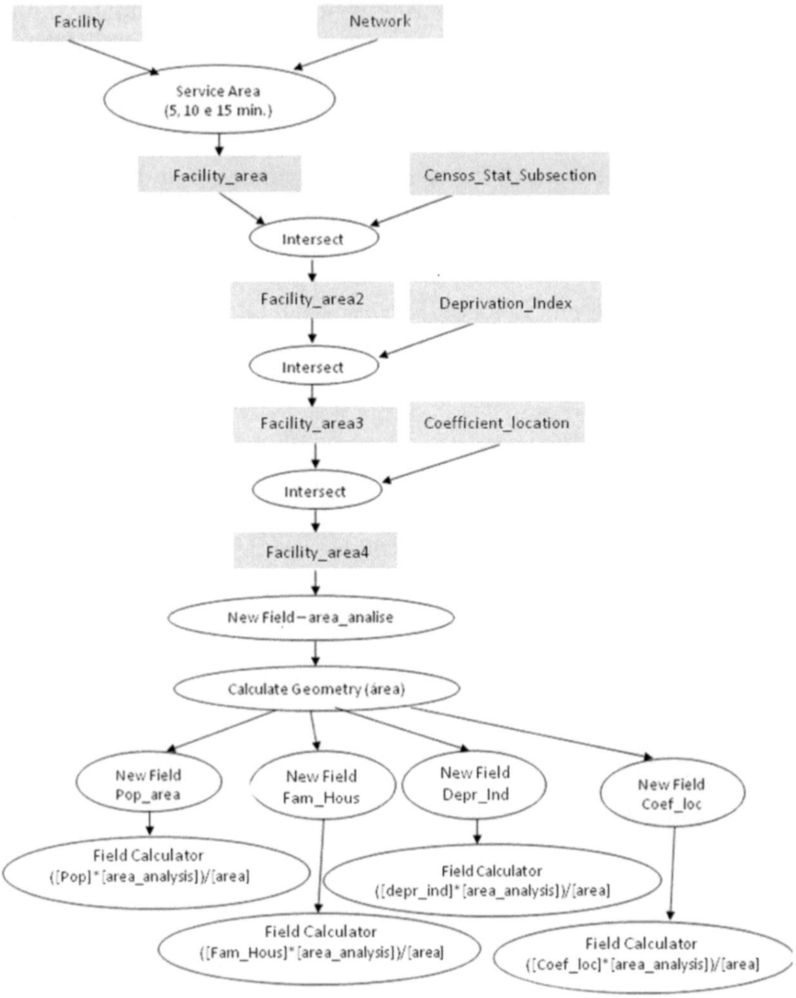

Fig. 2. Diagram of Method Adopted

Then, using the density of the accessible areas, the population and respective housing was calculated. For this analysis, the Location Coefficient[11] (Municipal Property Tax) was divided into classes based on the dis-

[11] The Location Coefficient is one of the indicators that influences the Municipal Property Tax, and refers to the location of the property. The criteria for the construction of this coefficient are: accessibility (quality and variety), proximity to social facilities, public transport services and location in areas of high property value (Nos. 3 and 4 of Article 42 of the Code of Municipal Property Tax (CIMI), approved by Decree-Law No. 287/2003, of 12th November. The coeffi-

tribution of the sample into quintiles. The percentage of the area accessible to each class (greater or lesser coefficient) was then gauged. Similar calculations were done for the percentage of accessible area with the greatest and least sociomaterial deprivation[12] (Santana *et al.*, 2007).

Supply was gauged using AMC databases showing the present distribution of these facilities and plans for future constructions. Different sources were used to identify this capacity. In the case of nursery schools, we used information available on line at the site of the social charter[13], while for kindergartens, this information was collected from Amadora Council's educational charter. This source also provided information about planned facilities of these types. Then, the ratio of capacity to catchment population (i.e. those within their accessibility areas) was calculated.

As the data collected from the 2001 Census did not enable the target population for preschool facilities to be established for the years 2008 and 2010, a method was developed to assess the relation between supply and potential demand based upon the 2008 capacity (supply) and the number of children born in the Fernando da Fonseca Hospital (FFH) between 2002 and 2007 and resident in the municipality (potential demand)[14]. Of the 7356 babies born between 2002 and 2007 whose mother resided in the municipality of Amadora, it was possible to locate 87.9% through their address, using the *geocoding* extension of the platform *ArcGis 9.2*.

---

cients relative to housing to be applied in each homogeneous zone may vary between 0.4 and 2, and may, in some situations of housing dispersed in a rural environment, be reduced to 0.35, and in areas of high property value, be as high as 3. The minimum and maximum values defined for Amadora were 1.00 and 2.16, respectively (Administrative Rule No. 1426, of 25th November) (Santana *et al.*, 2007).

[12] The Deprivation Index was constructed in accordance with the method used by Carstairs & Morris (1991), using selected variables collected from the NIS: proportion of individuals without basic schooling, proportion of unemployed (looking for first job or new job) and families whose main residence has no toilet (Censos, 2001 – Statistical subsection). The variables were standardized (*z-score* method), taking as effect the fact that each variable has the same influence on the final result. The deprivation índex is the sum of all variables, after standardization (Santana *et al.*, 2008).

[13] www.cartasocial.pt

[14] Births at the Fernando da Fonseca Hospital represent 71.1% of the total births in the municipality, according to a study presented by Machado *et al.*, 2007.

3 Pre-School Facilities in the Municipality of Amadora

3.1 Profile of the existing supply

Amadora, a municipality within the Lisbon Metropolitan Area, is a densely occupied urban territory, with 172,110 recorded inhabitants in 2008. There is a great deficit of facilities and services, particularly in the area of education and childcare, as a result of decades of population growth, without the provision of necessary measures to meet this increased demand (AMC, 2007).

As regards preschool facilities, there are in Amadora 31 nursery schools and 83 kindergartens. Within the sphere of the municipality's educational charter, the construction of a further 4 nursery schools and 9 kindergartens is planned, and 2 of the nursery schools are to be expanded (AMC, 2007:156).

Concerning nursery schools in particular, around 48% of the 0-4 year population and nuclear families with children under 6 have good access (less than 15 minutes' walk) to these facilities (Table 2). However, only a fifth (19.78%) are placed, due to the fact that the capacity of these establishments is insufficient to meet the needs of the resident population. These facilities tend not to be located centrally (Figure 3); rather, they are dispersed, near the areas of residence of the young and active population, which means that accessibility for areas further from the centre of the municipality could be improved. It should be noted, however, that the parishes in the north of the municipality, such as *São Brás*, *Brandoa* and *Mina*[15], have poorer geographic accessibility.

On the other hand, around 40% of the municipality's substandard housing and residences without toilets are concentrated in the areas of good and very good accessibility. That is to say, there is proximity (as regards travel on foot) between the nursery schools and the poorest housing in the municipality. This is because many rundown neighbourhoods are located in inner city areas, where various social security and charity organisations operate, providing nursery schools and kindergarten facilities to these communities. This is confirmed by the fact is that 19.4% of the areas that have good or very good accessibility also score high for sociomaterial deprivation. However, almost half (49.4%) of this area registers a Location Coefficient (IMI) above the average, motivated by the proximity of train stations and council-run green spaces.

[15] In figure 1 is possible see the location of this parishes.

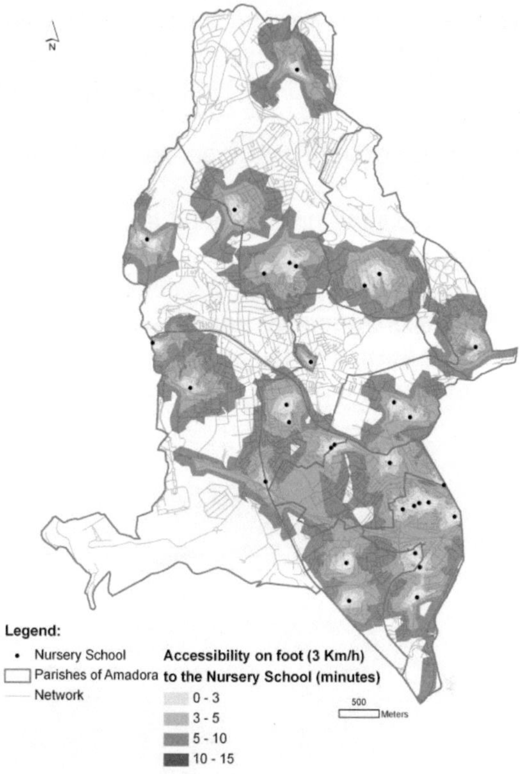

Fig. 3. Accessibility on foot (3km/h) of nursery schools. Source: Prepared using data from Amadora Municipal Council (AMC) and the 2001 Census (NIS)

Table 2. Accessibility on foot (3km/h) of nursery schools in the Municipality

Nursery schools: 31*	Potential population (%)		Potential residences (%)	
	0 to 4 year	Families with children under 6	Families in sub-standard accommodation	Families with no toilet in main residence
Total	*8662*	*8373*	*1438*	*1464*
0-3min	3.8	3.9	0.5	2.9
3-5min	4.1	4.1	0.9	3.1
5-10min	20.2	20.7	15.4	17.3
10-15min	19.8	19.8	19.9	18.1
<15min	48.1	48.6	36.7	41.5

Source: Prepared using data from Amadora Municipal Council (AMC) and the 2001 Census (NIS).

* From the 31 nursery schools, 17 are charity-runs and 14 are private.

As regards kindergartens, it was concluded that there are many such establishments across the whole municipality (Figure 4): In fact, all parishes have at least one facility of this type. As for the population covered, more than half (58.2%) of family groups with children under 6 are located within 15 minutes' walk of a kindergarten, and 30.7% are within 5 to 10 minutes' walk (Table 3).

However, despite the good or very good access to these facilities and the high municipal property tax practised in 49% of this area, around 25% of the population resides in areas of considerable sociomaterial deprivation. The greatest concentration of substandard housing and residences with no toilet are also located in the area of greatest accessibility to kindergarten, i.e. half of those existing in the municipality.

Fig. 4. Accessibility on foot (3km/h) of kindergartens. Source: Prepared using data from Amadora Municipal Council (AMC) and 2001 Census (NIS)

Table 3. Accessibility on foot (3km/h) of kindergarten in the Municipality

Nursery schools: 83*	Potential population (%)	Potential residences (%)	
	Families with children under 6	Families in sub-standard accommodation	Families with no toilet in main residence
Total	*8373*	*1438*	*1464*
0-3min	8,5	1,5	7,0
3-5min	7,4	1,9	6,5
5-10min	30,7	29,6	31,5
10-15min	1,2	9,3	11,5
<15min	58,2	50,4	56,4

Source: Prepared using data from Amadora Municipal Council (AMC) and the 2001 Census (NIS).
* From the 31 kindergartens, 27 are public, 25 are charity-runs and 31 are private.

3.2 Analysis of capacity and need for new facilities

By assessing the relation between the supply of facilities and potential demand, we attempted to identify whether further facilities were required.

As regards nursery schools, 2995 children were identified as having been born at the Fernando Fonseca Hospital and aged between 3 months and 3 yea in 2008, and all facilities existing in the municipality (i.e. public, private and charity-run) were included. Then geographical accessibility was assessed.

It was found that 24.5% of children reside more than 15 minutes' walk away from nursery schools. Moreover, the municipality does not have the capacity to receive all children that were born in the period; only 30.5% of the potential demand is catered for (Figures 5). When the analysis was limited to public and social security or charity-run facilities (in order to take account of the economic constraints suffered by some families), it was found that 37.3% of families with children born in the FFH live in neighbourhoods that are more than 15 minutes' walk from this type of facility. Moreover, these institutions cater for only 22.9% of the children analysed.

As regards the supply of kindergartens and the potential demand of 3427 children (born between 2002-2004 at the FFH, and aged between 4 and 6 years in 2008), it was found that these facilities were well distributed around the municipality, and that geographical accessibility was very good. 95.7% of families with children in this age band live within 15 minutes' walk of a facility and the supply exceeds the potential demand by

41.5% (Figure 6). When private kindergarten are excluded from the study, the results are different. Now 7.7% of children have accessibility problems (residing more than 15 minutes' walk away), though the capacity of public and social security or charity-run institutions to respond to demand covers almost 100% of potential users (97.8%).

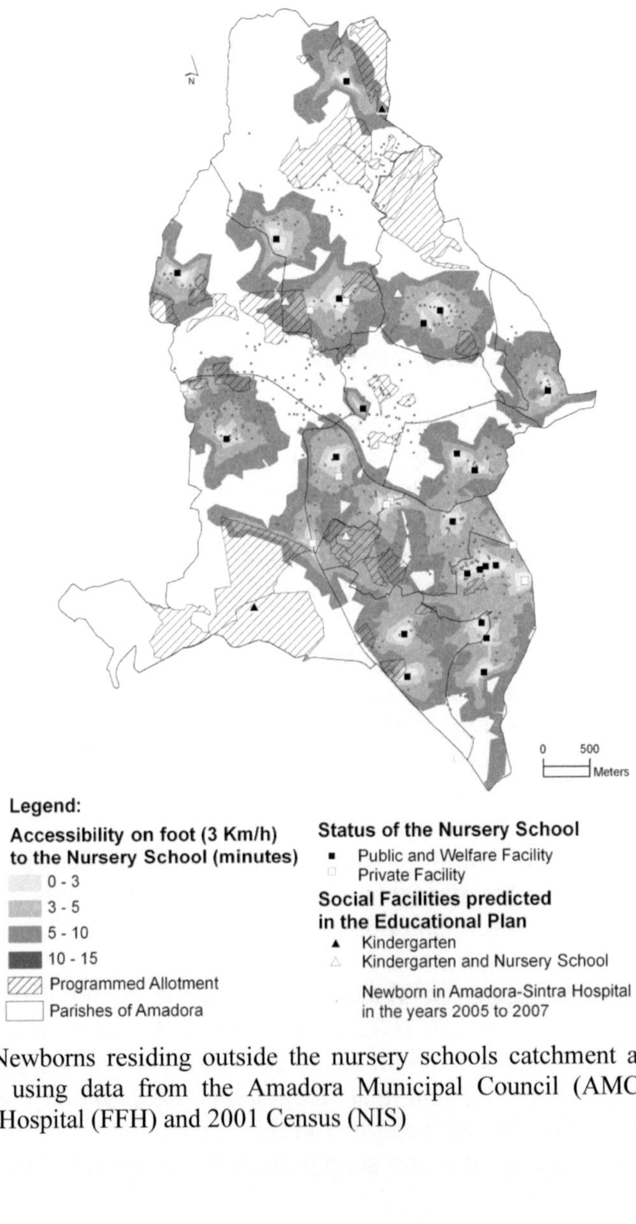

Fig. 5. Newborns residing outside the nursery schools catchment area. Source: Prepared using data from the Amadora Municipal Council (AMC), Fernando Fonseca Hospital (FFH) and 2001 Census (NIS)

Fig. 6. Newborns residing outside the kindergarten catchment area. Source: Prepared using data from the Amadora Municipal Council (AMC), Fernando Fonseca Hospital (FFH) and 2001 Census (NIS)

In a prospective analysis (considering the first group of children, born between 2005 and 2007: 2995 newborns (NB) and keeping the same residential locations of the families of the NB and the total supply on offer), the scenario relative to kindergartens indicates that 3.8% of those children will reside (in 2010) more than 15 minutes' walk distance away and that the capacity will be exceeded by 61.9% the needs predicted for 2010. Considering only public and social security or charity-run kindergartens, there will be less geographic accessibility to facilities (7.7% of children will be outside the 15 minutes accessibility area). However, there will continue to be a surplus of supply (11.9% places). However, it was found that 41.6% of the active population with 15 or more years of working or studying in the municipality of Amadora lives outside it (Census, 2001).

However, given the locations earmarked for the new areas of residential expansion in the municipality of Amadora, planned by the council, proximity to preschool facilities will be reduced. Around 72% of these new estates will be located more than 15 minutes' walk away from a nursery school, and 46% will be outside that distance for a kindergarten. Thus, there are infrastructure requirements that need to be attended to on the level of planning/establishment of new nursery schools and kindergartens (particularly the former). Thus, the analysis reveals that the Amadora Council's proposals for new nursery schools and kindergarten facilities, as laid down in the municipality's Educational Charter (2007), are adequate in number and location for previous needs.

4 Conclusion

Territorial profiling is and always will be conditioned by the data used and their scale. In fact, if we use only statistical data from the NIS, restricted to the level of the municipality or parish, as unique spatial attributes, the analysis may be inaccurate, leading to wrong decision taking. The same happens if we fail to analyse context indicators gathered from sources other than the NIS, such the location coefficient, urban space distribution, land use or the urban fabric.

The use of walking accessibility areas and of the number of children born in the *Fernando da Fonseca* Hospital (FFH) are two different ways not to fall into a MAUP problem: by using travel time on foot and accessibility areas of a facility it is possible to better know which population will use it and better to program the size of the facility; and by using hospital databases of newborns it is possible to get data about targeting groups of certain facilities which is not collected in the 10—years period between Census tracks.

On the other hand, these solutions are not applicable to all the planning projects. There are facilities whose catchment area may not be obtained and characterised only with recourse to data from statistical subsections or parishes. An example of this is the planning of a shopping centre, a hospital or an airport, since these are more regional than community in nature. Thus, this method may only be applied to facilities whose catchment area is the local community and which provides services for that population.

Other such examples are the planning of green spaces, schools, welfare and social security institutions, and sporting and cultural facilities.

As Barton & Tsourou (2000) have pointed out, the urban space should be organised in order to respond to real or current needs of human groups (safety, neighbourhood effect, social construction of place, etc) and not the reverse. Thus, the cost/benefit ratio of urban planning practices needs to be made evident, with the main objective of improving the health and quality of life of populations, reinforcing the role of the community and the effects of neighbourhood or proximity. Simultaneously, the attributes of place need to be rethought (public and private facilities, leisure spaces, etc) in order to tailor them to the needs of the potential user population.

References

Austroads. (1995). Guide to Traffic Engineering Practice. In: Austroads, *Pedestrians.* Sydney.

Ávila, R. P., & Monasterio, L. M. (s.d.). O MAUP e a análise espacial: um estudo de caso para o Rio Grande do Sul (1991-2001). 25.

Bailey, T., & Gattrel, A. (1995). *Interactive Spatial Data Analysis.* Londres: Longman.

Barton, H., & Tsourou, C. (2000). *Healthy Urban Planning. A WHO guide to planning for people.* Londres: Spon Press.

Carstairs, V., & Morris, R. (1991). *Deprivation and health in Scotland*. Aberdeen: Aberdeen University Press.

Corburn, J. (2004), Confronting the Challenges in Reconnecting Urban Planning and Public Health. In: *American Journal of Public Health*, Vol. 94, No. 4, Abril,

Costa Lobo, M., Pardal, S., Correia, P., & Lobo, M. (1995). *Normas Urbanísticas - Principios e Conceitos fundamentais* (2ª ed.). Lisboa: DGOTDU - UTL.

Dias, C. (2002). A componente geográfica nas estatísticas oficiais. In:E-SIG'2002 - VII encontro de utilizadores de Informação Geográfica, Oeiras, 13-15 November.

DGOTDU - Direcção-Geral do Ordenamento do Território e Desenvolvimento Urbano. (2002). *Normas para a programação e caracterização de equipamentos colectivos.* Lisboa: Direcção de Serviços de Estudos e Planeamento Estratégico - Divisão de Normas.

Geirinhas, J. (2001). *BGRI - Base Geográfica de Referenciação de Informação. Conceitos e Metodologias.* Acesso em 24 de Outubro de 2008, disponível em INE e Direcção Regional de Lisboa e Vale do Tejo: www.ine.pt/ngt_server/attachfileu.jsp?parentBoui=107197&attdisplay=n&attdownload=y

George, P. (1989). *O Homem na Terra - A Geografia em Acção.* Lisboa: Edições 70.

Green, M., & Flowerdew, R. (1996). New evidence on the Modifiable Areal Unit Problem. In: P. Longley, & M. Batty, *Spatial Analysis: Modelling in a GIS Environment* (pp. 41-55). Nova Iorque: John Wiley & Sons.

Haining, R. (2003). *Spatial Data Analysis: theory and practice.* Londres: Cambridge University.

Júnior, R., & Costa, S. (2007). Metodologia para caracterização sócio-economica do espaçoconstruído utilizando Geotecnologias. *Anais XIII Simpósio Brasileiro de Sensoreamento Remoto*, (pp. 5435-5442). Fiorlanópolis.

Koga, D. (2002). Cidades entre territórios de vida e territórios vividos. *Serviço Social & Sociedade , 72*, 22-52.

Loureiro, A. (2007). *Rede de Apoio Social - a cidade das pessoas e para as pessoas.* Tese de Licenciatura em Geografia, Ordenamento do Território e Desenvolvimento, Faculdade de Letras da Universidade de Coimbra.

Machado, M. C., Santana, P., Carreiro, M. H., Nogueira, H., Barroso, M. R., & Dias, A. (2007). *Iguais ou Diferentes? Cuidados de Saúde materno-infantil a uma população de imigrantes.* Laboratórios BIAL.

Martin, D. (2000). Census 2001: Making the best of zonal geographies. In: U. d. Manchester (Ed.), *The Census of Population: 2000 andbeyond.*

Martin, D. (1995). *Geographic Information Systems: Socioeconomic Applications.* Londres: Routledge.

Openshaw, S. (1978). An empirical study of some zone-design critiria. *Environment and Planning, 10*, 781-794.

Openshaw, S. (1996). Developing GIS-relevant zone-based spatial analysis methods. In: P. Longley, & M. Batty, *Spatial analysis: modelling in a GIS environment* (pp. 55-73). Cambridge: GeoInformation Internationnal.

Openshaw, S. (1984). Ecological fallacies and the analysis of areal census data. *Environment and Planning, 16*, 17-31.

Partidário, M. (1999). *Introdução ao Ordenamento do Território.* Lisboa: Universidade Aberta.

Ramos, F. (2002). *Análise espacial de estruturas intra-urbanas: o caso de São Paulo.* Tese de Mestrado, Instituto Nacional de Pesquisas Espaciais, São José dos Campos.

Ramos, F. (2004). Cartografias sociais como instrumentos de gestão social: a tecnologia a serviço da inclusão social. *IX Congresso Internacional del CLAD sobre la Reforma del Estado y de la Administration Pública*, (pp. 1-10). Madrid.

Ramos, F. R. (Outubro de 2000). Medidas Territoriais: Bairro, Distrito, Zona, Interdistrital, Intradistrital, Intermunicipal e outros Recortes do Espaço Urbano. São José dos Campos.

Rathener, H. (1974). *Planejamento Urbano e Regional.* São Paulo: Editora Nacional.

Santana, P. (2005). *Geografias da Saúde e do Desenvolvimento. Evolução e Tendências em Portugal,* Coimbra: Almedina

Santana, P., Nogueira, H., Costa, C., & Santos, R. (2007). Identificação das vulnerabilidades do ambiente físico e social na Construção da Cidade Saudável. In:

Paula Santana (coord.), *A Cidade e a Saúde* (pp. 165-179). Coimbra: Almedina.

Santana, P., Nogueira, H., Costa, C., & Santos, R. (2007). Avaliação da qualidade ambiental dos espaços verdes urbanos no bem-estar e na saúde. In: Paula Santana (coord.), *A Cidade e a Saúde* (pp. 165-179). Coimbra: Almedina.

Santana, P., Santos, R., Costa, C., & Loureiro, L. (2008). *Pensar Amadora Saudável e Activa.* 3º Prémio de Reconhecimento Científico da Rede Portuguesa de Cidades Saudáveis.

Santana, P. & Nogueira, H. (2008). Environment and Health: Place, sense of place and weight gain in urban areas. In: J. Eyles & Williams (eds.) *Place, sense of place and quality of life*, pp. 153-165.

Santana, P. (2009), Por uma Cidade Saudável, In: *JANUS, 2009 – Portugal no Mundo "Aspecto da Conjuntura Internacional. A Saúde no Mundo".* Jornal Público / Universidade Autónoma de Lisboa (p.83-84).

Santos, M. (2000). *Território e Sociedade: entrevista com Milton dos Santos.* São Paulo: Fundação Perceu Abramo.

Simões Lopes, A. (2001). *Desenvolvimento Regional* (5ª ed.). Lisboa: Serviços de Educação e Bolsas da Fundação Calouste Gulbenkian.

Teixeira, J. & Antunes, A. (2008). A hierarchical location model for public facility planning. In: *European Journal of Operational Research*, 185 (pp. 92-104).

Valle, J. (1993-94). Dinámica Demográfica y Planificación Urbana. *Cuadernos Geográficos* .

Wrigley, N., Holt, T., Steel, D., & Tranmer, M. (1996). Analysing, modelling, and resolving the ecological fallacy. In: P. Longley, & M. Batty, *Spatial analysis: modelling in a GIS environment* (pp. 25-40). Cambridge: GeoInformation International.

Weinstein, M. (1980). *Health in the city*, New York: Pergamon Press Inc.

A Geo-business Classification for London

Patrick Weber[1], Dave Chapman[2]

[1] Department of Computer Science, University College London, Gower Street London, WC1E 6BT
Tel: +44 (0)20 7718 5430 | Email: p.weber@ucl.ac.uk
[2] Department of Management Science and Innovation, University College London, Gower Street, London, WC1E 6BT
Tel: +44 (0)20 7679 0441 | Email: d.chapman@ucl.ac.uk

Abstract. This paper discusses the methodology and processes required to implement a geo-business classification to aid spatial decision making in the context of foreign direct investment promotion for London. This research is both timely and relevant since there is need for better decision support tools that will improve sub-regional location decision making ensuring London's diverse business neighbourhoods are presented effectively to potential investors.

The research methodology presented in this paper adopts principals and practices common place in consumer marketing in the form of geodemographic classification. The five key data domains associated with companies, working property stock, general living environment and accessibility were used to gather a range of input variables. These variables were then used as the input to a principal components analysis which simplified the data into 9 dimensions describing and contrasting London's diverse business neighbourhoods. These geo-business area profiles will form the basis for spatial decision support tools for business location decision making.

1 Introduction

The forces of globalisation have resulted in strong competition between countries, regions and cities to attract and retain increasingly mobile investment and talent. As one of the leading 'world cities', London is home

M. Painho et al., (eds.), *Geospatial Thinking*, Lecture Notes in Geoinformation and Cartography,
DOI 10.1007/978-3-642-12326-9_7,

to a highly internationalised workforce and is particularly reliant on these sources of foreign direct investment (FDI). In the face of increasing global competition and a very difficult economic climate, the capital must compete effectively to encourage and support such investors. As part of London's marketing and support infrastructure, its official foreign direct investment agency Think London seeks to develop a detailed understanding of the needs of inward investors in order to develop and deliver relevant support to inbound organisations.

Through a collaborative study with Think London, a geo-business classification relevant to business location decisions is presented here, based on an alternative geography of London's complex multi-nodal business landscape. The impetus behind such a classification is the need for geographical areas within London exhibiting similar characteristics to be ranked, visualised and compared to enhance the marketing of these areas for business location decision making.

The development of the geo-business classification draws upon the methods and practices common to many geospatial neighbourhood classifications. However our analysis is novel in that it is built upon a geography that divides the capital in meaningful socio-economic units before developing profiles for each region that enable investors to make meaningful comparisons.

2 Geodemographic technology and their relevance to geo-business classifications

Geodemographic classifications, also known as neighbourhood classifications or typologies, cluster together small areas at a consistent geographic scale according to the socio-economic similarity of residential populations (Harris et al. 2005). The underpinning notion is that similar people live in similar types of neighbourhoods, go to similar places, do similar things and behave in a similar manner as in the old adage; "Birds of a feather flock together". It is a term most commonly attached to the analysis of people according to where they live (P. Sleight 1997).

The methodology for developing a geodemographic classification can be described in the following steps (Harris et al. 2005, chap.6):

1. Evaluating potential data sources, their reliability, robustness and appropriateness.
2. Normalisation of absolute values to a base count.
3. Transforming variables through standardisation (z-scores) to make them comparable.

4. Identifying highly correlated variables to enable deletion of superfluous duplicate variables. Principal Components Analysis identifies the main differentiating components of a group of correlated variables.
5. Selecting weights to be attached to the different variables.
6. Classifying neighbourhoods into a set of clusters, through an iterative allocation-reallocation process such as K-means clustering. Individual clusters are grouped into types according to similarities.
7. Developing profiles for each cluster through the presentation of summary labels, portraits, indicative photographs, descriptive prose, charts and maps.

Contemporary commercial classifications have become commonplace and widely used to provide operational, tactical and strategic context to decisions related to questions of "where?" (Longley & Clarke 1995), for example to understand retail location choices in terms of access to consumers and demand for retail locations (Harris et al. 2005, p.4). Today in the UK, different types of geodemographic classifications exist, both commercial such as MOSAIC (Experian 2009), and public such as 2001 Output Area Classification (Office of National Statistics 2009).

It is apparent that companies also behave in a similar fashion by flocking together, see for example Porter (2000). The drive to be competitive and tap into existing information, social and knowledge networks leads businesses to locate near to each other, leading to an inherent spatial autocorrelation in the pattern of business clusters. See for example Saxenian (1991) excellent contribution to detailing co-location benefits for Computer Systems firms in Silicon Valley, California. Therefore it makes sense that the development process for a geo-business classification of London neighbourhoods follows closely the methodology outlined for geodemographic classifications.

3 A spatial data framework for a geo-business classification

The review of geodemographic classifications highlighted two important components that need to be established prior to the development of a geo-business classification: (1) appropriate input data variables need to be selected and integrated into a coherent database and (2) these variables need to be measured using a consistent geographical unit of analysis. The following section will address how this can be achieved for business location decision making in the context of foreign direct investment into London; identifying the most important variables and how they can be integrated into a coherent and relevant geographical framework able to form the basis on which to develop London area profiles.

4 Data framework

This research builds on previous work carried out by Weber & Chapman (2009), which highlighted the potential of geographical analysis in the collection, analysis and presentation of data to support foreign direct investment business location decision making, A comprehensive review of published literature, industry surveys, user-needs analyses and live enquiries allowed the authors to identify the critical factors and processes driving company decisions regarding location selection. The database developed in this case study supports the five major domains of locational knowledge needed for FDI promotion activities in London:

1. the discovery, quantification and qualification of industry sector clusters (*Companies*),
2. the characterisation of the available talent pool and daytime population (*Working Population*),
3. quantity and quality of the *Property Stock*,
4. a more general appreciation of the *Living Environment* of London neighbourhoods.

The fifth domain in the study is *accessibility* which is excluded in this research paper. In location decision making, accessibility is considered to be heavily dependent on individual requirements of investors, based on their sectorial provenance and functional activity. Both attraction, defined as collocation with potential suppliers, partners, customers, as well as repulsion, the desire to be located away from competitors, are relevant location variables. Accessibility to certain areas of London needs to be considered separately in a later stage of the decision making process, and is excluded from the spatial database informing the development of geo-business classification.

A detailed presentation of the 4 remaining domains of relevant location variables is given below, and also summarised in Table 1:

Companies

The discovery, quantification and qualification of collocation, or clusters, of similar firms is frequently requested in the context of FDI promotion activities, and ties in with a desire of potential investors to obtain an overview of economic activity patterns across London.

The Annual Business Inquiry (ABI) is a government statistical data set built from a sample of the Inter-Departmental Business Register, the comprehensive UK business register used for the generation of statistical data on companies and economic activity (Office of National Statistics 2008). It is generated through the aggregation of individual units to Standard Industrial Classification (SIC), employment size and local units. The variables available include number of enterprises, total turnover, approximate Gross Value Added, purchases, as well as number of employees. For this study, ABI employment statistics aggregated at Lower Super Output Area level and significant target industry sectors are used (LDA 2003). These target sectors highlight the most important industry sectors to London's economy and offer a relevant and easy to understand framework to analyse economic activity patterns in London.

Unfortunately, standard industrial classifications only record the industrial sectors in which companies operate, but do not encode any functional division of labour. It is impossible with the standard industrial classification to distinguish company headquarters and production facilities. Although this information would be very useful for a better understanding of the economic landscape of London such data is not available in any current London-wide dataset.

Following the methodology for geodemographics, the normalization of absolute counts is needed to allow the evaluation and comparison of different areas of London. Location Quotients (Leigh 1970), represent a ratio between the employment proportion locally and the national average of this proportion and are derived from the following formula:

$$LQ_i = \frac{e_i/e}{E_i/E}$$

where e_i represents the local area employment in ICT, e the total employment at the local area, E_i employment in ICT for London and E total employment in London.

Employee counts for each Lower Super Output Area were transformed to a Location Quotient which computes the relative concentration of jobs in a particular sector, identifying if a sector is over or underrepresented relative to the overall London average. Location Quotients enable the development of local activity profiles which inform analyses of the relative attractiveness of a given area with respect to particular industries.

A second measure of the business environment is the average size in terms of number of employees of firms in a given area. The ABI data both publishes data on the number of workplaces, which can be equated to a company, and the number of people working in a specific area. The ratio of employees divided by workplaces represents the average size of a workplace, as a proxy for average company size.

Characterising the workforce

The second data framework domain corresponds to the understanding of workforce differences across different areas of London. The definitive data source for population statistics is the UK Census from 2001. Using data from the Special Workplace statistics 2001, it is possible to access commuting flows, i.e. the travel to work patterns, at Census Ward level.

This removes a serious limitation from the usual Census dataset, namely that census data normally refers to the resident population of an area. Through access to the Special Workplace Statistics, a link is established between employees at their place of work or study and broader socio-economic data captured in the Census relating to their place of domicile. From this link, geo-demographic characterisations of the day-time population of sub-regions and neighbourhoods of London are possible.

The National Statistics Socio-Economic Classification measures employment relations and conditions of occupations and represents the best approximation of the qualification level of the local workforce. Given that the place of work is of interest, the count of people in the category "*Never worked and long term unemployed*" by definition is nil for the whole dataset, and the variable is excluded from the database.

Normalisation of the data through the generation of Location Quotients enables a better appreciation of relative over or underrepresentation of specific workforce classes in an area, irrespective of its geographic size.

Property stock

Another important facet of the data domains and of the qualitative makeup of London localities is the property stock available to potential investors. Investors looking to setup a new business location in London need data on both the quantity, and quality of the property stock.

The commercial FOCUS database collects information on the commercial property stock across London recording property transactions such as sales and lettings. Unfortunately upon closer inspection, FOCUS presents several fundamental drawbacks. First, the quality and completeness of the data records regarding property transactions is less than satisfactory, specifically regarding property rental or sale prices. Also, access to historical transaction data is only possible on a request basis involving significant consultancy charges.

Given these drawbacks, data gathered by the government through the work of the Valuation Office Agency is another source of property data. The Valuation Office Agency is charged with the collection of business rates, a tax on the occupation of non-domestic property based on certain variables of the property. These include quality, size, location and consider the economic conditions prevalent at the time of the estimation. Rateable Value can then be considered a proxy variable for commercial property quality, and by extension price. The Rateable Value statistics are complete, updated annually, publicly available and contain both indicators of quality/price of business premises, as well as information on the total floorspace for a set of business premises classes, such as factories, warehouses, retail premises and offices.

To make areas comparable, Location Quotients for these variables are generated and used in the spatial database instead of the source data.

Liveability

The work in developing a better understanding of location factors relevant to FDI also highlights the need to qualify a more informal living environment quality indicator. Such an indicator highlights the quality of the working environment, influencing the attractiveness to the workforce of the business location environment.

The Index of Multiple Deprivation (IMD) 2007 is a multiple indicator of deprivation structured into a series of domains, provided as a basis for policy making (DCLG 2009). The IMD is composed from data collected on persons or households in receipt of various government benefits made up of 7 domains.

The IMD allows the addition to the spatial database of a small scale indicator of liveability, or general attractiveness as a place of work for companies looking to setup in London.

Table 1. Summary table of initial data framework

Class	Source	Variables		Geography
Companies	Annual Business Inquiry 2007	Creative industries	Environmental	LSOA
		Higher Education & Research	Construction	
		Health	Retail	
		Social work	Transport & logistics	
		Tourism & leisure	Charity & voluntary	
		Utilities	Life sciences	
		Professional services	Pharmaceuticals	
		Financial services	Medical equipment	
		Food & drink	Manufacturing	
		ICT	Real estate	
		Ratio of Workplaces over Employees		
Working Population	Census 2001: Special Workplace Statistics	Higher managerial and professional Occupations: Large employers and higher managerial Occupations Higher professional Occupations		Census Wards
		Lower managerial and professional Occupations		
		Intermediate Occupations		
		Small employers and own account workers		
		Lower supervisory and technical occupations		
		Semi-routine occupations		
		Routine Occupations		
		Never worked and long-term unemployed		
Property Stock	Rateable Value Statistics 2007	Rateable Value per square meter - Offices		MSOA
		Rateable Value per square meter - Premises		
		Rateable Value per square meter - Factories		
		Rateable Value per square meter - Warehouses		
		Total Floorspace - Offices		
		Total Floorspace - Retail Premises		
		Total Floorspace - Factories		
		Total Floorspace – Warehouses		
Living Quality	Index of Multiple Deprivation 2007	Overall Score		LSOA

5 Defining a geographic framework

Looking back at the history of London and its urban development, Greater London has come into existence as a construction of many individual towns and cities (Hebbert 1998; Thurstain-Goodwin & Unwin 2000; Ackroyd 2001; URBED 2002). These nuclei of urban development still survive today and have separate identities and differing characteristics, in an economic sense, as well as socially, demographically and culturally.

The Town Centre Statistics Project satisfies the condition of not being based on pre-existing delineations, as it aimed to define town centre boundaries along with attached economic and social statistical indicators, for all towns falling within the M25. The result of the research was the production of an Index of Town Centeredness – which form thresholds which define town centre boundaries constrained by size and functional activity (Thurstain et al. 2001; Lloyd 2004).

Fig. 1. Overview map of the Town Centres inside the Greater London Authority Area

The Town Centre boundaries for London (see Figure 1) specifically define 208 boundaries representing the quantitative expression of the nucleus constituting London “Towns” described in history and defined previously through anecdotal evidence. Town Centre boundaries provide the geographical framework for the provision of quantifiable and comparable indicators, fit to constitute the basis of a set of Area Profiles for Greater London.

5.1 Transforming variables into a unified geographic framework

Given that the Town centre boundaries are derived from a set of surfaces contoured to a specified threshold, they are not coincidental with any existing administrative, statistical or political boundaries. In Figure 2, one can see an example of the problem for the exemplar Town Centre of Canary Wharf. Although the Town Centre boundary describes the area of principal town centre activity, the boundaries for the individual wards holding data from the spatial database are not coinciding. In fact, the town centre boundary intersects more than one ward boundary.

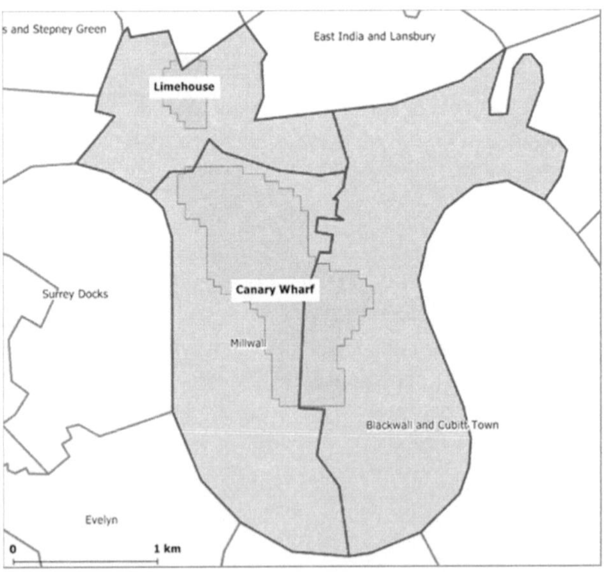

Fig. 2. Canary Wharf Town Centre Polygon and the selected touching Wards used for analysis

To transform the spatial database of location variables, data is aggregated for each Town Centre from the intersection between each Town Centre and respective spatial data units. The Town Centre statistics inherit an average value of the surrounding spatial data units. The inherent overestimation of the data collection area, allows the qualification of London neighbourhoods on a wider scale than the narrow view set by the Town Centre boundaries.

6 Developing a geo-business classification

The spatial database joined up into a coherent and relevant geographical framework, developed in the previous section, contains 50 variables deemed relevant to the FDI location decision making process.

Decision making, through comparison of Town Centres along each of these 50 variables, is possible but remains cumbersome due to the number of variables and amount of data involved. A data reduction and aggregation method is needed, satisfying the conditions of the variables included in the new system to sufficiently, although parsimoniously, tap the domain of the constructs in question, while the constructs, in turn, sufficiently, although parsimoniously, model the phenomena in question. (Bacharach 1989, p.506)

The following section is concerned with the development of a new classification system for Town Centres, reducing the number of variables involved to a manageable level, identifying and eliminating highly correlated variables, obfuscating judgements on location options. The classification is brought to life through the development of profiles detailing the different geo-business dimensions, giving a detailed account of the main distinctive characteristics of each class. Such a parsimonious classification allows for the meaningful characterisation and comparison between FDI location options.

This classification does not attempt to incorporate measures of accessibility, which are dealt with separately in the SDSS framework, instead it focuses on the social and economic characteristics of each town centre.

7 Principal Components Analysis

In exploratory analysis models of highly dimensional datasets, one of the techniques commonly used to uncover a variable structure is a Principal Components Analysis (PCA). In the context of geodemographic classifications, PCA is used to isolate the main differentiating factor or components of a group of correlated variables. The structure of these principal components is such that the first component accounts for as much data variability as possible and each succeeding component there after accounts for a decreasing amount of variability. For a detailed discussion of the methodology of PCA, refer to Robinson (1998, p.121).

PCA is used to reduce the spatial database of 50 input variables to a smaller number of components characterising both the common and unique variance of the original dataset.

Discussion of Outputs

The principal output from the PCA is contained in Table 2, showing generated components in decreasing order of contribution to the overall variance of the constituting variables. The eigenvalue of a given component measures the variance in all the variables which is accounted for by that component.

The plotted eigenvalues of the individual components help make an informed decision on the threshold beyond which supplemental components only add little to the model. Apart from subjective appreciations by the researcher of the interpretability of a given component, i.e. does a component make any sense and can be interpreted, a commonly accepted cut-off criterion is when the eigenvalue of a component drops below a value of 1, according to the Kaiser criterion (Kaiser 1960). There are numerous other methods to determine the component number threshold, for a detailed discussion and evaluation see Jackson (1993).

Table 2. Principal Components Analysis Output of Eigenvalues and contribution to overall variance of dataset

Comp	**Name**	**Initial Eigenvalues**			**Rotation Sums of Squared Loadings**		
		Total	**% of Variance**	**Cumulative %**	**Total**	**% of Variance**	**Cumulative %**
1	Urban professionals	7.943	20.368	20.368	4.602	11.800	11.800
2	Blue Collar Industry	4.606	11.810	32.178	4.069	10.433	22.233
3	Blue Chip Finance	2.942	7.543	39.721	3.324	8.523	30.756
4	Third Sector Centres	2.407	6.171	45.892	2.832	7.262	38.018
5	Big Sheds and Trucks	2.174	5.575	51.468	2.508	6.431	44.450
6	High (End) Streets	1.756	4.503	55.970	2.405	6.168	50.617

Table 2. (cont.)

Comp	Name	Initial Eigenvalues			Rotation Sums of Squared Loadings		
		Total	% of Variance	Cumulative %	Total	% of Variance	Cumulative %
7	Creative & Green Minds	1.460	3.742	59.713	2.298	5.893	56.511
8	Sights of London	1.397	3.582	63.294	1.678	4.302	60.812
9	Ivory Towers	1.215	3.115	66.410	1.602	4.108	64.921
10		1.130	2.898	69.307	1.476	3.784	68.705
11		1.098	2.816	72.124	1.333	3.419	72.124

Another possible indicator for the choice of component threshold is the Cattell Scree Plot (see Figure 3). This plot draws the components on the X axis, and the corresponding eigenvalues on the Y axis. As a rule, when the drop in the curve almost ceases, as to make an "elbow", Cattell advises to drop all further components after the one starting the elbow.

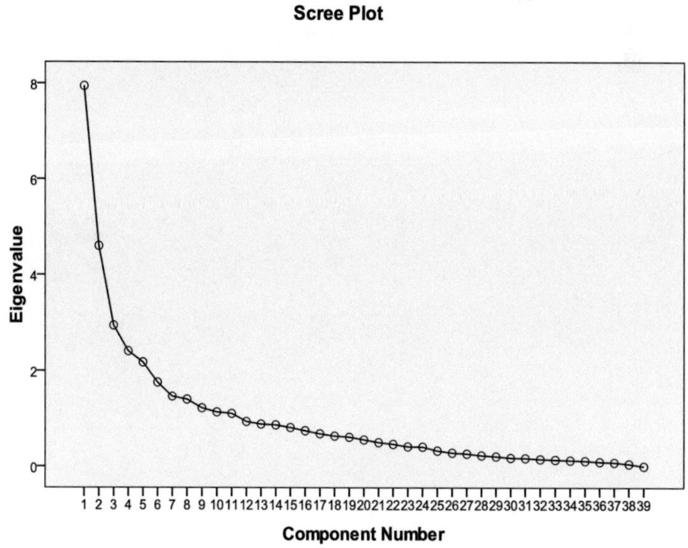

Fig. 3. Scree Plot of PCA output

Although these indicators are widely recognized and valid, the final decision on the number of components to retain rests on a careful consideration of the components and their contribution to the model in terms of explanatory value. Looking at the component loadings, which relate individual variables to the individual components, the interpretability of the components is tested to see if each component and its constituting variables add to the explanatory power of a model of London Town Centre characteristics.

To relate the components back to the individual variables, PCA generates a new correlation matrix (see Table 3). This loadings matrix represents the correlation coefficients between variables (rows, in our case the individual variables), and the components (columns). The squared sum of loadings of a given component is equal to the variance among the variables explained by the given component.

For example a component loading of 0.7 confirms that the variable is very well represented by a particular factor as half of the variance of a variable is explained by that component. The component loadings matrix allows the interpretation of the components to make an informed final decision on the number of components to retain.

Table 3: Component Loadings matrix.

Category	Variable	Components 1	2	3	4	5	6	7	8	9
Business	Retail	-0.85	-0.07	-0.07	-0.12	-0.03	0.23	0.11	0.16	-0.11
Workforce	Full Time Students	-0.84	-0.19	0.04	-0.04	-0.08	0.05	0.08	0.04	0.01
Workforce	Semi routine occupations	-0.80	0.32	-0.06	0.06	-0.07	-0.16	-0.22	-0.01	-0.06
Property	Floorspace - Retail Premises	-0.65	-0.10	-0.26	-0.02	-0.32	0.06	-0.10	0.01	-0.02
Workforce	Higher Professional Occupations	0.62	-0.37	0.27	-0.18	-0.18	0.22	0.12	0.04	0.35
Workforce	Large Empl/ & H. Manag. Occ.	0.56	-0.14	0.47	-0.25	0.07	0.14	0.38	0.04	-0.03
Workforce	Lower Manag. & Prof. Occ.	0.53	-0.50	-0.11	0.07	-0.02	0.48	0.23	0.02	0.03
Property	Floorspace - Offices	0.48	-0.39	0.47	-0.24	0.02	0.19	0.24	0.24	-0.11
Business	Professional Services	0.45	-0.35	0.02	-0.18	-0.22	0.11	0.24	0.23	-0.08
Property	RV per sqm - Offices	0.42	-0.30	0.17	-0.36	0.10	0.38	0.26	0.32	-0.06
Workforce	Routine Occupations	-0.18	0.81	-0.15	0.08	0.17	-0.17	-0.22	-0.13	-0.03
Business	Manufacturing	0.10	0.79	-0.03	-0.06	-0.02	-0.16	0.26	-0.05	-0.01
Workforce	Lower Superv. & Techn. Occ.	-0.24	0.74	0.02	-0.09	0.27	-0.23	-0.14	0.01	-0.11
Business	Food and Drink	0.01	0.56	-0.21	0.05	-0.10	0.03	-0.01	-0.01	0.05
Workforce	Intermediate Occupations	0.02	-0.23	0.83	0.02	0.04	-0.07	-0.09	-0.03	-0.06
Workforce	Small Empl. & Own Acc. Wrk.	-0.25	-0.05	-0.73	0.18	-0.08	-0.33	-0.14	0.01	-0.13
Business	Ratio Employes to Workplace	0.08	-0.04	0.73	-0.15	0.29	-0.03	-0.14	-0.04	0.03
Business	Financial Services	0.41	-0.25	0.42	-0.22	-0.28	0.08	-0.04	0.31	-0.11
Business	Charity and Voluntary	-0.03	0.02	-0.09	0.87	0.10	0.01	0.10	0.12	-0.05
Business	Social Work	-0.03	-0.11	-0.21	0.82	0.01	-0.04	-0.08	0.07	-0.05
Liveability	IMD 2007 Score	0.04	0.45	0.21	0.59	0.18	0.18	-0.02	-0.04	-0.10
Property	RV per sqm - Factories	0.07	-0.10	-0.03	0.29	0.69	-0.08	0.11	-0.11	-0.13
Business	Transport and Logistics	0.06	0.25	0.22	-0.19	0.67	-0.08	-0.04	-0.28	-0.06
Property	RV per sqm - Warehouse	0.06	0.01	0.31	-0.01	0.63	0.12	-0.04	0.31	0.12
Property	Floorspace - Warehouses	0.09	0.48	0.06	0.19	0.55	-0.05	-0.07	-0.02	0.15
Property	Floorspace - Factories	0.15	0.33	-0.13	0.28	0.46	-0.16	-0.10	-0.10	-0.13
Business	Construction	-0.02	0.22	-0.17	-0.05	-0.01	-0.71	-0.03	0.02	-0.24
Property	RV per sqm - Retail Premises	0.08	-0.17	0.20	-0.40	-0.06	0.58	0.14	0.39	-0.13
Business	Real Estate	-0.02	-0.22	-0.32	0.05	-0.08	0.53	0.01	-0.06	-0.11
Business	ICT	0.22	-0.11	0.01	-0.25	0.25	-0.44	0.31	0.17	0.04
Business	Creative Industries	0.08	-0.17	-0.23	0.00	0.05	0.21	0.82	0.09	-0.06
Business	Environmental	0.12	0.05	0.07	0.06	-0.12	-0.12	0.74	-0.10	0.18
Business	Healthcare	0.08	-0.21	-0.12	0.04	-0.11	0.00	-0.45	-0.38	0.18
Business	Pharmaceuticals	0.05	0.07	0.05	-0.22	0.09	0.05	0.00	-0.74	-0.07
Business	Tourism and Leisure	-0.14	-0.14	-0.42	-0.23	0.05	0.33	0.02	0.42	-0.02
Business	Life Sciences	0.01	0.05	0.02	-0.08	0.05	-0.05	0.15	0.03	0.79
Business	Higher Education & Research	0.14	-0.08	-0.04	-0.02	-0.06	0.15	-0.13	0.00	0.71
Business	Medical Equipment	0.02	0.10	-0.11	0.07	-0.07	0.04	-0.12	-0.15	-0.05
Business	Utilities	-0.17	-0.06	0.08	-0.17	0.03	0.01	-0.09	0.01	-0.02

Previous information for thresholds from the Kaiser criterion (11 components), and the Scree Plot (8 components) means that a informed decision has to be reached by the investigator. Given that component 9 contains significant correlations with the Life Sciences and Higher Education industry sectors (refer to Table 3), this component seems to contain information related to potential clusters of biotech companies, a sector of great interest to FDI, and thus is retained in the final PCA output.

The final selected set of 9 components retained for further analysis retain 64 percent of the original variance, giving a reasonable basis on which to proceed.

8 Development of Component Profiles

Through the application of the Principal Components Analysis, discussed in the previous section, the 9 most significant components were identified. The characterisation of the Town Centres according to these 9 dimensions achieves the initial goal set to reduce the original spatial database of variables down to a much reduced set of components.

To make the results of the PCA analysis accessible to end users, and as a visualisation and exploration tool, PCA components profiles are developed. These profiles describe, quantify and qualify each component according to the most significant variables attached to each component. Through the component scores, Town Centre rankings for each component are also computed indicating how representative of a given component each Town Centre is. Component profiles contain the following elements:

1. **Most Positively/Negatively correlated variables:** The identification of both the strength and direction of correlation to the original variables.
2. **Component Name/Label:** A short, memorable and distinctive component name allows the classification user to grasp the overall characteristic of a component, and to help memorisation.
3. **Keywords and/or Narrative:** Using the component loadings, an expanded narrative description of representative characteristics is developed in terms of of economic activity, liveability, socio-economic makeup of the workforce and property stock.
4. **Most/least representative Town Centres:** Through the component scores, a ranked list allows the identification of the most "typical" or "atypical" Town Centres.

5. **Example streetviews of most/least representative Town Centres:** A pictorial narrative is developed, showing typical street views, shops and other pictures giving a general context of the Town Centre profile.

The following section presents the profiles of the 9 components, in order of decreasing contribution to the overall variance of the spatial database. These profiles form the final geo-business classification retained to support business location decision-making.

Urban Professionals

This component characterises typical Central London activities in the knowledge and professional services industry, from big accountancy and professional services firms, across to major financial institutions. The property stock is mainly devoted to high quality office spaces and limited retail space of less than average value. Given that this component mainly represents the knowledge economy, there is an above average concentration of a highly skilled workforce. The Component is also marked by a relative scarcity of semi routine occupations, or full-time students.

The most representative Town Centres for this component thus are the individual Town Centres that make up the City of London global financial centre, along with Canary Wharf and Holborn in Central London.

Blue Collar Industry

"Blue Collar Industry" is mainly concerned with manufacturing activities, including the food and drink industry. Interestingly, logistics and distribution make a less significant contribution to the component characterisation (at a factor loading of 0.25). The workforce is mainly composed of supervisory, routine and technical employees. The property stock is characterised by an abundance of warehousing, with a relative scarcity of office space.

The geographic distribution of the most representative Town Centres brings up Dagenham, famous for its manufacturing tradition in the automotive sector such as the Ford motor factory, but also locations outside Central London in the East and North of London such as Bow, Tottenham, Edmonton, and Kenton in Harrow.

Blue Chip Finance

Although at first glance the component presents similar characteristics to "Urban Professionals", there are some marked differences. In this component, there is a significant correlation with the presence of larger firms, which is not the case for "Urban Professionals". Financial services are the only significantly positively correlated business sector, but not Professional Services (loading of 0.02). Workforce qualification levels are lower when compared to "Urban Professionals" as well. There is a marked negative correlation of self employed and small employers. Similar to "Urban Professionals", there is comparatively less retail space, and not a lot of tourism.

Looking at the geographical distribution of the most representative Town Centres for the component, again, the City of London Town Centres and Canary Wharf score highly, although it is worth noting that Croydon Retail Core comes in fourth.

Third Sector Centres

The "Third sector Centres" component is characterised by presence of both Charity and Voluntary as well as Social Work industry sectors, located in deprived areas, with sparse local work opportunities apart from social work and third sector charity work. Although there is retail activity, the property stock is of low value, with some small offices, again of low value. There is evidence of some factory stock.

The geographic distribution of the most representative "Third Sector Centres" is composed mainly of deprived inner and outer London Town Centres such as North Kensington, Brixton, Norbury, Maida Hill (see Figure 4). It is interesting to note that North Kensington is one of the most representative Town Centres, versus South Kensington as one of the least representative areas. Deprivation plays an important role in the characterisation of this component, and steep social gradients between neighbouring areas of London explain these extreme contrasts, an issue explored in more detail in Harris & Longley (2002).

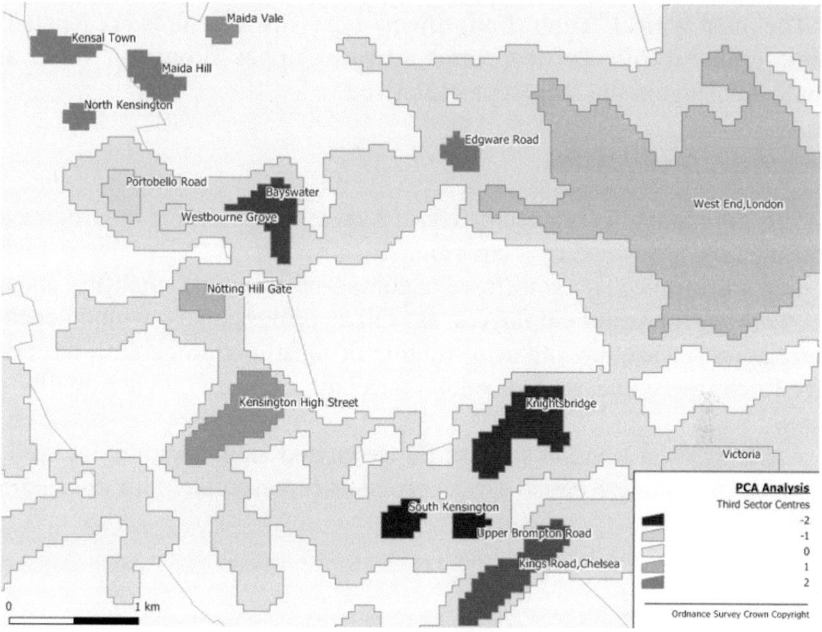

Fig. 4. Example map of least and most representative neighbouring Town Centres in West London according to component "Third Sector Centres"

Big Sheds and Trucks

"Big sheds and trucks" is marked by a significantly above average concentration of transport and logistics businesses, the property stock is mainly composed of both warehousing and factories, with above average value. The workforce is mainly composed of lower skilled workers, and there is a clear negative correlation, and thus lack of financial service businesses.

The most representative Town Centres are in West London focused around Heathrow as one of the most important international transportation hubs of London.

High (End) Streets

The label of the sector "High (End) Streets" is deduced from the concentration of correlated variables related to high end retail sector activities, including the real estate industry. High quality retail and office premises are significantly correlated, along with a workforce mainly comprised of lower managerial and professional occupations. This component is also marked by lack of ICT and Construction companies.

The most typical "High (End) Street" type Town Centres are located in West London with Town Centres such as Upper Brompton Road and South Kensington the most representative.

Creative & Green Minds

This component is representative of a creative sector and environmental consultancy type business environment, also characterised by a highly skilled workforce. The workforce is composed of highly qualified managers working for larger employers, as well as professionals, with a negative correlation for routine and semi-routine occupations. Regarding the property stock, there is a weak positive correlation towards higher quality offices.

Areas in West London such as Battersea and Hammersmith, as well as Camden (including Kentish Town) are most representative for this component.

Sights of London

"Sights of London" is named after the concentration of tourism and leisure, as well as retail activities, together with significant correlations with a property stock of high value offices. Geographically, these conditions are mostly found in Central and West London, which also are the main hubs of tourism activity. There are significant negative correlations with the Pharmaceutical and Healthcare industries, as well as with Transport and Logistics.

Ivory Towers

Given the significant correlation of this component with Life Sciences industries and Higher Education Institutions, the component is deemed useful to FDI promotion. Apart from these two sectors, the workforce is qualified as higher professional occupations. Given this combination of significant variables, it makes sense to relate this component to a dimension of collocation of universities and life science firms, possibly university spinouts. Due to the narrow definition of the component, the geographical distribution of the component scores highlights individual Town Centres where significant life sciences companies are located. This is the case for example in Mill Hill which hosts the National Centre for Medical Research, the largest UK medical research centre.

9 Concluding remarks

Given the methodology of Principal Components Analysis, each Town Centre is made up of several components in varying degrees, reflected by the component scores. For example, when comparing Canary Wharf and Cheapside (red squares and green triangles in Figure 5, both Town Centres have significant component scores for both "Urban Professionals" as well as "Blue Chip Finance". Dagenham, but contrast, is dominated by "Blue Collar Industry".

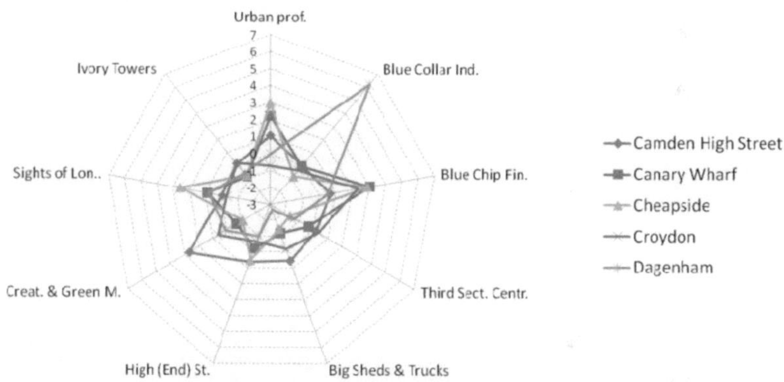

Fig. 5. Radar plot of factor loadings for a selection of London Town Centres

Comparing the methodology in this research to classical geodemographic classifications methodology, differences are clear. Although most steps have been observed, the geo-business classification at this point has not been run through a clustering algorithm to assign neighbourhoods to a set of discrete and exclusive clusters, to be characterised.

In the context of decision support for business location decision making, the aim isn't clustering and definition of exclusive labels attached to a set of areas. Rather it is to simplify and aggregate the spatial database of location variables. The components presented here are such a simplification, while still retaining the majority of the original datasets variance.

Through the collection of and processing of a set of variables relevant to business location making, a geo-business classification characterising London's diverse neighbourhoods emerged. The process, summarized in Figure 6, entailed the selection of relevant location variables as well as the standardisation and aggregation of these input variables into a spatial database within a consistent geographical framework. From this spatial database, through a Principal Components Analysis, 9 typical and distinct

London business neighbourhoods emerged, described in more detail through the generation of rich profiles describing each component, as well as scores relating each component back to its constituent variables.

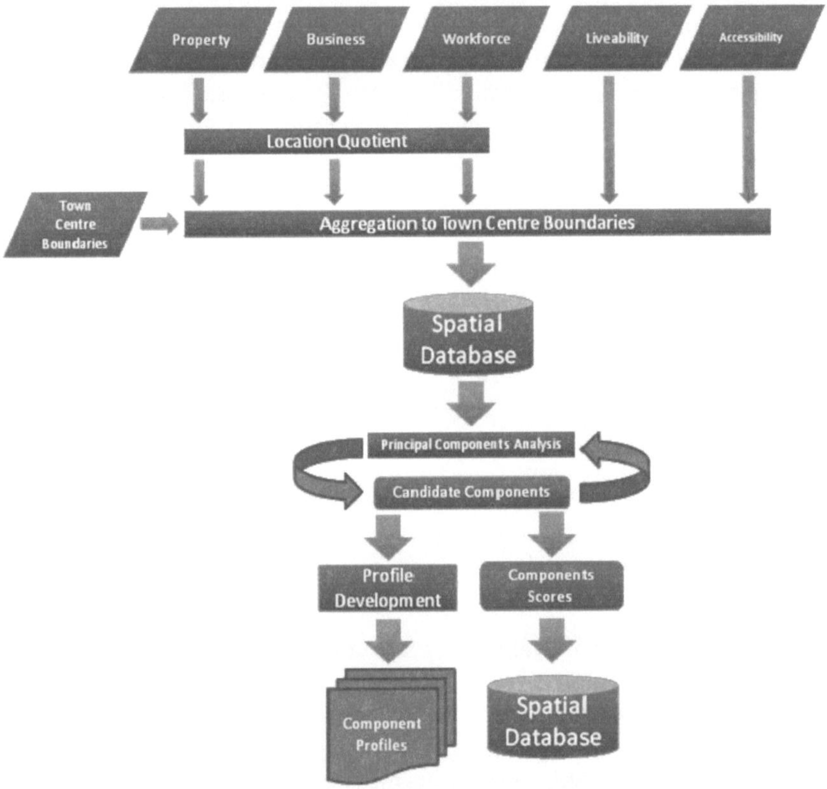

Fig. 6. Development Process for a geo-business classification

These profiles, built for each London town centre, can on their own be used to inform and improve FDI marketing through better location intelligence and identification of strengths and weaknesses of different areas.

At present, the work presented here is being integrated into the wider scope of the research project to develop a decision support tool to explore, characterise and compare different London locations. Users will be able to assign their own weights to different area characteristics, such as the area profiles presented here, but also other variables such as accessibility to markets, transport hubs or live property market data. The system allows the computation of a suitability index, used to rank Town Centres according to the investor's individual needs and presenting the user with a rich view to explore, compare and rank London's business neighbourhoods along with ancillary data.

References

Ackroyd, P., 2001. *London: The Biography* New edition., Vintage.

Bacharach, S.B., 1989. Organizational Theories: Some Criteria for Evaluation. *The Academy of Management Review*, 14(4), 496-515.

Charlton, M., Openshaw, S. & Wymer, C., 1985. Some new classifications of census enumeration districts in Britain: a poor man's ACORN. *Journal of Economic and Social Measurement*, 13(1), 69–96.

Department of Communities and Local Government, 2009. Indices of Deprivation 2007. *Indices of Deprivation 2007*. Available at: http://www.communities.gov.uk/communities/neighbourhoodrenewal/deprivation/deprivation07/ [Accessed July 21, 2009].

Experian, 2009. MOSAIC UK 2009. Available at: http://www.business-strategies.co.uk/ [Accessed July 20, 2009].

Harris, R., Webber, R. & Sleight, P., 2005. *Geodemographics, GIS and neighbourhood targeting*, John Wiley & Sons Inc.

Harris, R.J. & Longley, P.A., 2002. Creating small area measures of urban deprivation. *Environment and Planning A*, 34(6), 1073 – 1093.

Hebbert, M., 1998. *London: More by Fortune than Design*, London: John Wiley.

Jackson, D.A., 1993. Stopping Rules in Principal Components Analysis: A Comparison of Heuristical and Statistical Approaches. *Ecology*, 74(8), 2204-2214.

Johnston, R.J., 2000. *The dictionary of human geography*, Blackwell Publishers.

Kaiser, H.F., 1960. The Application of Electronic Computers to Factor Analysis. *Educational and Psychological Measurement*, 20(1), 141-151.

LDA, 2003. *Understanding London's Sectors*, London: London Development Agency.

Leigh, 1970. The use of location quotients in urban economic base studies. *Land economics*, 46(2), 202.

Lloyd, D., 2004. *Uncertainty in Town Centre Definition*. PhD Thesis. University College London.

Longley, P. & Clarke, G., 1995. *GIS for business and service planning*, John Wiley and Sons.

Office of National Statistics, 2009. National Statistics 2001 Area Classification. *Office of National Statistics Website*. Available at: http://www.statistics.gov.uk/about/methodology_by_theme/area_classification/default.asp [Accessed July 20, 2009].

Office of National Statistics, 2008. The IDBR Inter-Departmental Business Register; a key survey source. *The IDBR Inter-Departmental Business Register; a key survey source*. Available at: http://www.statistics.gov.uk/CCI/nugget.asp?ID=195 [Accessed July 21, 2009].

Pearson, K., 1901. On lines and planes of closest fit to systems of points in space. *Philosophical magazine*, 2(6), 559–572.

Porter, M.E., 2000. Location, Competition, and Economic Development: Local Clusters in a Global Economy. *Economic Development Quarterly*, 14(1), 15.

Robinson, G.M., 1998. *Methods and techniques in human geography*, London: J. Wiley.

Saxenian, A., 1991. The origins and dynamics of production networks in Silicon Valley. *Research Policy*, 20(5), 423-437.

Sleight, P., 1997. *Targeting Customers: how to use geodemographic and lifestyle data in your business*, NTC Publications Henley on Thames, UK.

Thurstain, M. et al., 2001. *Producing Boundaries and Statistics for Town Centres*, CASA, Center for Advanced Spatial Analysis, London.

Thurstain-Goodwin, M. & Unwin, D., 2000. Defining and Delineating the Central Areas of Towns for Statistical Monitoring Using Continuous Surface Representations. *Transactions in GIS*, 4(4), 305-317.

URBED, 2002. *A City of Villages: Promotion a sustainable future for London's suburbs*, London: Greater London Authority. Available at: http://www.london.gov.uk/mayor/planning/docs/tr11_villages.pdf.

Weber, P. & Chapman, D., 2009. Investing in geography: A GIS to support inward investment. *Computers, Environment and Urban Systems*, 33(1), 1-14.

Suggestive Geo-Tagging Assistance for Geo-Collaboration Tools

Marius Austerschulte, Carsten Keßler

Institute for Geoinformatics, University of Münster, Germany
m.austerschulte@gmail.com, carsten.kessler@uni-muenster.de

Abstract. An argumentation map is an online discussion forum for spatially related topics that combines the forum with an interactive map. The utility of an argumentation mapping tool highly depends on the accuracy and quantity of the geo-tags that link the discussion contributions to geographic locations. These geo-tags can be created manually by the users of the argumentation map or automatically by a geo-parsing application. However, in the case of manual geo-tagging users often do not create geo-tags as extensively as desired. In contrast, automatic geo-parsers have difficulties with the informal language often used in user-generated content and with resolving small-scale features.

This paper proposes a hybrid approach for geo-tagging user-generated content which involves the users in the process but supports them with an automatic geo-parser which suggests locations. The implementation of a prototype as well as a human participants test are presented in order to analyze the geo-tagging performance of this approach. It turns out that it is possible to reduce the number of geo-tagging errors but keep the recall rate approximately constant, compared to automatic geo-parsers.

1 Introduction

With the increasing importance of the web as a medium for global communication, new applications for asynchronous discussions have been developed. Rinner (2001) describes the concept of *argumentation maps* to support map-based discussions in online planning. Argumentation maps

M. Painho et al., (eds.), *Geospatial Thinking*, Lecture Notes in Geoinformation and Cartography,
DOI 10.1007/978-3-642-12326-9_8, © Springer-Verlag Berlin Heidelberg 2010

combine an online discussion forum (ODF) with an interactive geographic mapping component. Their web-based design implicates several advantages compared to traditional offline means of discussion, such as a lower inhibition threshold for participation in a discussion (Kingston et al. 1999), the ability to attend a discussion from any computer with online access, or the possibility to share information among many attendees (Laurini 2004).

While conventional forums without a mapping component only allow the structuring and retrieval of contributions by keywords, argumentation maps in principle also provide the possibility to retrieve messages by the geographic area they are referring to. Discussion participants are instantly aware of what geographic area a specific discussion is about. Locations which would be difficult or laborious to explain in words (e.g., locations in uninhabited areas) can easily be marked in a map and ambiguous place names can be disambiguated with little effort.

Although the general concept of argumentation maps has been around for several years now, practical implementations are still relatively rare. Keßler (2004) created a prototypical discussion forum for spatial decision-making and geo-collaboration called "Argumap". Its successor is called "ArgooMap". Both combine an online discussion forum with an interactive map. Forum contributions can be *geo-tagged*, i.e., linked to geographic locations by clicking locations in the map.

A usability test was carried out by Sidlar and Rinner (2007). They attested the *Argumap* prototype a generally high usability but revealed several issues that could be improved. According to Rinner et al. (2008) users often do not create references to every location that appear in their contributions, or do not even set any references at all.

Another important limitation of both Argumap and ArgooMap is that entire contributions instead of single geographic names mentioned in the texts are linked to geographic locations (Rinner et al. 2008). This means that the match between place name and location is lost. The resulting complexity could lead to insufficient geo-tagging activity since the discussion participants cannot see a benefit from adding more references to the map.

In this paper we will present an argumentation mapping prototype that allows users to assign geo-tags to single terms instead of entire contributions.

With the term *user-generated text content* we denote text resources that are created on publicly accessible web sites by end-users and are not direct subjects to an editorial authority. Such text resources are sometimes written in colloquial or not well-authored language and may contain slang words and misspellings, especially in terms of upper and lower case.

Important examples for web sites that are characterized by user-generated text content are online discussion forums, wikis, or blogs. Geo-tagging of text is often done *a posteriori*, i.e., it is not done by the authors of the texts themselves (which is *a priori* geo-tagging). A human or, in most cases, a software program tries to infer the intended geographic locations from the place names mentioned in the text and their context. This is in principle inexact in many cases, especially when performed by software programs that cannot understand the context in which a place name is mentioned. Automatic geo-tagging software[1], so-called "geo-parsers" or "geo-taggers", already achieves relatively good recognition rates on corpora containing well-authored texts, such as news stories (Silva et al. 2006). However, these geo-parsers have problems with resolving ambiguous place names. Locations that are not listed in the geo-parser's gazetteer, such as single buildings or locations in open land, cannot be geo-tagged at all[2]. User-generated text as encountered in discussion forums is often written in an informal style, contains typos and slang expressions and thus makes it even harder for geo-parsers to achieve high recognition rates. Therefore, a further motivation for this work is to find a method that allows effective geo-tagging of user-generated content. It should outperform automatic geo-parsers in this task.

2 Related Work

Online discussion forums[3] (ODF) are a well-established and popular application on the web. Besides the use of ODFs for personal informal discussions and information exchange, the concept also got into the focus of researchers who are investigating its benefits for professional applications. The idea of discussing things by using ODFs is gaining increasing attention in areas like e-Learning (Wu 2004), public participation[4], and the theory of deliberative democracy (Wright 2007). Laurini (2004) depicts the advantages of online discussion forums in public participation. He states that web-based discussions do not need fixed appointments and are more convenient and relaxed than conventional participation procedures. For people who feel uncomfortable when

[1] For example MetaCarta, http://www.metacarta.com

[2] Except by entering the according geographical coordinates directly. However, this is cumbersome and not practically feasible for a large number of users.

[3] One of the first online discussion forums, namely UBB.classic, emerged in 1996 (according to http://www.tomrell.com/

[4] http://www.e-participation.net/taxonomy/term/32

speaking in front of large groups, online discussion forums are a good way of making themselves heard (Kingston 1999).

Argumentation mapping tools are based on the combination of an ODF and an online mapping component. Rinner (2001) defines the argumentation maps concept as an object-based model for geographically referenced discussions. Argumentation maps aim at “supporting any argumentative process that has a spatial component and can benefit from explicit links between arguments and the corresponding places they refer to“ (Rinner et al. 2008, p. 6).

In the argumentation map model, Rinner (2006) distinguishes argumentation elements, geographic reference objects, and graphic reference objects (see Fig. 1). Argumentation elements are the formal representations of arguments expressed by the participants of a discussion. In a spatially related discussion these argumentation elements potentially refer to one or more geographic reference objects, which are part of the map. At the same time they refer to one or more graphic reference objects, which are created by the discussion participants. Graphic reference objects are markers in the map that highlight a point or an area. Between all three kinds of elements and objects several kinds of relations are defined that represent the structure of the corresponding discussion. For instance, multiple argumentation elements may be linked to each other through logical relations, which may indicate a response to a certain argument.

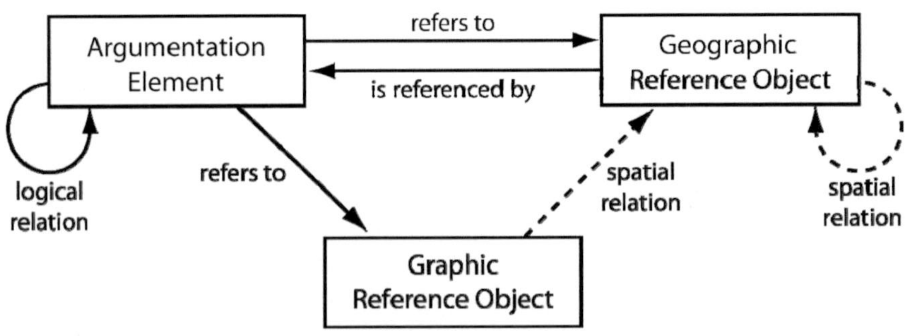

Fig. 1. Conceptual model for argumentation maps. Source: Rinner (2006)

3 Suggestive Geo-Tagging

The usefulness of an argumentation mapping tool highly depends on the accuracy and quantity of the geo-tags that link the discussion contributions to geographic locations. Generally we can assume the following: The more tedious it is for a user to create geo-tags and the less the user can see a benefit from these geo-tags, the less geo-tags will be created by her. Geo-tagging resources such as photographs and video is a common task on the web which can be completed easily by clicking a point in a map. In contrast, geo-tagging of texts generally requires more effort since text resources, unlike photographs, often refer to many different locations. Thus making this process easy and convenient is a key aspect of improving the overall quality of the geo-tags.

Two specific factors are crucial for a geo-parser's recognition rate on text content from online discussion forums: the authoring quality, including the grammatical and orthographic correctness of the given text, and the general prominence of the geographic features mentioned in this text.

Authoring quality of a text. Geo-parsers make use of Natural Language Processing and Named Entity Recognition to spot possible geographic identifiers and therefore require well-authored texts with only few grammatical mistakes and typos to achieve good results. In contrast, discussion forum contributions contain user-generated text content that is written spontaneously in many cases. Often, no great store is being set on grammatically and orthographically correct writing. As shown by Amitay et al. (2004), this negatively influences the recognition rate of geo-parsers to a great extent.

Prominence of a feature. The prominence of a feature is the general importance of a feature. Features with a high prominence, e.g., cities with a high population, are usually more likely to be mentioned than features with a low prominence. Geo-parsers utilize this assumption to disambiguate between equally-named features. However, spatial topics are often discussed at large scales (Rinner et al. 2008). Features at this scale, e.g., single buildings, that discussion participants might refer to, normally have a low prominence and are difficult for a geo-parser to recognize and resolve correctly.

To shift the geo-tagging process from an a posteriori to an a priori approach, the contribution authors have to be involved in the geo-tagging process to some extent. In the following we present a prototype of an argumentation mapping tool that supports the authors by suggesting locations based on the content of the contributions but leaves the final decision whether to create a new geo-tag in the hands of the authors (*suggestive geo-tagging*).

3.1 Geo-Parsing and Suggestive Geo-Tagging

A geo-parser tries to find terms in a text that denote location names and maps found entity names to the geographic coordinates of the intended real-world counterpart:

`Washington` → (38.895, -77.037)

Geo-parsing and suggestive geo-tagging are technically for the most part identical. However, while geo-parsing aims at geo-tagging a given text mainly automatically, a suggestive geo-tagger suggests features for names in a text. It generates a list of feature candidates, ranks them according to their computed relevance, and lets a human make the final decision of linking a geographic location to a name in the text or not. The key objective here is to support the users with a sufficient number of suggestions. We assume that a higher number of false positives (i.e. terms that are falsely regarded as geographic names) delivered by the geo-parser is tolerable, since the final geo-tagging decision is up to the users. However, too many false positives might confuse users and may therefore result in a lower number of geo-tags created by the users.

3.1.1 Disambiguation of Geographic Names

By far the biggest challenge that a geo-parser is confronted with is *ambiguity*: In many cases one word not only has exactly one meaning but is the name of different geographic features and denotes other completely different non-geographic things. Amitay et al. (2004) distinguish two kinds of ambiguities:

geo/non-geo ambiguity occurs if a word or a sequence of words is the name of a geographic place, but also has a different meaning that is not referring to any geographic place. For example, *Turkey* is the name of a country but also denotes a bird species.

geo/geo ambiguity occurs if there are several geographic places with the same name, e.g., *Paris* (France) and *Paris* (Texas).

The prototype presented in this paper utilizes its own geo-parser that has been implemented for the task of suggestive geo-tagging. This geo-parser is based on the Geonames gazetteer[5], whose database contains several types of features like cities, countries, mountains, lakes, parks, etc. In the following, the geo-parsers' disambiguation techniques are introduced.

[5] http://www.geonames.org/

In most cases humans can easily resolve ambiguity from linguistic and extra-linguistic context (Leidner et al. 2003). In contrast, geo-parsers do not truly understand a text. They try to resolve ambiguity by combining several different methods, rules, and heuristics (Overell et al. 2006; Amitay et al. 2004; Leidner 2004; Rauch et al. 2003; Blessing et al. 2007; Leidner et al. 2003). To make use of implicit context information, two special *minimality heuristics* can be applied (Gardent and Webber 2001) to disambiguate place names:

One sense per discourse. If one geographic place name is mentioned several times, it is assumed that it refers to the same location throughout the text (Gale et al. 1992).

Minimal spanning region defines interpretation. If there are more than one geographic names occurring in a text, the location candidates (interpretations) which span the smallest region are chosen (Leidner et al. 2003). For example, if the cities *Bedford* and *Everett* are mentioned, it is assumed that Bedford and Everett, Pennsylvania, are intended (since they only lie 10 km apart) and not Everett, Pennsylvania, and Bedford, UK.

The geo-parser makes use of several other methods used to disambiguate place names. These include Part-of-Speech tagging (information about the grammatical structure of the user-generated text), population data (places with higher population are more likely to be mentioned), and the "focus score". The focus score method calculates a score based on the map area that is currently visible in the prototype and an inverse distance weighting. The map area can be adjusted by the users independently and should specify the approximate geographic scope of the discussion. Feature suggestions lying inside this area are assigned a higher score than those lying outside.

4 Design and Implementation of the Prototype

In order to overcome the shortcomings of the previous versions of Argoomap the prototype has been redesigned, away from a contribution-based towards a word-based geo-tagging mechanism. This should clarify the relations between the words in the contributions and the map and increase the benefit the users gain from adding new geo-tags.

The new prototype is called "ArgooMap 2". It is a web page based on standard web technologies (HTML, CSS, JavaScript) that can be used with any common web browser. No additional plug-ins or software installations are required to run it.

The prototype layout is horizontally divided into two panels (see Fig.2). The left panel contains the textual content where forum contributions can be displayed and new topics and replies can be written. The right panel is dedicated to an interactive map that can be dynamically panned and zoomed. Users can choose between four different map types: a street map, satellite imagery, satellite imagery with an overlaying street map, or OpenStreetMap data. ArgooMap 2 makes use of the Google Maps API[6]. Due to its prominence, many users may already be familiar with the map controls.

Fig. 2. Front page of the ArgooMap 2 prototype website

The left panel lists the posted forum topics in chronological order. As the map is panned and zoomed to a different geographic area the list of topics is dynamically updated with posts referring to the current map extent.

Clicking the header of a topic brings up the according discussion including all the replies that have been posted. It automatically adjusts the map pane so that all geographic locations mentioned in the discussion are shown. Words that have been linked to geographic locations in the map are highlighted in blue. The associated points are flagged with blue markers in the map. When the mouse is moved over a highlighted word in the text, the corresponding markers in the map change their color to red. The other way around, when a marker in the map is clicked, an info window opens that lists all words in the discussion that refer to this marker. Moving the

[6] See http://code.google.com/apis/maps/documentation/reference.html

mouse over an entry in the info window immediately highlights the associated word in the discussion.

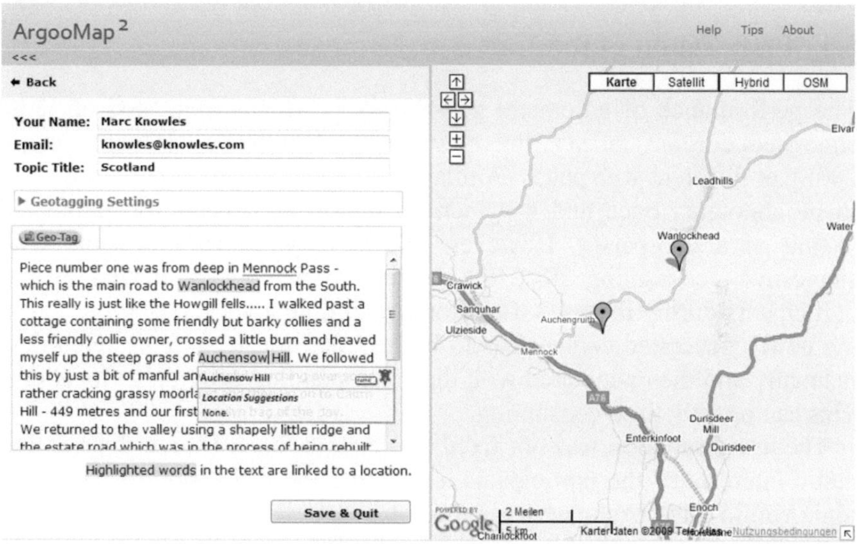

Fig. 3. Composing a new contribution. The two terms in the text highlighted in blue are each linked to one of the two blue markers in the map

There are two different possibilities to link words or terms in the contribution text to one or more geographic locations: automatic geo-tagging and manual geo-tagging. The easiest way is to start the automatic geo-tagging process by clicking the “Geo-Tag“ button. Words that are recognized as geographic names are underlined. Clicking such a word brings up the location suggestions that the author can choose from.

If none of the automatically retrieved suggestions matches the intended location, or if a word has not been recognized as a geographic name at all it can still be geo-tagged manually. This is done by simply clicking the according location in the map. Until the contribution has been saved the authors can create, edit, and remove references at any time.

5 Human Participants Test

To analyze the applicability of the prototype for geo-tagging user-generated content, a group of people was asked to take part in an experimental online discussion by using ArgooMap 2. The achieved geo-tagging effectiveness of ArgooMap 2 has been compared to the performance of the automatic geo-parsers Yahoo! Placemaker and MetaCarta GeoTagger.

5.1 Preparation of the Test

The performance of automatic geo-parsers is usually evaluated on large annotated corpora taken from newspapers, Wikipedia (Overell et al. 2006), or different web pages (Amitay et al. 2004). Such corpora only have to be annotated once and can then be reused as a basis for automated testing of a geo-parser. However, this is only partially possible with suggestive geo-tagging since the content to geo-tag is created by the participants during the test[7]. Therefore, each time after carrying out a test, the newly generated content has to be searched for geographic references manually and then annotated with the corresponding geographic locations. This can be very time-consuming.

The test discussion was not fixed to a certain topic. Due to diverse thematic interests of the participants, restricting the discussion to a specific topic would have excluded a number of people from taking part. Instead, some initial questions were posted in the forum acting as conversation starters.

5.2 Execution of the Test

For the test, 41 people were invited via email to participate. The email explained the objective of the test and the capabilities of the prototype but did not *prompt* the readers to make use of the geo-tagging function. The participants should not have the feeling of having to place geo-tags extensively for the purpose of this test. In contrast, it was left up to them to find out the advantages of well geo-tagged content.

When using the prototype the first time the participants were presented an introductory video of approximately four minutes length. This video demonstrated the basic concepts and the operation of the program.

Although the discussion started sluggishly it became more vital with an increasing number of contributions. Travel reports emerged as a favorite topic. To keep the discussion going, additional contributions and replies were occasionally posted by the moderator.

[7] Nevertheless, the geo-parser integrated in ArgooMap 2 could be trained on an annotated corpus containing user-generated text content.

5.3 Evaluation

Table 1 depicts the participation statistics for the entire test. Until the end of the testing period, 20 of the 41 invited persons (49%) had written at least one contribution. The total number of contributions (excluding those that were created by the author) was 33 which corresponds to an average number of 1.7 contributions per participant. There were 19 threads with a total number of 17 replies, i.e., approx. 1 reply per thread on average.

Table 1. Discussion participation statistics. Note: Contributions posted by the researchers are not included

Number of invitations	41
Number of participants	20
Participation rate	49%
Number of contributions	33
Average number of contributions per person	1.7
Number of threads	19
Total number of replies	17

To determine the geo-tagging performance, all contributions had to be scanned for geographic references manually and then annotated with the corresponding geographic locations. There were some cases where the correct annotation was not completely clear. For instance, one thread was about getting from Münster to Coventry and several airplane routes were discussed. The participants provided advice such as "fly from Dortmund to Stansted". In these cases it was defined that the locations of the corresponding airports were referenced and not the cities of Dortmund and Stansted themselves. Other ambiguities frequently encountered referred to the correct tagging of multi-word units like "Münster central station" or "Cologne cathedral". Here often only the city name ("Münster") was geo-tagged. In this case it was defined that the correct tagging applies to the full multi-word unit "Münster central station". Phrases like "Dresden, Germany" (where "Germany" acts as a specifier) were defined to be treated as references to one single location ("Dresden") and not separately ("Dresden" and "Germany").

The annotation results are summarized in Table 2. 192 geographic references have been made in the contributions which refer to 152 distinct places. This means that there were 4.7 distinct places mentioned per contribution on average. This number is relatively high compared to the two discussions from the case study that was carried by Rinner et al. (2008) (2.94 and 3.13, respectively).

Table 2. Geographic references statistics

Total number of geographic references	192
Number of referenced places	156
Contributions without geographic references	5 (15%)
Contributions with one geographic reference	4 (12%)
Contributions with multiple geographic references	24 (73%)
Average number of geographic references per contribution	5.8
Average number of referenced places per contribution	4.7

5.4 Evaluation Measures

In information retrieval, the most frequently used measures for the effectiveness of a system are recall and precision (Manning et al. 2008). These measures can easily be adapted for evaluating a geo-tagging application:

$$R = \frac{\#(\text{correct geo-tags})}{\#(\text{geographic references})}$$

$$P = \frac{\#(\text{correct geo-tags})}{\#(\text{created geo-tags})}$$

where R and P denote recall and precision, respectively.

Depending on the application, one of the two measures might be more important than the other. As a trade off, the *F measure* is applied (Manning et al. 2008):

$$F = \frac{1}{\alpha\frac{1}{P} + (1-\alpha)\frac{1}{R}} = \frac{(\beta^2+1)PR}{\beta^2 P + R} \text{ where } \beta^2 = \frac{1-\alpha}{\alpha}$$

where $\alpha \in [0,1]$ and $\beta \in [0,\infty]$

With $\beta = 1$ (written $F_{\beta=1}$), recall and precision are equally weighted. If $\beta < 1$ precision is emphasized whereas values of $\beta > 1$ emphasize recall. In the case of maps more emphasis should be put on precision, since incorrect geo-tags might easily confuse people during a discussion and might adulterate the geographic scope. A value of 0.5 for β is chosen here. It is considered to accentuate precision and decrease the influence of recall in a reasonable magnitude. Accordingly, we will also look at $F_{\beta=0.5}$, besides recall and precision.

5.5 Geo-Parsers

The ArgooMap 2 prototype is evaluated against the two automatic geo-parsers Yahoo! Placemaker[8] and MetaCarta GeoTagger[9]. These are freely available geo-parsing web services which are accessed via HTTP POST. After posting the text content the services return an XML document containing the found places and the corresponding geographic coordinates as well as information on where in the text the geographic names have been found.

5.6 Test Results

To determine recall and precision of ArgooMap 2, Yahoo! Placemaker, and MetaCarta GeoTagger, for each contribution the correct geo-tags, the geo/geo tagging errors, the geo/non-geo tagging errors, and the geographic references that were not geo-tagged were counted. Table 3 summarizes the results achieved by each of the three applications.

From the 192 geographic references mentioned in all contributions, 91 where geo-tagged correctly with ArgooMap 2. There were 13 geo/geo tagging errors and 88 geographic references were not geo-tagged. Geo/non-geo tagging errors were not encountered. About one third of the geo-tags was created manually by clicking a location in the map and two thirds were inserted after selecting a feature suggestion. It is remarkable that all of the 13 geo/geo tagging errors occurred after selecting a suggestion but none after marking a location manually in the map. Although ArgooMap 2 provides the possibility to reference multiple locations to single geo-tags, this function was not utilized by any of the participants.

The number of geographic references tagged correctly by Yahoo! Placemaker was 90, and therefore slightly lower than the value achieved by ArgooMap 2. However, the number of tagging errors was more than twice as high (27), most of them geo/geo errors and only 2 geo/non-geo errors.

The MetaCarta geo-parser achieved the highest number of correct geo-tags (105) but also made the highest number of mistakes (34). Thus, with

[8] http://developer.yahoo.com/geo/placemaker

[9] http://ondemand.metacarta.com/?method=GeoTagger

54.7% the MetaCarta geo-parser attained by far the highest recall value, followed by ArgooMap 2 with 47.4% and Yahoo! Placemaker with 46.7%. However, this is achieved at the price of the highest error rate and the lowest precision. ArgooMap 2 clearly gained the highest precision value, namely 87.5%. Yahoo! Placemaker's precision is 10.6 percentage points lower (76.9%), followed by MetaCarta having a precision of 75.5%.

Table 3. Geo-tagging effectiveness of ArgooMap 2 compared to Yahoo! Placemaker and MetaCarta

	ArgooMap 2	Yahoo!	MetaCarta
Correct	91	90	105
Geo/geo errors	13	25	31
Geo/non-geo errors	0	2	3
Not geo-tagged	88	77	56
Recall	47.4%	46.9%	54.7%
Precision	87.5%	76.9%	75.5%
$F_{\beta=0.5}$	0.75	0.68	0.70

Table 4. ArgooMap 2 specific geo-tagging effectiveness statistics

Suggested geo-tags	71 (68.3%)
Manual geo-tags	33 (31.7%)
Geo/geo errors in suggested tags	13 (100%)
Geo/geo errors in manual tags	0 (0%)
Geo-tags referring to multiple locations	0

Table 4 shows some additional ArgooMap 2 specific statistics. More than two thirds of the geo-tags created in ArgooMap 2 were created by selecting an automatically generated location suggestion. Manual geo-tags were mainly created for places for which no appropriate suggestion was available. The corresponding error rates show that all geo-tagging errors occurred together with suggested tags. In contrast, the manually created geo-tags were correct in all cases.

6 Discussion

The experimental discussion has shown that the recall rate of the implemented prototype was slightly higher than the one achieved by Yahoo! Placemaker, but did not reach the recall rate of the MetaCarta geo-parser. On the other hand only every eighth geographic reference was tagged falsely by the users of the ArgooMap 2 prototype, while MetaCarta and Yahoo! Placemaker tagged about one quarter of the geographic references incorrectly. Hence, the geo-tagging precision of ArgooMap 2 was clearly the highest among all geo-taggers. However, the prototype's recall rate is not completely satisfying. Some issues became apparent during the human participants test that negatively influence the recall rate. These issues as well as possible solutions are discussed in the following.

The task of geo-tagging was well understood by the participants. There were only 3 contributions that did contain geographic references, but no geo-tags. In such cases, the reason was mostly that the geo-parser did not provide any suggestions for the mentioned geographic names. Although it was possible to geo-tag these names manually, the participants did not, either to avoid extra work or because they were not aware of this possibility. As the feature of manually creating geo-tags is a key functionality of the prototype which also separates it from fully automatic geo-parsers, its usage should be made clearer in the user interface. This could be achieved, by adding a prominent button that explicitly provides a simple option to manually create geo-tags. Additionally, the introduction presented to first time users of ArgooMap 2 should put more emphasis on this feature.

The human participants test has shown that the majority (68%) of the referenced geographic names were geo-tagged by confirming a suggestion made by the built-in geo-parser. This illustrates the importance of the geo-parser's performance for achieving a high recall. However, especially two shortcomings of the geo-parser had a negative impact on the recall during the test. First, generating location suggestions for a given text may take, depending on the text length, up to 20 seconds. This is clearly too long and it may discourage people to utilize this functionality. Hence, this fact very likely leads to a decrease in recall. Special focus should therefore be put on accelerating this process. The second drawback concerns the different names of features in various languages. The ArgooMap 2 geo-parser gazetteer only contains the local names of geographic features, e.g., it contains *Köln* but not *Cologne*, and it contains *Roma* but not *Rome*. As a result, mentions of *Cologne* were not recognized by the geo-parser as geographic names, not highlighted, and for this reason also not geo-tagged by some users.

In many contributions, specific geographic features were mentioned several times. Even if there are correct suggestions available for each instance of this name, users clearly tended to reference only one instance of this name. For example, one user mentioned "Coventry" four times but only geo-tagged the first occurrence of it, presumably for convenience. Here the *one sense per discourse* heuristic could be applied to relieve the user from the work of confirming the same suggestion for every single geographic name instance. When one instance of such a name is geo-tagged by the user, he should be asked whether all remaining occurrences of this name should be automatically referenced as well.

A problem that was encountered several times was that participants forgot to adjust the map area of interest before initiating the geo-parsing process. Hence, places lying further outside the map area were not recognized. Since the adjustment of the map area by the participants turned out to be an important means for disambiguation, it should be preserved as such. However, a simple solution to this problem might be to just remind the users of adjusting the map area each time they start the automatic geo-parsing process. A balance has to be struck here in order to not annoy the users unnecessarily.

6.1 Transfer to Other Application Areas

The concept of suggestive geo-tagging is not limited to the field of argumentation maps and online discussion forums. Potential other application areas include all kinds of public platforms that deal with user-generated text content, such as wikis (Wikipedia) or blogs. Figure 4 shows a workflow diagram that abstracts the general principle of suggestive geo-tagging.

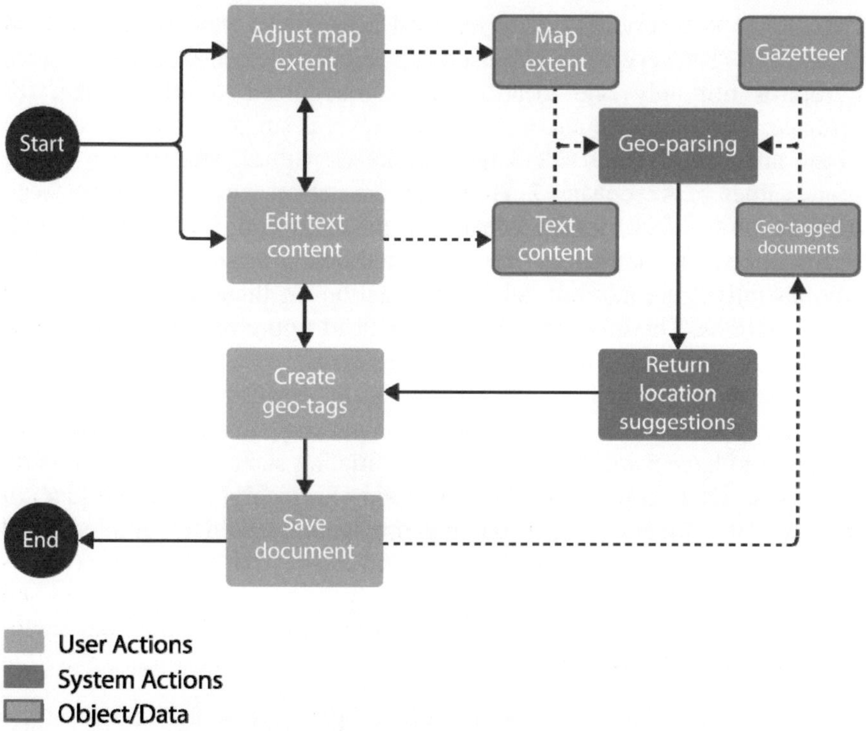

Fig. 4. Suggestive geo-tagging workflow diagram

7 Conclusion

This work has shown that geo-tagging user-generated text content can be done with acceptable recall and with an especially high precision if the geo-tagging is delegated to the users, i.e., is done a priori. An important prerequisite is a geo-tagging software that actively suggests locations. It should support the users in the geo-tagging work as much as possible to achieve a better recall rate. On the other hand it should still give them full control over what is being geo-tagged in order to keep the precision rate high.

7.1 Future Work

In the medium term it should be considered whether the geo-parser implemented for ArgooMap 2 can be replaced by the MetaCarta

GeoTagger web service, due to its good geo-tagging performance and its high speed. The service can be used as a suggestive geo-parser, too, since it returns not only one location for a recognized geo-term (like, for instance, Yahoo! Placemaker does), but several suggestions, if available. These are weighted by calculated *confidence* values, analogously to the score values in ArgooMap 2. However, the reason why this has not been done yet is the fact that the MetaCarta results do not inhere information about the actual names of the suggested features and their according administrative regions, but solely information on their feature types and their locations. Therefore, it can be difficult to figure out which suggested location actually refers to the intended feature.

Currently both Yahoo! Placemaker and MetaCarta GeoTagger use proprietary formats to access their geo-parsing services. It would be desirable to have a standardized way of initiating such services, preferably through an OGC compliant Web Processing Service[10] (WPS) interface. In this case the suggestive geo-tagging application could be implemented independently of the associated geo-parser.

7.2 Geographic Scope

One of the most important advantages of geo-tagged documents is the possibility to retrieve them according to geographic criteria. However, not all referenced geographic locations can be considered equally important for the subject matter of the document. Therefore, to figure out the most relevant geo-tags and thus the most relevant documents for a specific region, the *geographic scope* of a document is attempted to be determined. The geographic scope is defined, if it exists, as the region where more people than average would find that document relevant (Silva et al. 2006). There are different approaches of how to calculate the geographic scope (Martins et al. 2005; Silva et al. 2006). Amitay et al. (2004) propose the *focus scoring algorithm* to calculate a geographic scope (they call it *page focus*) based on the existing geo-tags in the document. Silva et al. (2006) describe a number of heuristics that they rely on for their approach to compute a geographic scope for web pages.

In the context of suggestive geo-tagging it should be investigated how the geographic scope can be inferred from user-generated content that has been geo-tagged with the help of the ArgooMap 2 prototype. The human participants test posed the assumption that, albeit the recall rate is moderate, users tended to geo-tag mostly those geographic names that they

[10] OGC Web Processing Service, http://www.opengeospatial.org/standards/wps

thought were most important for the subject matter of their contribution. Furthermore, the map extent that was set at the time of geo-tagging might prove useful at this point.

Acknowledgments

This work has been funded by the SimCat project granted by the German Research Foundation (DFG Ra1062/2-1 and Ja1709/2-2, see http://sim-dl.sourceforge.net). Special thanks go to Claus Rinner and Werner Kuhn for valuable comments.

References

Amitay E, Har'el N, Sivan R, Soffer A (2004). Web-a-where: Geotagging web content. In: Proceedings of the 27th annual international ACM SIGIR conference on Research and development in information retrieval, ACM Press, pp 273-280

Blessing A, Kuntz R, Schütze H (2007). Towards a context model driven German geo-tagging system. In: GIR '07: Proceedings of the 4th ACM workshop on Geographical information retrieval, ACM, pp 25-30

Chaves M, Silva M J, Martins B (2005), A Geographic Knowledge Base for Semantic Web Applications. In: Proceedings of SBBD-05, the 20th Brazilian Symposium on Databases, UFU, pp 40-54

Gale W, Church K, Yarowsky D (1992). One sense per discourse. In: Proceedings of the workshop on Speech and Natural Language, HLT '91, Association for Computational Linguistics, pp 233-237

Gardent C, Webber B (2001). Towards the Use of Automated Reasoning in Discourse Disambiguation. In: Journal of Logic, Lang. and Inf. 10(4): 487-509, Kluwer Academic Publishers, Hingham, USA

Keßler C (2004), Design and Implementation of Argumentation Maps, Diploma thesis, Westfälische Wilhelms-Universität Münster, Germany.

Kingston R, Carver S, Evans A, Turton I (1999), A GIS for the Public: Enhancing Participation in Local Decision Making. GIS Research UK.

Laurini R (2004), Computer Systems for Public Participation. Laboratoire d'Ingénierie des Systèmes d'Information, University of Lyon, France. http://www.gisig.it/VPC_sommet/CD_Sommet/ws3/articololaurini.pdf

Leidner J, Sinclair G, Webber B (2003). Grounding spatial named entities for information extraction and question answering. In: Proceedings of the HLT-NAACL 2003 workshop on Analysis of geographic references, Association for Computational Linguistics, pp 31-38

Leidner J (2004). Toponym Resolution in Text: Which Sheffield is it? In: Proceedings of the SIGIR 2004 conference on Research and Development in Information Retrieval, Sheffield, UK, July 25th -29th 2004. ACM Press, New York, USA

Mahemoff M (2006), Ajax Design Patterns, O'Reilly Media, Sebastopol, USA

Manning C, Raghavan P, Schütze H (2008). Introduction to Information Retrieval, Cambridge University Press, New York, USA

Martins B, Chaves M, Silva M J (2005), Assigning Geographical Scopes To Web Pages. In: Advances in Information Retrieval, Springer, pp 564-567

Overell S, Magalhaes J, Ruger S (2006). Place disambiguation with co-occurrence models. In: A. Nardi, C. Peters, and J. L. Vicedo, editors, CLEF 2006 Workshop, Working notes, September 2006

Rauch E, Bukatin M, Baker K (2003). A confidence-based framework for disambiguating geographic terms. In: Proceedings of the HLT-NAACL 2003 workshop on Analysis of geographic references, Association for Computational Linguistics, pp 50-54

Rinner C (2001), Argumentation maps: GIS-based discussion support for on-line planning. Environment and Planning B: Planning and Design 30(6): 847-863

Rinner C (2006). Argumentation Mapping in Collaborative Spatial Decision Making. In: S. Dragicevic, S. Balram, Collaborative GIS, Idea Group Publishing, pp 85-102.

Rinner C, Keßler C, Andrulis S (2008). The use of Web 2.0 concepts to support deliberation in spatial decision-making. In: Computers, Environment and Urban Systems 32(5): 386-395

Sidlar C, Rinner C (2007), Analyzing the Usability of an Argumentation Map as a Participatory Spatial Decision Support Tool. URISA Journal 19(1): 47-55

Silva M J, Martins B, Chaves M, Afonso A P, Cardoso N (2006). Adding Geographic Scopes to Web Resources. In: Computers, Environment and Urban Systems 30(4): 378-399

Wright S, Street J (2007). Democracy, deliberation and design: the case of online discussion forums. In: New Media Society 9(5): 849-869

Wu D, Hiltz S R (2004). Predicting Learning from Asynchronous Online Discussions. In: JALN 8(2), April 2004

A Novel Approach to Mining Travel Sequences Using Collections of Geotagged Photos

Slava Kisilevich[1], Daniel Keim[1], Lior Rokach[2]

[1] University of Konstanz, Konstanz, Germany
{slaks, keim}@dbvis.inf.uni-konstanz.de
[2] Department of Information Systems Engineering and
The Deutsche Telekom Laboratories
Ben-Gurion University of the Negev, Beer-Sheva, Israel
liorrk@bgu.ac.il

Abstract. In this paper we present a novel approach for analyzing the trajectories of moving objects and of people in particular. The mined data from these sequences can provide valuable information for understanding the surrounding locations, discovering attractive place or mining frequent sequences of visited places. Based on geotagged photos, our framework mines semantically annotated sequences. Our framework is capable of mining semantically annotated sequences of any length to discover patterns that are not necessarily immediate antecedents. The approach consists of four main steps. In the first step, every photo location is semantically annotated by assigning it to a known nearby point of interest. In the second step, a density-based clustering algorithm is applied to all unassigned photos, creating regions of unknown points of interest. In the third step, a travel sequence of every individual is built. In the final step, travel sequence patterns are mined using the semantics that were obtained from the first two steps. Case studies of Guimarães, Portugal (where the conference takes place) and Berlin, Germany demonstrate the capabilities of the proposed framework.

Keywords: sequence mining, trajectories, sequence patterns

M. Painho et al., (eds.), *Geospatial Thinking*, Lecture Notes in Geoinformation and Cartography,
DOI 10.1007/978-3-642-12326-9_9,

1 Introduction

Location acquisition technologies and web-centric information sharing are ubiquitous in today's world and have become a focus for research in a variety of fields, data mining in particular, due to the vast quantity of data involved. Existing works on analyzing people's mobility mainly concentrate on the trajectories obtained by GPS-enabled devices. Such trajectories usually consist of many space-and-time referenced points measured at a constant interval where the foremost nontrivial task is to extract (semantically) important parts or stay points.

Several approaches exist to find the important elements: (1) applying density functions to find regions where intersections of trajectories are high or (2) finding parts of a trajectory where the object stayed for a significant period of time. After the stay points are found, data mining algorithms can be applied to mine frequent sequences.

These approaches involve several issues. (1) Important intersection sites for various individuals may seem to be the same but in fact correspond to different sites that were visited. For example, one person visited a bank and another entered a shop. The bank and the shop are situated close to each other and these regions were defined as one stay point in the trajectory of these two persons. (2) Since the stay points are defined mainly using characteristics of the trajectory, without any background knowledge, there is a need to interpret the obtained sequences. The first issue can be tackled by assuming that the regions visited by people are important, making no distinction between the sites visited in these regions. The second issue can be resolved by using external databases of points of interest (POIs) to explain the important places. However, a POI database may be unavailable, inapplicable to the data (shopping, work) or incomplete.

Large-scale, GPS-based datasets of people's trajectories are still unavailable partly because of data acquisition problems. For example, Zhen et al. (2009) reported that a *large* GPS dataset was created from data collected by 107 users carrying GPS-enabled devices with them for one year. The regions that these users covered included 36 cities in China and various areas in the USA, South Korea, and Japan. Without regard to the difficulty of data acquisition, the question of whether 107 users are enough to mine travel sequences in different parts of the world remains open. We will also try to answer this question in this paper.

A recent trend in analyzing people's activity and travel behavior is the use of geotagged photos shared by people and publicly available on such photo-sharing sites like Flickr[1] or Panoramio[2]. The data from these geo-

[1] http://www.flickr.com

tagged photos differ technically and semantically from raw GPS-based type trajectories. Unlike trajectories recorded by GPS devices and measured at a constant time, photo data can be regarded as a private case of raw trajectories in which an individual is capturing an important event. Using time and location of photos taken by a person, it is possible to construct event-based trajectories, which can then be used to analyze travel activity. The act of sharing the photo with others through photo-sharing sites reveals important information, including time, location, title, tags and the photo itself. Therefore, this data can be directly used in retrieving interesting places, providing us with the opportunity to discover travel sequences and understand in what order people visit such places.

In this paper, we address the problem of automatically finding semantically annotated sequences. For instance, consider the following sequences:

1. A → B → C
2. A → * → D

The first sequence can be interpreted as a route followed by people from place A to place B and from place B to place C. It is important to note that those who reached C from B are the same persons as those who reached B from A. In the second sequence, those who started from A and reach D, did not necessarily visit a particular place, rather they may have visited any possible place before visiting D.

Our approach to mining travel sequences consists of four main parts. In the first part, we automatically assign every geotagged photo to a nearby POI using an external POI database. Since we do not perform any image analysis, we cannot really know what was photographed. However, the fact that the photo was taken near some known POI assumes the presence of the photographer in that place. After step one, there are photos that were not assigned to any POI. There are two reasons for this. (1) A photo was taken in an area where there are no POIs (for example a forest or parking lot near someone's house). (2) A photo was taken in an attractive place but a POI is missing in the database. Therefore, there is still a need to analyze these locations and artificially create points of interest using several constraints. For this purpose, we apply a density-based clustering algorithm in order to find dense areas (Rokach and Maimom 2005). This allows us to filter out outliers – sparse areas, where the number of people who took photos is less than a predefined threshold. The dense regions that are obtained are new, unknown points of interest which are added to the areas acquired in the first step. The automated process annotates these areas with symbolic names and stores the boundaries of these regions for future access. In the third step, the travel sequence of each person is con-

[2] http://www.panoramio.com

structed using the notion of a session: a time frame in which a person takes photos in a particular area. In the fourth step, travel sequence patterns are mined using semantics obtained from the first two steps.

The goal of this paper is to suggest an automatic approach for mining semantically annotated travel sequences using geotagged photos by searching for sequence patterns of any length. The sequences obtained may contain patterns that are not necessarily the immediate antecedents. Moreover, the approach that we propose can examine sequences in which the same pattern is repeated more than once in the same sequence.

The main contribution of this paper is the development of a new data mining process that employs concepts that have been developed in various other fields such as bioinformatics and artificial intelligence.

2 Related work

The mining of frequent sequential patterns in databases of customer transactions was first presented by Agrawal and Srikant (1995). The method adopts an a-priori-like approach (Agrawal and Srikant, 1994) where the idea is to find subsets that are common to at least a minimum number of sequences, termed itemsets. The method uses the following observation: if the sequence of length k is not frequent, then neither can the sequence of length $k+1$ ever be frequent. The algorithm can be applied to generic items provided they can be sorted using transaction time. Time, however, is not considered in pattern mining. The limitation of the approach is that it cannot find sequences with repeating patterns and sequences in which patterns are not necessarily immediate antecedents.

There are application domains where time duration between adjacent events is also important. This issue was addressed in MiSTA, a generic algorithm for mining temporally annotated sequences, where frequent patterns are mined using sequence and temporal similarity (Giannotti et al., 2006). As an extension to MiSTA, Giannotti et al. (2007) presented three different approaches to mining trajectory sequences that are reflecting site visits at approximately the same time. These two approaches share the idea that the transformation of a trajectory into a sequence of significant parts and the application of semantic meaning are done as a preprocessing step prior to mining the sequence patterns. Since the trajectory is transformed into a sequence of generic events, the MiSTA algorithm can be directly applied to them.

The MiSTA authors suggested two general methods for performing preprocessing. In the first case, background knowledge should be applied to trajectories. To perform this task may require an additional database of POIs or a domain expert. In the second case, significant parts are found without using background knowledge, only the properties of the trajectories themselves. Specifically, the authors proposed to divide the area of investigation into grids and to count the density of trajectories in every grid. Thus, the significant places are defined in terms of frequency of visits by different persons. In contrast to temporal annotated sequences, we define sequences as a frequent move from one place to another without regard to time similarity.

Alvares et al. (2007a) proposed a generic model for semantically annotating trajectories and representing a moving pattern in the geographic database. This approach has two main parts. In the first part, the significant places in a trajectory are found by identifying moves and stops (Spaccapietra et al., 2007). Stops are significant places that are also called *stay points,* sites where a person stayed for a certain period of time. The extraction of stops depends on time and distance thresholds. Moves are transitions between consecutive stops. In the second part, stops are integrated into the database along with geographic data like POIs. This makes it possible to perform spatial queries on stop regions by annotating them with semantically meaningful information. They demonstrated this approach for mining frequent trajectory patterns between two stops of conference attendees (Alvares et al., 2007b).

Zheng et al. (2009) mine travel sequences by inferring interesting places from trajectories and the person's experience. The method is based on calculating probabilities that a person will take a specific path using information about how many people move from one place to another. The most interesting sequences of length n can be found by summing the probabilities of every two-length sequence comprising the larger sequence and selecting sequences with high score. However, the notion of such sequences differs from classical sequences based on the frequency of patterns. The authors report that finding sequences of length n is possible but a time consuming process and hypothesize that people would not likely visit many places in a trip. Thus, two-length sequences were only considered in their paper.

The main differences between our work and existing state-of-the-art methods can be summarized as follows:

1. We work with trajectories on the semantic level instead of trajectories as raw points.
2. We introduce a concrete approach for semantically annotating points within a trajectory.

3. Interesting places are found before mining sequence patterns in contrast to existing approaches where interesting places are found using characteristics of trajectories such as density, frequency, stay time, stop points.
4. The sequence patterns can be of any length in which patterns are not necessarily immediate antecedents.
5. We evaluate our algorithm on a real-world database obtained from Flickr.

3 Framework

Fig. 1 presents the proposed framework. First, we try to match photo coordinates with known POIs. Then the remaining unassigned photos are clustered and new POIs are identified. This is followed by converting the individual's trajectory into sequences of POIs. These sequences are analyzed and new sequence patterns are discovered. The following subsections describe each step in more detail.

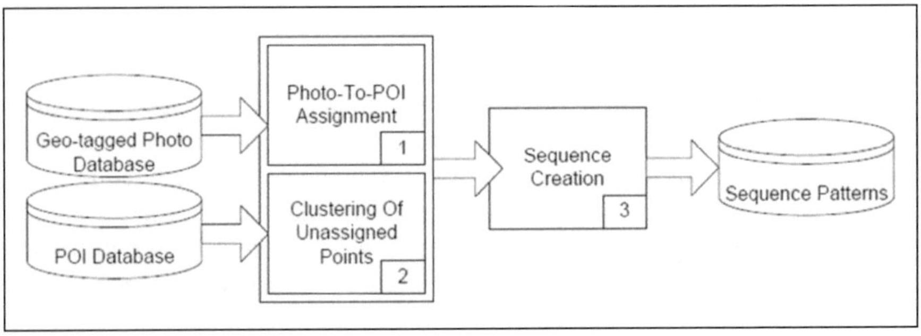

Fig. 1. Steps of the framework

3.1 Datasets

We collected metadata of geotagged photos from the Flickr photo-sharing Web site using its publicly available API. The Flickr API does not allow downloading metadata by providing exact boundaries of the area of interest. Therefore, we used an approach similar to Web crawling. We downloaded all the photo metadata of arbitrarily selected subjects and obtained the list of their contacts as well as the list of groups their photos belonged to. The same procedure was iteratively applied on other retrieved users.

We began collecting the data from the beginning of June. By the end of October 2009, we managed to collect 64,975,609 entries from 2,617,271 users. In the preprocessing step, we converted coordinates expressed in degrees into meters based on Universal Transverse Mercator (UTM) coordinates. This was done in order to enable us to apply distance-related functions. In total, 56,305 entries with wrong or missing dates were removed. These entries included 6,229 with incorrect dates; 50,076 photos were taken after October 1, 2009,

We used the Wikipedia database[3] as a source for POI data. This database is an on-going community project aimed at applying geographic annotation to articles describing interesting sites around the world. The database we obtained contains 450,637 entries of various geotagged sites such as cities, landmarks, monuments, buildings, towers, etc. For our purposes, the most important information that the entries contained were *id, title,* and *coordinates.*

3.2 Photo to POI assignment

In this step, every geotagged photo from the database is matched to a nearby POI using a distance threshold called *photo-to-POI.* If the distance between the photo and a POI is not longer than the *photo-to-POI* distance threshold, the photo is assigned to that POI. If there are several POIs within the distance threshold, the photo is assigned to the closest POI.

3.3 Discovery of new POIs based on unassigned photos

In this step, we use a clustering algorithm to create regions of unknown POIs using photos that were not assigned during the previous step. In our previous work (Kisilevich et al., 2010), we showed that density-based clustering can be used in finding attractive areas. In general, the density based clustering algorithms has several advantages over other types of clustering algorithms: Density based clustering algorithms require minimum domain knowledge to determine the input parameters and can discover clusters with arbitrary shape. In addition, density-based clustering algorithms can filter outliers and work effectively when applied to large databases.

---

[3] http://de.wikipedia.org/wiki/Wikipedia:WikiProjekt_Georeferenzierung/Wikipedia-World/en

3.4 Sequence Creation

In this step, we assemble the POIs visited by a person into a sequence of places using the time stamp of the photo. If two consecutive photos are assigned to the same POI, only one photo is taken into consideration. We discard sequences that have only one POI since they do not contribute to discovering new sequence patterns. In general, sequences of any length can be built in this step. However, sequence creation can be constrained using such criteria as a time interval between every two consecutive photos or a total time interval between first and last photo. For example, Girardin et al. (2009) applied a 30-day interval threshold to differentiate between tourists, whose photo sessions lasted less than 30 days and locals whose sessions were longer. This heuristic approach can be used for differentiating between travel patterns of various groups of visitors. In our experiments, we implemented the same idea.

3.5 Sequence Patterns

The term "sequence pattern" usually refers to a set of short sequences that is precisely specified by some formalism. As is the practice in bioinformatics research, we are also adopting a regular expression in order to represent sequence patterns. A pattern is defined as any string consisting of a letter of the alphabet and the wild-card character '*'. The wild-card (also known as the "don't care" character) is used to denote a position that can be occupied by any letter of the alphabet.

In this paper, we consider the Teiresias algorithm (Rigoutsos and Floratos, 1998) which was originally developed as a combinatorial pattern discovery algorithm in bioinformatics for analyzing DNA sequences. The algorithm identifies recurrent maximal patterns within sequences. Although the method is combinatorial in nature and able to produce all patterns that appear in at least a (user-defined) minimum number of sequences, it achieves a high degree of efficiency by avoiding the enumeration of the entire pattern space. The algorithm, which has also been successfully used for information retrieval and intelligent manufacturing (Rokach el al., 2008A and Rokach et al., 2008B), performs a well-organized exhaustive search. In the worst case, the algorithm is exponential, but works very well for usual inputs. Furthermore, the reported patterns are maximal; any reported pattern cannot be made more specific and still keep on appearing at the exact same positions within the input sequences. Teiresias searches for patterns that satisfy certain density constraints, limiting the number of wild-cards occurring in any stretch of pattern. More specifically, Teiresias looks for maximal <L,W> patterns with support of at least K (i.e. in the corpus there are at least K distinct sequences that match this pattern). A pattern P is called <L,W> pattern if every sub-pattern of P with length of at least W operations (combination of specific operations and "." wild-card operations) contains at least L specific operations. For example, given the following corpus of 6 trajectory sequences:

```
1. Reichstag → Der Bevölkerung → Brandenburg Gate →
Memorial to the Roma and Sinti Holocaust Victims → Pariser
Platz

2. Reichstag → Marienviertel → Memorial to the Murdered Jews
of Europe → Brandenburg Gate

3. Reichstag → Berliner Dom → Liebknecht Bridge →
Checkpoint Charlie → Brandenburg Gate → Treptower Park →
Pariser Platz

4. Reichstag → 18th of March Square → Brandenburg Gate

5. Potsdamer Platz → Zoological Garden → Marienviertel →
Reichstag → 18th of March Square → Brandenburg Gate

6. Sony Center → Pleasure Garden → Reichstag → Der
Bevölkerung → Unter den Linden → Memorial to the Murdered
Jews of Europe → Brandenburg Gate
```

The Teiresias program (L=K=2 and W=3) discovers 5 recurring patterns shown in the following table. The first column represents the support of the pattern.

Table 1. Illustrative results of the Teiresias algorithm

#	Sequence patterns
2	Reichstag → 18th of March Square → Brandenburg Gate
2	Reichstag → Der Bevölkerung
2	Memorial to the Murdered Jews of Europe → Brandenburg Gate
3	Reichstag → * → Brandenburg Gate
2	Brandenburg Gate → * → Pariser Platz

4 Experimental Evaluation

In this section, we present an experimental evaluation using two case studies of areas in Guimaraes, Portugal (where the conference takes place) and Berlin, Germany. In particular, this experimental study has the following goals:

1. To examine whether the proposed method can be applied to regions with different scales, number of persons and their photos, and several points of interest.
2. To examine the effect on travel patterns of such parameters as the photo-to-POI threshold (Sect. 3.2), the distance threshold for density-based clustering and the minimum number of people in a cluster (Sect. 3.3), and session length (Sect. 3.4), .

Throughout the entire experimental process, we observed a constant session time of ten days, a cluster threshold of three people and a minimum support K=3 of sequence patterns . We used session time as a heuristic for classifying people into locals and tourists. We classified a person as a tourist if she took photos during a period of no more than ten days. Otherwise, she was considered as a local resident and her sequences were discarded. The following subsections describe the experimental study in detail.

Case 1. Guimarães, Portugal

Guimaraes is a relatively small city with historical roots going back to the 9th century. The city was the first capital of Portugal and is often called “the birthplace of the Portuguese nationality”. UNESCO declared its historical section as a World Heritage site. In spite of its historical importance, only a very small number of people shared their photos on Flickr compared to the sharing of photos that is generally derived from other cities.

We defined an area of approximately 8.5 square kilometers around the center of Guimaraes with the following boundaries: longitude = 8.318° West and 8.276° East; latitude = 41.435° South and 41.457° North. From 2005 until October 2009, we were able to obtain only 391 photos from 152 people. The Wiki database contains only 11 POIs in the defined area: *Nossa Senhora da Oliveira, Guimaraes Castle, Palace of the Dukes of Braganza, Church of Sao Miguel do Castelo, Guimaraes Historical Center, Sao Paio, Dom-Afonso-Henriques-Stadion, Azurem University, Sao Sebastiao, Pousada de Santa Marinha, Oliveira do Castelo.*

We used 200 and 400 meters as a distance for a photo-to-POI assignment (Sect. 3.2) in order to obtain the sequence patterns. We applied DBSCAN (Ester et al., 1996) on unassigned photos using a distance threshold of 100 meters and identified unknown POIs (Sect. 3.3). These new POIs were added to the existing POIs. A total of 342 photos from 127 individuals were assigned to existing and unknown POIs. Fig. 2 shows regions of existing and unknown POIs using a photo-to-POI threshold of 200 (Fig. 2a) and 400 meters respectively (Fig. 2b).

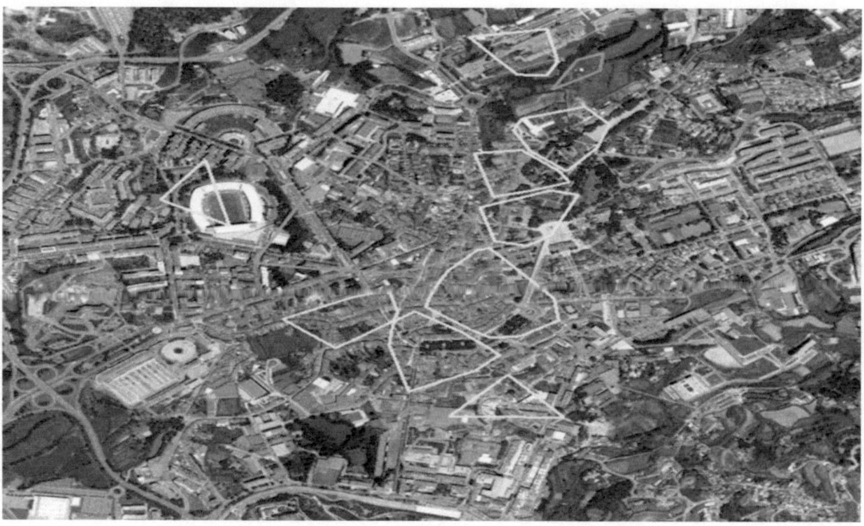

(a)

(b)

Fig. 2. Guimaraes, Portugal. Cluster boundaries of photos assigned to existing POIs (yellow) using a photo-to-POI distance threshold of (a) 200 meters and (b) 400 meters. Cluster boundaries forming new areas of POIs were obtained using a distance threshold of 100 meters and a density threshold of three people in a cluster (green)

The Teiresias algorithm discovered frequent sequence patterns of length two only. The general statistics pertaining to sequences and patterns are presented in Table 2. It can be seen that only 18 out of 127 sequences for a photo-to-POI threshold of 200 meters and 24 out of 138 sequences for a photo-to-POI threshold of 400 meters were created. There are two reasons for this. Firstly, the majority of people took photos in only one place. Secondly, some of the sequences were discarded because their length exceeded the 10-day threshold. Teiresias discovered 8 patterns using a photo-to-POI threshold of 200 and 7 patterns using 400 meters respectively. Table 3 shows five most frequent sequence patterns for every photo-to-POI threshold, where three generated sequences do not differ in two cases. The sequences that are different for 200 and 400-meter threshold are marked in bold.

Table 2. Guimaraes, Portugal. General statistics

Photo-to-POI threshold	<L,W>	# of people in sequences	# of valid sequences	# of sequence patterns
200	<2,3>	127	18	8
400	<2,3>	138	24	7

Table 3. Guimaraes, Portugal. Sequence patterns using L=2, W=3

Photo-to-POI threshold	# of input sequences	Sequence patterns
200	5	Guimaraes Historical Center → Nossa Senhora da Oliveira
	3	Guimaraes Castle → Church of Sao Miguel do Castelo
	3	**Nossa Senhora da Oliveira → Church of Sao Miguel do Castelo**
	3	Church of Sao Miguel do Castelo → Nossa Senhora da Oliveira
	2	**Church of Sao Miguel do Castelo → Nossa Senhora da Oliveira**
400	4	Guimaraes Historical Center → Nossa Senhora da Oliveira
	4	**Guimaraes Castle → Nossa Senhora da Oliveira**
	3	**Nossa Senhora da Oliveira → * → Nossa Senhora da Oliveira**
	2	Church of Sao Miguel do Castelo → Nossa Senhora da Oliveira
	3	Guimaraes Castle → Church of Sao Miguel do Castelo

Case 2. Berlin, Germany

Berlin is the capital of Germany and its largest city. It is one of the most popular tourist destinations in the EU. In 2008, a total of 17,758,591 persons visited Berlin according to European Cities Tourism Site[4]. Of this total, 7,033,593 people were classified as foreign visitors.

We defined an area of approximately 46.7 square kilometers around the center of Berlin with the following boundaries: longitude = 13.341° West, 13.483° East; latitude = 52.495° South and 52.537° North. We retrieved 71,821 photos from 9,505 people between 2005 and October 2009. The Wiki database contains 857 POIs in the defined area.

We used 200 and 400 meters as a threshold for a photo-to-POI assignment. These new POIs were added to the existing POIs. A total of 68,624 photos from 8,952 users were assigned to existing and unknown POIs. Fig. 3 shows regions of existing and unknown POIs using a photo-to-POI distance threshold of 200 (Fig. 3a) and 400 meters (Fig. 3b) respectively. When the photo-to-POI threshold was 400 meters, almost all the photos were assigned to existing POIs and only two clusters of unknown POIs were created (see Fig. 3b).

[4] http://www.europeancitiestourism.com/

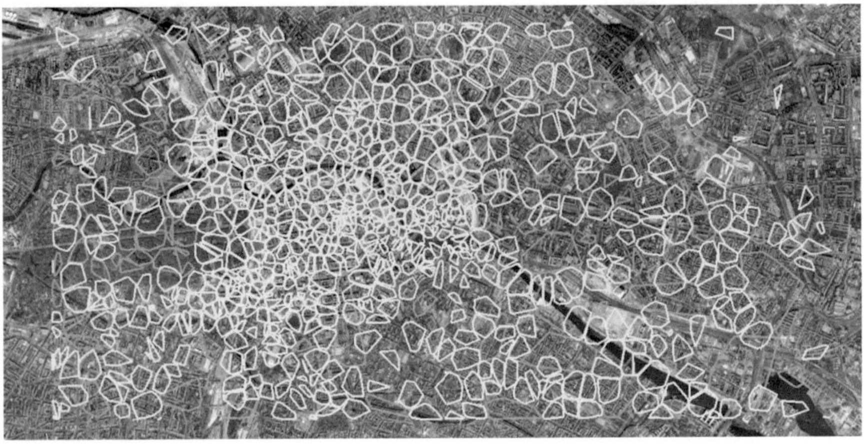

(a)

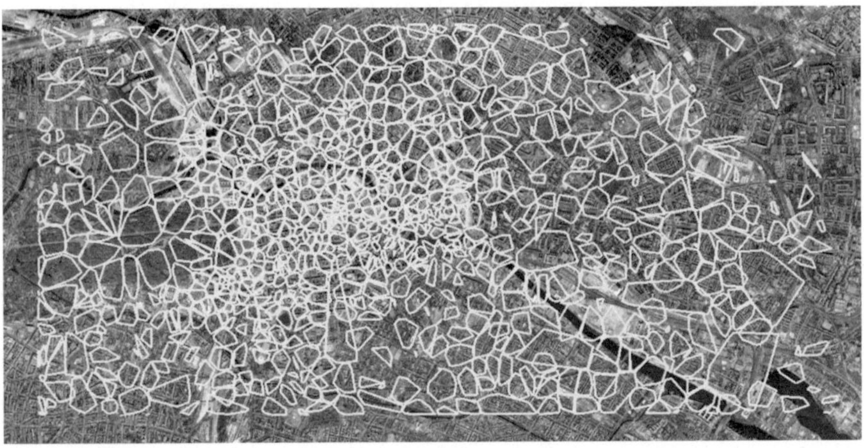

(b)

Fig. 3. Berlin, Germany. Cluster boundaries of photos assigned to existing POIs (yellow) using a photo-to-POI distance threshold of (a) 200 meters and (b) 400 meters. Cluster boundaries forming new areas of POIs were obtained using a distance threshold of 100 meters and a density threshold of three people in a cluster (green)

While the Teiresias algorithm has the potential for discovering patterns of up to length four if applied on Berlin data, we only present patterns of length 2 and 3 in keeping with the editorial limitations of this paper. The general statistics pertaining to sequences and patterns are presented in Table 4. Tables 5 and 6 present the five most frequent patterns discovered by the Teiresias algorithm.

Table 4. Berlin, Germany. General statistics

Photo-to-POI threshold	<L,W>	# of people in sequences	# of valid sequences	# of sequence patterns
200	<2,3>	8952	2844	2047
	<3,4>			186
	<4,5>			9
400	<2,3>	8968	2845	2086
	<3,4>			195
	<4,5>			11

From Table 3 we can see that using 2,844 sequences from a total of 8,952 sequences and a photo-to-POI distance threshold of 200 meters, the algorithm discovered 2,047 patterns of length 2; 186 patterns of length 3; and 9 patterns of length 4. Using 2,845 sequences from a total of 8,968 sequences with a photo-to-POI distance threshold of 400 meters, the algorithm discovered 2,086 patterns of length 2; 195 patterns of length 3; and 11 patterns of length 4. The first four sequence patterns of length 2 and 3 are identical for two photo-to-POI distance thresholds (Tables 4-5). The first three sequence patterns of length 2 (Table 4) suggest that people began photographing at Brandenburg Gate and then continued to other places. The third sequence pattern in Table 4 contains a wild character indicating that that people started from Brandenburg Gate, then visited any POI and finished at the Reichstag.

We should also note that unknown POIs created by applying density-based clustering (Sect. 3.3) are not part of the most frequent sequence patterns.

Table 5. Berlin, Germany. Sequence patterns using L=2, W=3

Photo-to-POI threshold	# of input sequences	Sequence patterns
200	74	Brandenburg Gate → Reichstag
	53	Brandenburg Gate → Memorial to the Murdered Jews of Europe
	46	Brandenburg Gate → * → Reichstag
	41	Reichstag → Brandenburg Gate
	36	**Pariser Platz → Brandenburg Gate**
400	71	Brandenburg Gate → Reichstag
	51	Brandenburg Gate → Memorial to the Murdered Jews of Europe
	47	Brandenburg Gate → * → Reichstag
	43	Reichstag → Brandenburg Gate
	34	**Reichstag → * → Reichstag**

Table 6. Berlin, Germany. Sequence patterns using L=3, W=4

Photo-to-POI threshold	# of input sequences	Sequence patterns
200	13	Reichstag → Der Bevölkerung → Reichstag
	10	Brandenburg Gate → Memorial to the Roma and Sinti Holocaust Victims → Reichstag
	8	Pariser Platz → Brandenburg Gate → 18th of March Square
	8	Reichstag → Brandenburg Gate → Memorial to the Murdered Jews of Europe
	7	**Der Bevölkerung → Reichstag → Der Bevölkerung**
400	14	Reichstag → Der Bevölkerung → Reichstag
	10	Brandenburg Gate → Memorial to the Roma and Sinti Holocaust Victims → Reichstag
	9	Pariser Platz → Brandenburg Gate → 18th of March Square
	8	Reichstag → Brandenburg Gate → Memorial to the Murdered Jews of Europe
	7	**Zeughaus →Alte Kommandantur → Lustgarten**

5 Discussion

We demonstrated how an automatic data mining process could be used in finding travel patterns from a collection of geotagged photos. However, geographical data mining is far more complex process than its "classical" counterpart. There are several reasons for this:
(1) Data quality, spatial precision and uncertainty play a crucial role in a spatio-temporal analysis.
(2) Many spatial problems are ill-defined. This makes it impossible to apply fully automatic data-mining process to solving particular problems (Andrienko et al., 2007).
(3) The geographical analysis is very sensitive to the length or area over which an attribute is distributed (Miller and Hand, 2009).

Data quality (spatial and temporal) and precision depends on the way the data is generated and should be taken into consideration during analysis and validation of results. Movement data is usually collected using GPS-enabled devices attached to an object or by geotagging images shared on the Web. For example, when a person enters a building a GPS signal can be lost or the positioning may be inaccurate due to a weak connection to satellites. These concerns are valid for geotagged photo data as well. Specifically, there are two ways to geotag a photo and upload it on the Web. One way involves attaching a GPS to a camera. In this case, the geotagging is performed automatically and the person can face the same problems as with conventional GPS devices described above. Alternative solution would be to manually annotate a photo during upload. In this case, several possibilities exist: the individual photographer may geo-annotate the object being photographed instead of the exact place where it was taken or the exact place could be geotagged with a different level of precision. In addition, the timestamp of a taken photo may not correspond to the correct time at which the photo was taken because of: (1) time zones differences between the user's country of origin and the visiting country (2) careless setting of the camera's clock to some unrealistic time or (3) a software failure reading the timestamp of a photo.

In regard to the second issue raised in this section, there are two basic approaches for discovery of interesting sequence patterns: user-driven and data-driven. The user-driven approach is based on an expert's knowledge. However, it is not always efficient when an expert is required to find interesting sequences from thousands of sequence patterns such as was the case of Berlin. In our examples, we used frequency of patterns as a selection measure. However, frequent sequences do not necessarily constitute the most interesting patterns. In fact, frequent sequences usually represent the obvious patterns. Therefore, different interestingness measures for ranking patterns (Piatetsky-Shapiro, 1991) can be combined with expert's knowledge to find some new unexpected patterns.

The difficulties associated with spatio-temporal data mining indicate that an analyst should select the parameter values very carefully. Unfortunately, we could not cover all the possible combinations of parameter values in our experiments. However, we demonstrated that changing only the distance threshold of the *photo-to-POI* while keeping all other parameters constant, may produce slightly different pattern sequences. Changing parameters at every step of our approach could lead to completely new sequence patterns. While background knowledge of an analyst or domain expert could help overcome the weakness of the automatic process, some degree of human involvement is necessary for inspecting the data, tuning the parameters, controlling the analysis process and revising the obtained results. For example, an unknown POI can be discovered using the procedure presented in Sect. 3.3. The newly discovered POI may be adjacent to the region of an existing POI. An automatic process treats these two regions as distinct. However, visual inspection might reveal that the unknown POI belongs to the existing POI and that the two regions should be merged into one. Therefore, the solution to this issue would be incorporation of data mining techniques into geovisual analytics systems.

6 Conclusion

In this paper, we presented a novel approach for mining travel sequences using geotagged photo data. We showed that the method is capable of mining semantically annotated sequences of any length with patterns that are not necessarily immediate antecedents. We demonstrated the feasibility of our approach on two different cities using real data. We showed that the approach could be applied to different spatial scales -- to places that have a great number of visitors (Berlin) and POIs, and to sites that have relatively few visitors (Guimaraes) and POIs.

In our future work, we intend to integrate our approach within a visual analytics framework. We shall investigate in detail sequence patterns based on: user profiles (locals/tourists); activity (night/day); and seasonal changes. We shall also concentrate on analyzing sequences with specific parameter settings, and then validating and comparing the resulting patterns to existing solutions. In addition, we shall apply different interestingness measures to help the analyst in discovering interesting sequence patterns.

References

Agrawal, R. and Srikant, R. (1995) Mining sequential patterns. Proceedings of International Conference on Data Engineering (ICDE'95), pp. 3-14, 1995.

Agrawal, R. and Srikant, R. (1994) Fast algorithms for mining association rules. Proceedings of International Conference Very Large Data Bases, pp. 487-499, 1994.

Alvares, L.O., Bogorny, V., Kuijpers, B., de Macedo, J.A.F., Moelans, B. and Vaisman, A. (2007a) A model for enriching trajectories with semantic geographical information. Proceedings of the 15th annual ACM international symposium on Advances in geographic information systems, pp. 22, 2007.

Alvares, L.O., Bogorny, V., de Macedo, J.A.F., Moelans, B. and Spaccapietra, S. (2007b) Dynamic modeling of trajectory patterns using data mining and reverse engineering. Tutorials, posters, panels and industrial contributions at the 26th international conference on Conceptual modeling, volume 83, pp. 149-154, 2007.

Andrienko, G., Andrienko, N., Jankowski, P., Keim D., Kraak M.-J., MacEahren A., Wrobel S. (2007) Geovisual analytics for spatial decision support: Setting the research agenda. International Journal of Geographical Information Science, 21/8, pp. 839–857.

Ester, M., Kriegel, H.-P., Sander, J., and Xu, X. (1996) A density-based algorithm for discovering clusters in large spatial databases with noise. Proceedings ACM KDD, pp. 226-231, 1996.

Giannotti, F., Nanni, M. and Pedreschi, D. (2006) Efficient mining of temporally annotated sequences. Proceedings of the 6th SIAM International Conference on Data Mining (SDM'06), pp. 346-357, 2006.

Giannotti, F., Nanni, M., Pinelli, F. and Pedreschi, D. (2007) Trajectory pattern mining. Proceedings of the 13th ACM SIGKDD international conference on Knowledge discovery and data mining, pp. 339, 2007.

Girardin, F., Vaccari, A., Gerber, A., Biderman A., and Carlo R. (2009) Quantifying urban attractiveness from the distribution and density of digital footprints. International Journal of Spatial Data Infrastructures Research, 4, pp. 175-200, 2009.

Kisilevich, S., Florian, M., Peter, Bak., Tchaikin, A., Keim, D. (2010) Where Would You Go on Your Next Vacation? A Framework for Visual Exploration of Attractive Places, accepted in Geoprocessing 2010.

Miller H. and Han J. (2009) Geographic data mining and knowledge discovery. Chapman & Hall/CRC.

Piatetsky-Shapiro G., (1991) Discovery, analysis and presentation of strong rules. In: G. Piatetsky-Shapiro and W.J. Frawley Editors, Knowledge Discovery in Databases AAAI (1991), p. 229

Rigoutsos, I. and Floratos, A. (1998) Combinatorial pattern discovery in biological sequences: The TEIRESIAS algorithm. Bioinformatics, 14/1, pp. 55-67.

Rokach, L. and Maimon, O. (2005), Clustering methods, In O. Maimon & L. Rokach (Eds.), Data Mining and Knowledge Discovery Handbook, Springer, pp. 321--352.

Rokach, L., Romano R. and Maimon O. (2008A), Negation recognition in medical narrative reports, Information Retrieval, 11(6):499-538.

Rokach, L., Romano R. and Maimon O. (2008B), Mining manufacturing databases to discover the effect of operation sequence on the product quality, Journal of Intelligent Manufacturing, 19(3):313-325.

Spaccapietra, S., Parent, C., Damiani, M.L., de Macedo, J.A., Porto, F. and Vangenot, C.(2007) A conceptual view on trajectories. Technical report, Ecole Polytechnique Federal de Lausanne, 2007.

Zheng, Y., Zhang, L., Xie, X. and Ma, W.Y. (2009) Mining interesting locations and travel sequences from GPS trajectories. Proceedings of the 18th international conference on World Wide Web, pp. 791-800, 2009

Exposing CSW Catalogues as Linked Data

Francisco J. Lopez-Pellicer, Aneta J. Florczyk, Javier Nogueras-Iso, Pedro R. Muro-Medrano, F. Javier Zarazaga-Soria

Department of Computer Science and Systems Engineering
Universidad de Zaragoza, Spain
{fjlopez,florczyk,jnog,prmuro,javy}@unizar.es

Abstract. The OpenGIS Catalogue Services (CS) specification defines a set of abstract interfaces for the discovery, access, maintenance and organization of metadata repositories of geospatial information and related resources in distributed computing scenarios, such as the Web. The CS specification also defines a HTTP protocol binding, which is called "Catalogue Services for the Web" or CSW. A fair description of CSW is a remote catalogue interface over the HTTP protocol, but not over the architecture of the mainstream Web where search engines are the users' gateway to information. This paper identifies some aspects of CSW that difficult the findability of metadata in the Web, and hence, the discovery of resources. This paper also presents a toolkit that exposes as Linked Data the content of metadata repositories offered through CSW with the purpose of improving the discovery of metadata records in search engines.

1 Introduction

A catalogue is a system that helps publish, query and retrieve items of information in a systematic way. The OpenGIS Catalogue Services (CS) specification provides discovery, access, maintenance and organization interfaces for metadata catalogues of geospatial information and related resources, and allows users to find information in distributed systems (Nebert et al. 2007). The CS specification defines a HTTP protocol binding named Catalogue Services for the Web (CSW). Spatial Data Infrastructures (SDIs) use CSW as one of the gateways to their geospatial resources.

M. Painho et al., (eds.), *Geospatial Thinking*, Lecture Notes in Geoinformation and Cartography,
DOI 10.1007/978-3-642-12326-9_10,

An example of the relevance of CSW is the recommendation issued in the context of INSPIRE by the INSPIRE Network Services Drafting Team (2008) to SDIs in European Union to derive the base functionality of discovery services from the ISO profile of CSW defined in Voges et al. (2007).

However, CSW is not properly prepared for the mainstream Web where search engines are the users' gateway to information. Some features of the infrastructure for the discovery of information in the mainstream Web are:

- **Search engines try to browse and index Deep Web databases**. Surfacing Deep Web content is a research problem that concerns the search engine community since its description by Bergman (2001). The term Deep Web refers to the database content that is behind Web forms and applications. From this point of view, SDI metadata repositories are hidden behind catalogue applications; therefore, SDI metadata is part of the Deep Web. Hence, the findability in search engines depends on the success of crawling processes that require the analysis of the Web interface, and then the automatic generation of queries.

- **Applications ask for Linked Data**. The Linked Data community, which has blossomed in the last three years, promotes a Web of data based on the architectural principles of the Web (Bizer et al., 2008). Linked Data is a set of best practices to publish, share and connect data, information and knowledge using URIs that are resolved to Resource Description Framework (RDF) documents. RDF is a W3C recommendation for modelling and exchanging metadata (Miller et al., 2004). According to Bizer et al. (2009), in May 2009, the approximate amount of information released under this practice amounts to 4,700 million of RDF statements connected by 142 million of links and a growing number of relevant nodes.

- **Evolution of metadata vocabularies.** Well known metadata vocabularies have evolved to models based on RDF with an emphasis in the linking of metadata descriptions. The abstract data models of Dublin Core Metadata Initiative (DCMI) and the Open Archive Initiative (OAI) have evolved side of the RDF data model. This process has resulted in abstract models based on the RDF data model (Nilsson et al. 2008; Lagoze et al. 2008) that empathizes the use (and reuse) of entities rather than plain literals as the value of properties. This evolution enables the effective hyperlink of metadata and traverse queries using query languages and protocols, such as SPARQL (Seaborne et al., 2008).

This paper identifies three drawbacks in CSW in relation with the depicted scenario:

- The protocol is hard to crawl by standard Deep Web crawlers.
- The remote procedure call style for accessing metadata is orthogonal to the linked data approach.
- The support of association links between metadata in queries is limited.

The most relevant consequence is that metadata published by SDIs have become part of the Deep Web content not surfaced by search engines. Therefore, the resources offered by SDIs are more difficult to be discovered in the mainstream gateway to information.

This paper proposes republishing CSW catalogues as Linked Data to make their content easily accessible for search engines and machine-to-machine applications aware of the Web of data. This paper also proposes the CSW2LD toolkit for republishing SDI metadata. The mission of the toolkit is to expose the content of standard CSW catalogues as Dublin Core metadata conform to the RDF data model and the principles of Linked Data.

The structure of this paper is as follows. Section 2 identifies related work. Section 3 presents CSW and the above drawbacks. Section 4 discusses the general approach that the CSW2LD toolkit follows to map catalogue information models to the RDF abstract data model. Section 5 describes the CSW2LD toolkit for publishing metadata. Finally, the conclusions review the ideas presented and sets the next research goals.

2 Related work

This section presents related work in the geographic information domain about the use of semantic descriptions and search engines in catalogue systems, and presents some publishing tools of the Linked Data community related with CSW2LD.

Egenhofer (2002) proposes the use of the Semantic Web to face problems of semantic heterogeneity in geo-resource discovery. Studies on geospatial catalogue usability, such as Larson et al. (2006), identify as a potential improvement the use of semantic techniques for knowledge description and discovery. Some authors have considered the use of search engines. For example, the approach of Oates et al. (2007) is to provide metadata encoded in KML about resources and make the KML files discoverable through Google search.

There is a variety of Linked Data publishing tools. Many of them are services that publish the content of relational databases as Linked Data

(Bizer et al., 2009). Large geographical information providers are investigating how Linked Data and other Semantic Web technologies can assist the diffusion of geographic data. For example, Ordnance Survey is developing datasets in RDF and publishing them using the Linked Data principles (Goodwin et al. 2009). The Linked Data community has an increasing interest in the geospatial databases. In particular, the LinkedGeoData project maps OpenStreetMap data into linked data (Auer et al., 2009), and the GeoNames ontology describes the content of the GeoNames database (Vatant et al., 2007).

Haslhofer et al., (2008) is the only directly related work found in the literature but it does not belong to the geo community. It proposes a server that wraps the metadata protocol for digital libraries OAI-PMH (Lagoze et al. 2002), exposes metadata as Linked Data and provides metadata access via a SPARQL endpoint.

3 Catalogue Services for the Web

CSW defines the interaction between a catalogue client and a CSW server that exposes the contents of an opaque catalogue. Request and response messages must conform to the CSW specification or to application profiles derived from it.

3.1 The context

CSW is the HTTP protocol binding of the OpenGIS Catalogue Services (CS) specification. The CS specification defines interfaces for the management, the discovery and the access to collections of metadata about geospatial information resources. The management interface supports the ability to administer and organize collections of metadata in the local storage device. The discovery interface allows users to search within a catalogue and provides a minimum query language. Finally, the access interface facilitates access to metadata items previously found with the discovery interface. The CS specification also defines an abstract information model that includes a core set of shared attributes, a common record format that defines metadata elements and sets, and a minimal query language called CQL.

Additionally to the HTTP protocol binding, the CS specification includes binding implementation guidance for the application protocols Z39.50, a pre-Web protocol widely used in digital libraries, and

CORBA/IIOP, a remote procedure call specification *in a niche of relative obscurity* (see Henning 2008).

3.2 Request and response example

CSW is quite complex. For example, the operation `GetRecordById` fetches representations of metadata records using the identifier of the metadata in the local metadata repository. The parameter `elementSetName`, if used, establishes the amount of detail of the representation of the source record. Each level of detail specifies a predefined set of record elements that should be present in the representation. The predefined set name `full` represents all the metadata record elements. By default, the operation `GetRecordById` returns a metadata record representation that validates against the information model of the metadata repository. The parameter `outputSchema` allows user agents to request for a response in a different information model, and the CSW implementations must support at least the representation of the common information schema defined in the CSW standard.

Figure 1 shows a sample `GetRecordById` request for a metadata record available in IDEE, the SDI of Spain, and the corresponding response. The request URI identifies the location of the CSW server, the operation, the identification of the metadata record (parameter `id`), the amount of detail of the representation (parameter `elementSetName`), and the output schema (parameter `outputSchema`). The XML response consists of a `<GetRecoredByIdResponse>` element that contains a record that conveys the information of the source metadata. When a `<SummaryRecord>` element is the conveyor, the retrieved representation contains a summary of the original metadata record. The value of the output schema identifies the subset that conforms to the common information schema defined in the CSW standard.

Request:

```
GET/csw/servlet/cswservlet?request=GetRecordById&id=
ESIGNMAPASRELIEVESERIE200701180000&elementSetName
full&outputSchema=http://www.opengis.net/cat/csw/
2.0.2 HTTP/1.1
Host: www.idee.es
```

Response:

```
HTTP/1.x 200 OK
Content-Type: application/xml;charset=ISO-8859-1
...

<?xml version = '1.0' encoding = 'ISO-8859-1'?>
<GetRecordByIdResponse
 xmlns="http://www.opengis.net/cat/csw/2.0.2"
 xmlns:dc="http://purl.org/dc/elements/1.1/"
 xmlns:dcterms="http://purl.org/dc/terms/"
 ...>
 <SummaryRecord>
  <dc:title>Mapas en Relieve</dc:title>
  <dc:identifier>
   ESIGNMAPASRELIEVESERIE200701180000
  </dc:identifier>
  ...
  <dc:format>PVC</dc:format>
  <dc:subject>elevation</dc:subject>
  <dc:subject>imageryBaseMapsEarthCover</dc:subject>
  <dc:type>dataset</dc:type>
  ...
  <dcterms:spatial>COUNTRIES.SPAIN</dcterms:spatial>
  ...
  <dcterms:spatial>
   northlimit=43.8;
   southlimit=37.83;
   westlimit=-9.32;
   eastlimit=0.72;
  </dcterms:spatial>
 </SummaryRecord>
</GetRecordByIdResponse>
```

Fig. 1. Sample CSW `GetRecordById` request and response

3.3 Identified drawbacks

CSW is undoubtedly useful to enable the discovery and access to geographic information resources within the geographic community (Nogueras et al, 2005). However, it presents the following drawbacks:

- **Mismatch with operational model of Deep Web crawlers**. The search engines have developed several techniques to extract information from Deep Web databases without previous knowledge of their interfaces.

The operational model for Web crawlers, described in Raghavan (2001), based on (1) form analysis, (2) query generation and (3) response analysis is widely accepted. It models queries as functions with *n* named inputs $X_1..X_n$. where the challenge is to discover the possible values of these named inputs that return most of the content of the database. This approach is suitable for CSW HTTP GET requests. However, the constraints are encoded in a single named input as a CQL string (see Nebert et al. 2007), or an XML Filter (Vretanos, 2004). This characteristic is incompatible with the query model of the Deep Web crawlers. Researchers working for search engines, such as Google (see Madhavan et al. 2008), discourage the alternative operational model that consists in the development of ad-hoc connectors as non-sustainable in production environments.

- **RPC approach to access metadata.** Metadata repositories are behind a proprietary RPC from the point of view of other communities. CSW does not define a simple Web API to query and retrieve metadata. Some communities that potentially can use CSW are accustomed to simple APIs and common formats. For example, many geo mashups and related data services (see Turner, 2006) use Web APIs to access and share data built following the REST architectural style (Fielding, 2000) and the vision of Berners-Lee et al. (2001) about the Semantic Web. These APIs are characterized by the identification of resources by opaque URIs, semantic descriptions of resources, stateless and cacheable communication, and uniform interface based on the verbs of the HTTP protocol in opposition to the RPC style.

- **Queries limited to same record properties**. The field based query model of the CS specification does not define the support for associations in the CQL or Filter syntax. CSW application profiles may describe the support of associations. For example, the ISO application profile (Voges et al. 2007) supports the linkage between services and data instances. However, the linkage is based in the equality of literal values of properties such as `MD_Identifier.code`, and the profile does not extend the CQL and the XML Filter syntax. Hence, association queries require being decomposed in parts. For example, in a metadata repository where metadata records about data and services instances are linked, a query that returns the services that serves data created by a producer requires (1) to query initially about the data created by this producer, (2) to retrieve their identifiers, and then, (3) to query about servers that serve data with these identifiers.

4 Mapping SDI metadata to RDF

The annotation of geographic resources is based on the concept of metadata. Metadata are information and documentation that enable data to be understood, shared and exploited effectively by all users over time. As mentioned in Nebert (2004), the geographic metadata help geographic information users to find the data they need and determine how to use.

One of the main goals of the creation of geographic metadata is the reuse of organization's data by publishing its existence through catalogue metadata records that conveys information about how to access and use the data (FGDC, 2000). In the context of European SDI, the information is conveyed as ISO 19115 / ISO 19119 metadata records represented in XML. However, RDF is the lingua franca for the metadata interchange in the Semantic Web. The publication of SDI metadata in the Semantic Web requires a mapping from the metadata schema to the RDF data model.

4.1 The RDF data model

RDF is a metamodel for expressing metadata about resources. A resource may be an abstract concept, a real world concept or a digital asset such as an entire Web site. The RDF provides a simple model to describe relationships between resources in terms of properties associated with a name and a set of values. The RDF conceptual model is a graph-based model with directed labelled arcs. The nodes of the graph are resources, named or blank, and values, also known as literals. Each named node has an associated URI that uniquely identifies the node. The rules of the arcs, known as triples, are:

- The subject, that is, the origin of the arc, is a resource.
- The property or predicate, that is, the label of the arc, is a named resource.
- The object, that is, the target of the arc, is a resource or a literal

There are two kinds of literals: plain and typed. A plain literal is a character string that optionally has a tag that documents the language of the character string. A typed literal is a pair composed by a value encoded as a character string, and the data type, which defines both the semantics of the value and the syntax of the encoding. For the declaration and the interpretation of these properties, the RDF Schema (RDFS) provides a language to define and restrict the interpretation of the RDF vocabularies

Table 1. CWA 14857: Crosswalk ISO 19115 Core Metadata for Geographic Datasets – Dublin Core; the prefix *dct:* maps to the http://purl.org/dc/terms/ namespace; the entities *Agent*, *Location*, *MediaType*, *LinguisticSystem* and *RightsStatement* of RDF property range are DCMI terms classes

DC property	ISO 19115:2003 property mapping	RDF property	RDF property range
Contributor	MD_Metadata.identificationInfo. MD_DataIdentification.credit	dct:contributor	Agent
Coverage	MD_Metadata.identificationInfo. MD_DataIdentification.extent. EX_Extent.geographicElement. EX_GeographicBoundingBox	dct:spatial	Location
Creator	MD_Metadata.identificationInfo. MD_DataIdentification.citation. CI_Citation.CitedResponsibleParty. CI_ResponsibleParty. OrganisationName[role="originator"]	dct:creator	Agent
Date	MD_Metadata.identificationInfo. MD_DataIdentification.citation. CI_Citation.date.CI_Date	dct:modified	Typed literal (date)
Description	MD_Metadata.identificationInfo. MD_DataIdentification.abstract	dct:abstract	Plain literal
Format	MD_Metadata.distributionInfo. MD_Distribution.distributionFormat. MD_Format.name	dct:format	MediaType
Identifier	MD_Metadata. MD_Distribution. MD_DigitalTransferOption.onLine CI_OnlineResource.linkage.URL	dct:identifier	Plain literal
Language	MD_Metadata.identificationInfo. MD_DataIdentification.language	dct:language	LinguisticSystem
Publisher	MD_Metadata.identificationInfo. MD_DataIdentification.citation. CI_Citation.CitedResponsibleParty. CI_ResponsibleParty. OrganisationName. [role="publisher"]	dct:publisher	Agent
Relation	-	dct:relation	Resource
Rights	-	dct:rights	RightsStatement
Source	MD_Metadata.dataQualityInfo. DQ_DataQuality.lineage. LI_Lineage. source. LI_Source.description	dct:source	Resource
Subject	MD_Metadata.identificationInfo. MD_DataIdentification.topicCategory.	dct:subject	Resource
Title	MD_Metadata.identificationInfo. MD_DataIdentification.citation. CI_Citation.title	dct:title	Plain literal
Type	MD_Metadata.hierarchyLevel	rdf:type	Class

4.2 Expressing geographic metadata in RDF: the Dublin Core crosswalk approach

There are several geographic metadata crosswalks to the Dublin Core vocabulary. Table 1 describes the crosswalk of the geographic metadata ISO 19115 to the Dublin Core vocabulary defined in CWA 14857 (Zarazaga-Soria et al., 2003). We propose the use of well-known Dublin Core crosswalks to implement uniforms mappings from geographic metadata schemas to the RDF data model. This approach consists of three steps:

- Apply a metadata crosswalk from the original metadata schema to the Dublin Core vocabulary.
- Add additional metadata such as provenance of the record, original information model or crosswalk identification.
- Apply the profile for expressing as RDF the metadata terms.

The output of the crosswalk can be augmented by adding additional metadata descriptions that log the crosswalk and the provenance of the metadata. Then, this metadata description is transformed to the RDF data model by applying a profile for expressing the metadata terms as RDF. Table 1 also includes an example of a profile. This table includes for each Dublin Core term its mapping to a RDF properties and its range. The RDF Dublin Core profiles are different from the XML Dublin Core profiles. The DCMI abstract model (DCAM) has a reference model formalized in terms of the semantics of the RDF abstract model since 2005 (Powell et al., 2007). One of the changes is that properties may have a formal range. In the RDF data model, this range can be literal, for example the property title, or a resource, for example the property creator. With this approach, when the object of a property refers to an entity, it can be properly identified and described.

5 The CSW2LD toolkit

Our approach to solve the drawbacks of CSW is the CSW2LD toolkit. The ideas behind the design of the CSW2LD toolkit are presented below.

5.1 Conceptual model for re-publishing metadata

The conceptual model can be decomposed as follows:

- **CSW interface model**. A metadata repository contains metadata about resources. Client applications use CSW requests to query metadata repositories. The CSW requests may generate metadata snapshots that are subsets of metadata at the time of the request. The CSW request determines the amount of information (user defined, brief, summary or full records) and the information schema of the metadata snapshot. The CSW response contains the realization of the metadata snapshot in a supported media format. XML is the only media format that all CSW implementations must support.

- **Harvest model.** The harvest produces a set of metadata snapshots realized in XML representations. The harvest process asks for metadata records whose information model can crosswalk to Dublin Core. The CS specification defines a common group of metadata elements expressed using the Dublin Core vocabulary. CSW defines a default mapping of the common group of metadata elements to XML that all CSW implementations must support. The harvest process queries for the common representation if no crosswalk is applicable to the information model of the catalogue.

- **Semantic publication model**. The harvested representation of the metadata snapshot is mapped to the RDF data model and published following the Linked Data principles. The base of the mapping is the DCMI recommendation for expressing Dublin Core using RDF (Nilsson et al., 2008). The result is a semantic description about a resource that is a version of the metadata snapshot that describes the same resource. This semantic description is published according to the best practices to publish Linked Data on the Web (Bizer et al., 2007). The model assumes that a dereferenceable URI, the semantic URI, can identify the resource that the semantic description describes. This semantic URI is owned by the responsible of the semantic publication and redirects to an URI where user agents can get a RDF representation of the semantic description. The semantic description, in turn, has the semantic URI as subject in its assertions. If the mapping process discovers links between the resources, it may replace the original RDF mapping by these semantic URIs. For example, the description of a service may include a brief description of the data. Then, this brief description can be replaced with the URI that identifies the semantic description of the data. The semantic description may contain a link that encodes a CSW HTTP GET request equals to the CSW request done in the harvest. Semantic browsers and search engines, such as Tabulator (Berners-Lee et al. 2006) and Sindice (Tummarello et al., 2007) respectively, can browse and index the semantic descriptions, and use the links to navigate to other resources or to retrieve transparently the original metadata description.

- **Non-semantic publication model**. Given the semantic descriptions described in the previous point, the model assumes that an URI identifies the human-readable representation in HTLM format. The semantic URI of a resource may be resolved to this URI if the agent requests a human-readable representation of its semantic description. This representation uses the HTLM element `<link>` to provide information to navigate alternative representations. At least, it includes a link that points to the semantic URI and a link that encodes the CSW HTTP GET request. Web browsers and search engines can browse and index respectively these representations. In addition, they can use the links to navigate to the semantic representations and to retrieve transparently the original metadata description.

5.2 Algorithm for harvesting and publishing a CSW server

Figure 2 summarizes the process in the context of SDIs where the information model of many catalogues is ISO 19115 / ISO 19119. The steps of the harvesting process are:

- Analyze the capabilities of the CSW service to discover the information models served and the levels of amount of information.
- Fetch identifiers of new and updated records with the CSW `GetRecords` operation.
- Retrieve new and updated records using the `GetRecordById` operation; request ISO 19115 / ISO 19119 information models if they are available.
- Crosswalk to the Dublin Core vocabulary if the requested information model is not the common information model.
- Map the set of Dublin Core metadata terms to the RDF data model.
- Generate or update the human readable and machine-readable representations from the RDF graphs.

The `GetRecords` operation does a search and returns piggybacked metadata. The harvest process uses the `GetRecords` operation to determine the number of metadata records to retrieve, and to obtain piggybacked unique identifiers for retrieve metadata records. Optionally, along with the identifier, the harvester process can ask for the creation or update date of the record within the catalogue. The identifiers, and, if available, the creation or update date, are compared with the previous harvest of the same repository to detect new and updated records to retrieve. Deleted records may be kept for archiving reasons.

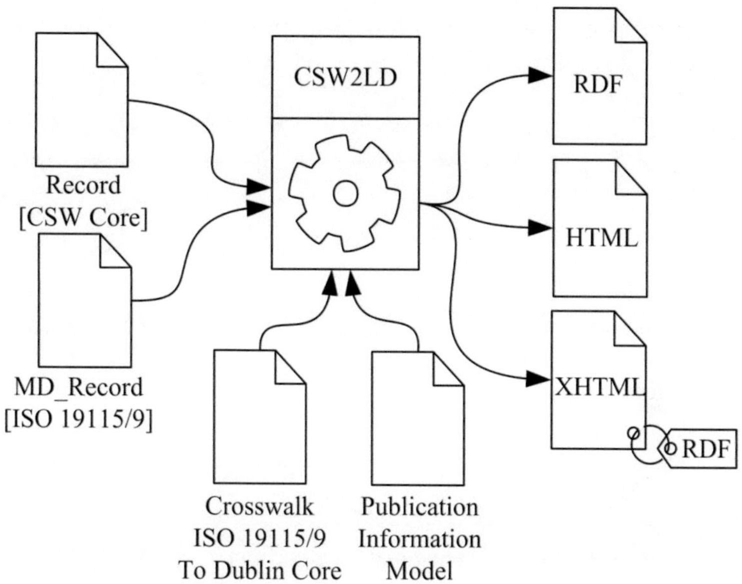

Fig. 2. Overview of the republish process of CSW served catalogues in terms of metadata representations

The current implementation implements a crosswalk described in Nogueras-Iso et al. (2004). The available formats for the machine readable and human readable representations are RDF/XML, N3, TURTLE and XHTML with RDF annotations (RDFa) for the former, and HTML and XHTML for the later.

The harvest process configures an Apache HTTP server for publishing the representations following the conventions of Linked Data (Berrueta et al, 2008). The configuration enables the server to publish machine-processable and human-readable representations. Figure 3 shows the core logic of the redirection and content negotiation implemented in the configuration. If the URI matches the web folder, the server returns a `303 See other` response that locates an HTML that informs the user agent about the metadata records exposed in the folder. If the URI matches with a resource contained in the web folder, the server identifies the resource and returns with a `303 See other` the location of the representation that matches the kind of representation requested. The harvest process also creates the index document. It contains hyperlinks to the URIs of the representations of the semantic descriptions with a summary of the information such as title and keywords. If the catalogue is large, the harvest process creates multiple index documents linked each other simulating pagination and modifies the redirection logic.

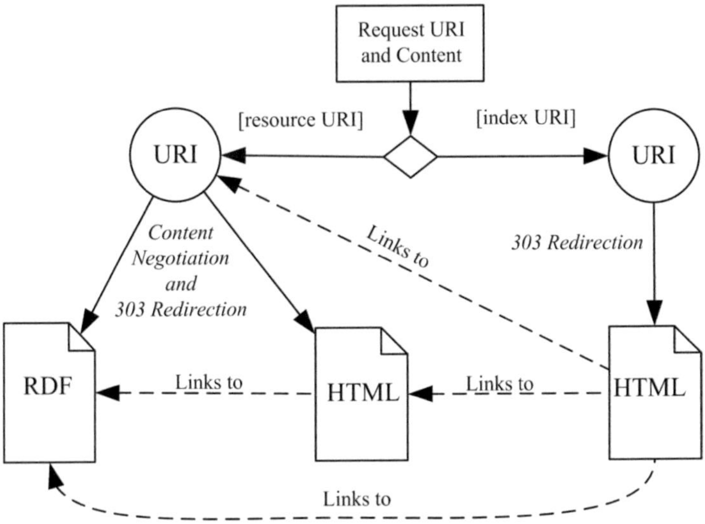

Fig. 3. Redirection and content negotiation algorithm

5.3 Enabling transparent access to the metadata repository

One on the goals of the CSW2LD toolkit is to provide transparent access to CSW served repositories. Transparent access and provenance metadata are related concepts in the CSW2LD toolkit. Each semantic description includes a simple provenance description as a triple with an `rdfs:seeAlso` predicate whose subject is the semantic URI and the object is a CSW HTTP GET request. The information content of the semantic description may include additional information about the metadata snapshot, for example, the retrieval date.

Figure 4 shows how semantic aware user agents can access transparently to the metadata repository. The user agent can discover the semantic URL of a resource in a semantic search engine. Then, it can retrieve its semantic description. After processing the content, the user agent can require additional information. The semantics of RDF says that this might be found traversing the `rdfs:seeAlso` property. As the target is a complete CSW HTTP GET request, the user agent can retrieve a XML representation of the original metadata.

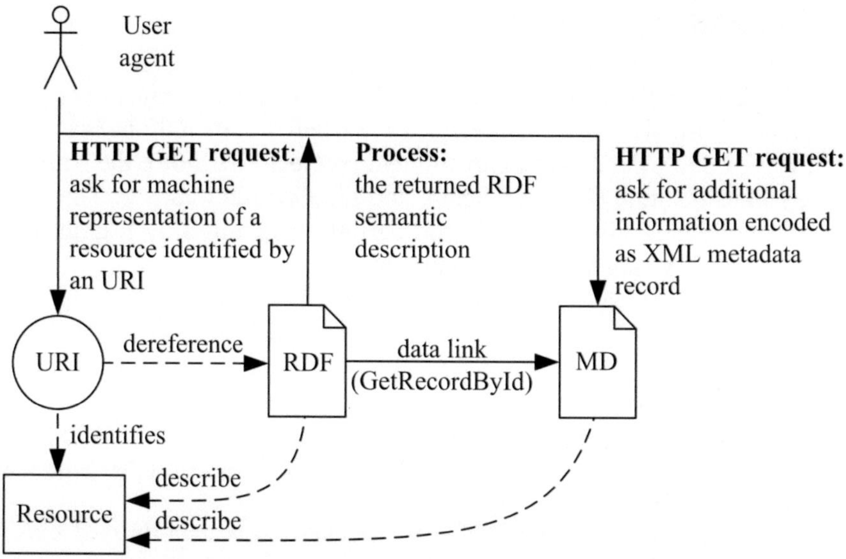

Fig. 4. A semantic user agent can access to the content of the metadata repository without previous knowledge of CSW

6 Conclusions

This paper presents the CSW2LD toolkit: a software component that republishes according to the Linked Data metadata from repositories accessible through CSW. Applied to SDI metadata catalogues, the CSW2LD toolkit exposes the description of SDI assets as dereferenceable Web resources, and allows search engines to index them. On the other hand, the published RDF description of metadata records and resources is not standard, and can be semantically inaccurate. The main reasons lie on the lack of standards mappings from geographic metadata schemas to the RDF model, and the heterogeneity of communities targeted by CSW.

Future versions of the CSW2LD toolkit should include additional technical features, such as additional crosswalks, and functional features, such as the generation of links between the metadata and existing thesauri and ontologies, augment the meta-metadata available about the provenance and quality of the exposed information, and describing the exposed data as aggregations.

Acknowledgements

This work has been partially supported by Spanish Government (projects "España Virtual" ref. CENIT 2008-1030, TIN2007-65341 and PET2008_0026), the Aragón Government (project PI075/08), the National Geographic Institute (IGN) of Spain, and GeoSpatiumLab S.L. The work of Aneta Jadwiga Florczyk has been partially supported by a grant (ref. AP2007-03275) from the Spanish government.

References

Auer, S.; Lehmann, J. and Hellmann, S. (2009) LinkedGeoData – Adding a spatial Dimension to the Web of Data Proceedings of 8th International Semantic Web Conference (ISWC)

Becker, C. and Bizer, C. (2008) DBpedia Mobile: A Location-Enabled Linked Data Browser Proceedings of the Linked Data on the Web Workshop, Beijing, China, April 22, 2008, CEUR Workshop Proceedings

Berners-Lee, T., Hendler, J. and Lassila, O (2001). The Semantic Web Scientific American , 284, 34-43

Berners-Lee, T., Chen, Y., Chilton, L., Connolly, D., Dhanaraj, R., Hollenbach, J., Lerer, A. and Sheets, D. (2006) Tabulator: Exploring and Analyzing linked data on the Semantic Web Proceedings of the 3rd International Semantic Web User Interaction

Berrueta, D. And Phipps, J (2008) Best Practice Recipes for Publishing RDF vocabularies [online]. W3C. Available from: http://www.w3.org/TR/swbp-vocab-pub/

Bizer, C., Cyganiak, R., and Heath, T. (2007) How to Publish Linked Data on the Web [online]. Freie Universität Berlin, Available from: http://www4.wiwiss.fu-berlin.de/bizer/pub/LinkedDataTutorial/

Bizer, C.; Heath, T.; Idehen, K. and Berners-Lee, T. (2008) Linked data on the web (LDOW2008) WWW '08: Proceeding of the 17th international conference on World Wide Web, ACM, 1265-1266

Bizer, C., Heath, T. and Berners-Lee (2009) Linked Data - The Story So Far International Journal on Semantic Web and Information Systems

Egenhofer, M. J. (2002) Toward the Semantic Geospatial Web. In GIS '02: Proceedings of the 10th ACM international symposium on Advances in geographic information systems, New York, NY, USA, ACM, pp. 1–4.

Federal Geographic Data Committee (2000) Content Standard for Digital Geospatial Metadata Workbook [online]. Technical report, Federal Geographic Data Committee, Washington, DC. Available from: http://www.fgdc.gov/metadata/documents/workbook_0501_bmk.pdf.

Fielding, R. T. (2000) REST: Architectural Styles and the Design of Network-based Software Architectures University of California, Irvine

Goodwin, J.; Dolbear, C. and Hart, G. (2009) Geographical Linked Data: The Administrative Geography of Great Britain on the Semantic Web Transactions in GIS, doi: 10.1111/j.1467-9671.2008.01133.x

Haslhofer, B. and Schandl, B.(2008) The OAI2LOD Server: Exposing OAI--PMH Metadata as Linked Data Proceedings of the Linked Data on the Web Workshop, Beijing, China, April 22, 2008, CEUR Workshop Proceedings.

Henning, M. (2008) The rise and fall of CORBA Communications of the ACM, ACM, 51, 52-57

Jacobs, I. & Walsh, N. (2004) Architecture of the World Wide Web, Volume One. W3C

Lagoze, C. and de Sompel, H. V. (2002) The Open Archives Initiative Protocol for Metadata Harvesting (OAI-PMH) – version 2.0. Available from: http://www.openarchives.org/OAI/openarchivesprotocol.html

Lagoze, C. and de Sompel, H. V. (2008) ORE User Guide – Resource Map Implementation in RDF/XML Open Archives Initiative. Available from: http://www.openarchives.org/ore/1.0/rdfxml

Madhavan, J., Ko, D., Kot, L., Ganapathy, V., Rasmussen, A. and Halevy, A. (2008) Google's Deep Web crawl, Proceedings of the VLDB Endowment, VLDB Endowment, 1, 1241-1252

Miller, E. and Manola, F. (2004) RDF Primer. W3C, Available from http://www.w3.org/TR/2004/REC-rdf-primer-20040210/

Nebert, D.D. (2004) Developing Spatial Data Infrastructures: The SDI Cookbook [online]. Technical report, Global Spatial Data Infrastructure Version 2.0. Available from: http://www.gsdi.org/docs2004/Cookbook/cookbookV2.0.pdf.

Nebert, D., Whiteside, A. and Vretanos, P. A. (2007). Open GIS Catalogue Services Specification. OpenGIS Publicy Available Standard, Open GIS Consortium Inc.

Network Services Drafing Team (2009) Technical Guidance Document for INSPIRE Discovery Services v 2.0 [online]. Available from: http://inspire.jrc.ec.europa.eu/documents/Network_Services/Technical%20Guidance%20Discovery%20Services%20v2.0.pdf

Nilsson, M.; Powell, A.; Johnston, P. and Naeve, A. (2008) Expressing Dublin Core metadata using the Resource Description Framework (RDF) [online] Dublin Core Metadata Initiative, DCMI Recommendation, Available from: http://dublincore.org/documents/dc-rdf/

Nogueras-Iso, J.; Zarazaga-Soria, F. J.; Lacasta, J.; Béjar, R. & Muro-Medrano, P. R. Metadata standard interoperability: application in the geographic information domain Computers, Environment and Urban Systems, 2004, 28, 611- - 634

Nogueras-Iso, J., Zarazaga-Soria, F.J., Muro-Medrano, P.R. (2005) Geographic Information Metadata for Spatial Data Infrastructures: Resources, Interoperability and Information Retrieval. Springer-Verlag New York, Inc., Secaucus, NJ, USA

Powell, A., Nilsson, M., Naeve, A., Johnston, P. and Baker, T. (2007) DCMI Abstract Model [online] Dublin Core Metadata Initiative, Available from: http://dublincore.org/documents/abstract-model/

Raghavan, S. and Garcia-Molina, H. (2001) Crawling the Hidden Web VLDB '01: Proceedings of the 27th International Conference on Very Large Data Bases, Morgan Kaufmann Publishers Inc., 129-138

Seaborne, A. & Prud'hommeaux, E. (2008) SPARQL Query Language for RDF W3C. W3C Recommendation. Available from: http://www.w3.org/TR/2008/REC-rdf-sparql-query-20080115/

Tummarello, G., Oren, E. and Delbru, R. (2007) Sindice.com: Weaving the Open Linked Data Proceedings of the 6th International Semantic Web Conference and 2nd Asian Semantic Web Conference (ISWC/ASWC2007), Busan, South Korea, Springer Verlag, LCNS 4825, 547-560

Turner, A. (2006) Introduction to neogeography O'Reilly Media, Inc.

Vatant, B. and Wick, M.(2007) GeoNames Ontology. GeoNames, Accessed 6 June 2009. Avilable from: http://www.geonames.org/ontology/

Voges, U. and Senkler, K. (2007) Open GIS Catalogue Services Specification 2.0.2: ISO Metadata Application profile. Open GIS Consortium Inc.

Vretanos, P. A. (2004) OpenGIS Filter Encoding Implementation Specification Open Geospatial Consortium Inc.

Zarazaga-Soria, FJ, Nogueras-Iso, J., Ford, M. (2003) Mapping between Dublin Core and ISO 19115, Geographic Information – Metadata. CWA 14857, CEN/ISSS Workshop - Metadata for Multimedia Information - Dublin Core

Applying Semantic Linkage in the Geospatial Web

Aneta J. Florczyk, Francisco J. Lopez-Pellicer, Rubén Béjar, Javier Nogueras-Iso and F. Javier Zarazaga-Soria

Computer Science and Systems Engineering Department
University of Zaragoza, Spain
{florczyk, fjlopez, rbejar, jnog, javy}@unizar.es

Abstract. The Semantic Web is an attempt to add meaningful annotations to Web resources, services and content that requires developing reference ontologies, which help to understand these annotations. The venue of the Web of Data makes the geographic information, which has become an important part of the current Web, widely usable.
This paper demonstrates how the Geospatial Web might take advantage from the Semantic Web. The show case is a services catalog dedicated to support the visualization applications based on on-the-fly data integration. The presented infrastructure for improving the catalog functionality applies an *administrative geography*, i.e. an ontology of political organization of the territory, published as Linked Data. The principal advantage of this approach is reflected by enhancing the functionality of the user application.

1 Introduction

From the very beginning, the growth of the Web has been uncontrolled and rapid, implying as a result the creation of disorganized content loosely connected. Soon, it was obvious that information searching and organizing techniques based on lexical analysis of the content were not satisfactory. A new approach has been needed to transform the Web from a document repository into an information resource which would improve reflection of

M. Painho et al., (eds.), *Geospatial Thinking*, Lecture Notes in Geoinformation and Cartography,
DOI 10.1007/978-3-642-12326-9_11,

human reasoning in sharing and processing information of Web resources by automated tools, such as search engines.

The machine-understandable Web resulted in the Semantic Web is based on the ideas of (1) semantic description of every resource available on Web and (2) knowledge representation for reasoning on the relation among concepts. Thus, the Semantic Web involves the idea of an ontology – "a formal, explicit specification of a shared conceptualisation" (Gruber, 1993). The semantics that capture the cognitive content of Web resources might be presented in different ways. The easiest way is to add simple *metadata*, e.g., specially designed tags in XML-based format. The semantics might be also represented as data models via other Web resources that provide conceptual structures, for example RDF (Klyne & Carroll, 2004). The most complex but also the most rich in meaning are ontology-based semantics expressed in the form of RDF+RDFS (Brickley and Guha, 2004) or OWL (McGuinness & Harmelen, 2004).

The combination of RDF documents and the HTTP protocol has gained recently considerable interest in the Semantic Web community, as it allows publishing structured data on the Web as *Linked Data* (Heath, 2009). This best practice from the Semantic Web offers logic references to any related resources. The potential of the created *Web of Data* consists of the identification of a concept via dereferenceable URI that permits retrieving the description of the concept from Web as the RDF document. This document may contain references to other documents about the same concept (i.e. identifies its instances) or states the logic relation with other concepts referenced via their URIs. In this simple manner it is possible to create a web of interlaced concepts.

Nowadays, geographic information has become an important part of the Web. The services, such as geocoding or map services, offered out-of-charge by commercial providers (e.g. Google, Yahoo) or open communities, such as OpenStreetMap (Haklay &Weber, 2008), have become essential elements of many Web applications. Historically, it has been difficult to integrate digital spatial data from different geoprocessing sources and integrate it to non-spatial information systems (McKee, 2004). The most important standardization bodies that deal with digital spatial data are (1) ISO/TC 211[1], which works on standardization in the field of digital geographic information; and (2) Open Geospatial Consortium (OGC)[2], a voluntary consensus standards organization for interoperability issues of geospatial and location based services. Also, there are many important initiatives from official institutions which aim to improve geographic in-

[1] http://www.isotc211.org/

[2] http://www.opengeospatial.org/

formation organization and accessibility via Spatial Data Infrastructures (SDI), e.g. Infrastructure for Spatial Information in the European Community (INSPIRE) established by the European Union directive (Inspire, 2007). The common Implementation Rules (IRs), adopted as Commission Decisions or Regulations, ensure that the spatial data infrastructures of the Member States are compatible and usable in a Community. The IRs are adopted in a number of specific areas (Metadata, Data Specifications, Netowrk Services, Data and Service Sharing, Monitoring and Reporting). The OGC services has been indicated as possible implementations of the INSPIRE compliant services, i.e., view services (OGC Web Map Service, WMS), download services (OGC Web Feature Service, WFS), invoke spatial service services (OGC Web Processing Service, WPS), transformation services (Application Profile of OGC WPS) and discovery services (OGC Web Catalogue Service, CSW). Following the Service Oriented Architecture approach the discovery service is the linking point responsible for the reusability of resources offered in SDI.

Geographic information has to be accompanied with a knowledge backbone to be appropriately handled due to its peculiarities (Egenhofer, 2002). Therefore, the proper semantic description of geographic information seems to be the first step in the improvement of its usage. The core of standards communities' efforts (e.g. ISO, OGC) focuses on *metadata* to describe services and their content, data models and spatial data itself. Recently, an OGC discussion paper has been published which proposes a methodology for referencing plain-text annotations to a backbone ontology (Maue, et.al., 2009). These additional annotations might be added on three levels, (1) resource meta-data (e.g., an OWS Capabilities Document), (2) data model (e.g. a GML Application Schema), and (3) data entities (e.g., a GML file). The formal specifications of concepts from the reference ontologies can be used then for tasks such as semantics-based information retrieval during workflow definition process.

As the contents of Spatial Data Infrastructures are concealed for common Web users, the Web community has created proper solutions applying successfully linking approaches, such as geographic Web platforms (e.g., semantic geocoding service of GeoNames[3]) and for publishing geo-data (e.g., LinkedGeoData[4]). On the other hand, the Linked Data has been applied successfully in spatial solutions (e.g., DBpedia Mobile (Becker & Bizer, 2008)). Both technologies, OGC services and Linked Data, might complement each other. There are many advantages of applying the best practices from the Semantic Web, such as Linked Data, in SDI. For exam-

[3] http://sws.geonames.org/

[4] http://linkedgeodata.org/About

ple, a created ontology of geographic features might link instances of the same geographic feature across different sources. This framework would provide an integrated view of geographic features rich in logic and spatial information. The richness of geographic feature description, direct or provided by linked sources, might be helpful in dealing with conflation for database integration, the well known problem from the database field (Dolbear & Hart, 2008). Additionally, such unified ontology might be used for defining and publishing complementary logic relations among geographic features (e.g., to represent different territorial organization) instead of creating new instances of OGC services. As for the Semantic Web community, the publication of geographic ontologies by official providers (i.e., a public administration organ) according to Linked Data principles might be a valuable source of references, for example the Administrative Geography of Great Britain (Goodwin, 2009).

One of the main advantages of applying the linking-based approaches to reference geographic features is the maintenance of the abstraction from their spatial representations. Usually, geographic features are characterized by blurriness of their footprints. Topological elements of physical world usually lack well defined conceptual boundaries and, consequently, this influences theirs spatial definition (e.g., rivers, chains of mountains). What is more, the computational representation of a geographic feature footprint is limited due to the resolution limits. Any geographic feature may be represented via different spatial objects which depend on the system application, the data model, and the applied technological solution. For example, a point is the best choice as a location reference while a polygon is for landscape visualization. Therefore the spatial object should be interpreted as one of the possible representations of an individual. Figure 1 presents the global view of spatial reference definition.

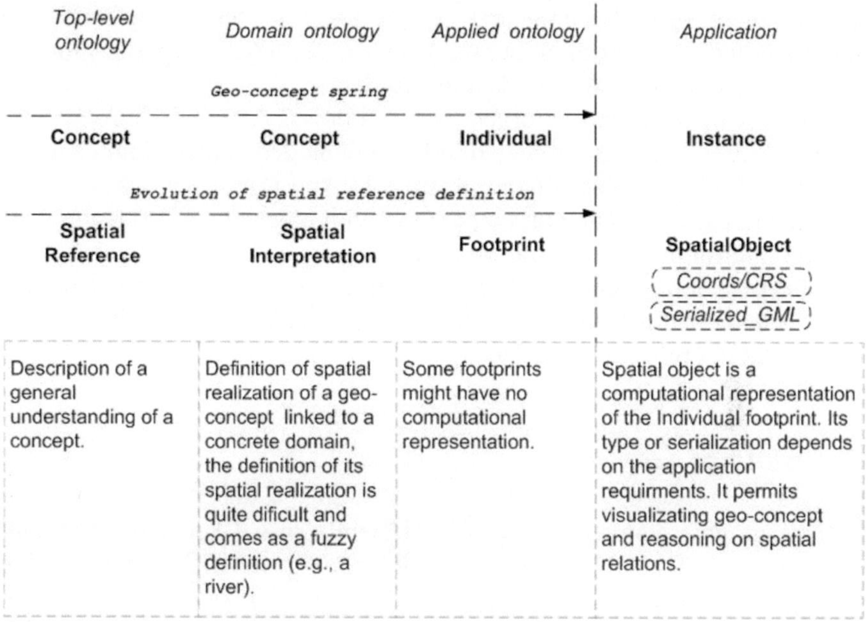

Fig. 1. Modelling the spatial representation of a geo-concept

The objective of this paper is to present how a SDI might take advantage of best practices from the Semantic Web. An *administrative geography* (AG), i.e. an ontology of political organization of the territory, published as Linked Data will be introduced as a relevant element of a SDI, since it permits (1) reasoning on logic relations among administrative units, and (2) accessing their footprints. In the use case, presented as an example later on, we will show how such ontology can improve functionality of the services catalog deployed within the Spatial Data Infrastructure of Spain (Nogueras-Iso, et.al., 2009) by applying geographic reasoning. One of the principal characteristics of a service deployed in a SDI is its geographic extent provided by the publisher as part of metadata description (i.e., *getCapabilities* response). According to the INSPIRE Metadata Implementating Rules (Craglia, 2009), the geographic location is defined as *minimum bounding box* (MMB). The descriptive metadata are used by the discovery service to answer the user requests and among the requestable parameters, there is the geographic location of the offered spatial data. Since services from SDI are provided by public administrations, frequently their geographic extent corresponds with the administrative area of the provider. The usage of MMB introduces false positive in the catalog response as the administrative areas are not rectangular. A catalog provided with knowledge about the hierarchy of administrative units and accurate spatial objects of each administrative area might increase considerably the precision and recall of the requests, which is especially important for applications based on on-the-fly data integration.

The rest of this paper is organized as follows. After this introduction, second section presents the state of art in linking geographic features in the Semantic Web community (starting from Linked Data approaches to deal with geographic information, and ending with examples of semantic geographic platforms existing on the current Web) and the Geospatial Web. The next section describes the infrastructure for the improvement of services catalog and the prototype application. Finally, some conclusions are drawn and future work is outlined.

2 State of art

The geographic feature is managed differently in the Semantic Web and the Geospatial Web. The first community treats geographic features as the additional contextual information, which might link to some other concepts. For the Geospatial Web, the concept of geographic feature is the core element of a geographic platform. However, the Geospatial Web has focused on the interoperability issues maintaining the boundaries among the concepts from different geographic sources.

2.1 Linking geographic features in the Semantic Web

Nowadays, deploying the data gathered in relational data bases as Linked Data on the Web is possible using integrated technological solutions such as OpenLink Virtuoso (Erling & Mikhailov, 2007) or D2R Server (Bizer & Seaborne, 2004). These approaches are based on mapping data base models onto a reference ontology and may provide a Semantic Web Browser and a SPARQL client. It is also possible to join RDF data from different endpoints providing a transparent on-the-fly view to the end user (Langegger, et al., 2008). The most important initiative related to creating and publishing interlinked contents on the Web is the Linked Open Data project. In May 2009, the amount of Linked Data datasets consist of over 4.7 billion RDF triples interlinked by around 142 million RDF links (Bizer, et al., 2009). The RDF links are navigable using Semantic Web browsers, and Semantic Web Search Engines can apply sophisticated queries over crawled data. The expressive semantic queries might be executed via SPARQL access points as well.

Since on-line contents involve geographic information, geographic features have become the part of Linked Data datasets, such as DBPedia, where geographic information has been extracted from Wikipedia (Auer, et al., 2008). Another example is the LinkedGeoData (Auer, et al., 2009b), which aims at adding geo-semantic meaning to the Web. It offers a Linked Geo Data Knowledge Base with RDF descriptions of more than 350 million spatial features from the OpenStreetMap database linked to DBPedia.

An example of applying Linked Data principles in location based solution is the DBpedia Mobile (Becker & Bizer, 2008), a location-aware client for the Semantic Web for mobile devices. The current user location is used to extract corresponding datasets from the underneath DBpedia database which are interlinked with various other location-related datasets.

An interesting proposal is Triplify (Auer, et al., 2009) which supports a kind of circular spatial requests. This system uses directly DB Views model as base for creating the RDF documents and URLs of published datasets, which facilitate the development. The underneath data base is responsible for processing spatial and semantic queries which are encoded explicitly in the request URL. The spatial query permits to retrieve the geographic features located in a circular region defined via a point and radius added to the request URL. The proposal enables limited spatial query (just a circular region) notwithstanding the Semantic Web techniques can not take advantage of this facility.

The idea of linking geographic features to create interlinked web influenced the appearance of geographic Web platforms, such as Yahoo! GeoPlanet[5] and GeoNames[6]. Both of them belong to a new branch of geocoders, the *semantic geocoders*, which return URIs to identify uniquely the named places instead of standardized textual description and location reference (e.g., a point). Although, they use the idea of linking, they are not following the pure Linked Data approach. GeoPlanet uses unique identifiers (URIs) to identify the named place which permits to retrieve its semantic description, however, the platform uses simple xml file instead of RDF. Additionally, the important spatial relations (e.g., "child", "neighbour", "siblings") are encapsulated into the URI definitions. GeoNames is *almost* Linked Data based. It also uses unique identifiers of concepts to identify the named places. The RDF description of features contains spatial relations defined in the published OWL reference ontology. However, GeoNames distinguishes the *concept* from the *descriptive document*. The feature (i.e. concept) is identified via an URI but the geonames server uses

[5] http://developer.yahoo.com/geo/
[6] http://www.geonames.org/

303 redirection to display its location on map. The RDF description is available by adding the "/about.rdf" at the end of the feature URI.

The spatial requests supported by the presented solutions of current Semantic Geo Web are based on a branch of predefined spatio-logic relations (e.g., *near-by*, *belongs-to*, *child*, *siblings*). Since it is impossible to express all spatial relations among geographic features via definition of logic relations, the Semantic Web needs to use spatial representation of features. Currently, the spatial objects usually used in the Semantic Web are limited to points or MMB. The complex spatial requests that mix spatial objects and spatial relations defined in a reach ontology still remain the open issue.

2.2 Linking in the Geospatial Web

The requirement of unique identifiers for geographic features, *geoidentifiers,* in the geospatial community has been present from its beginning. Any Geospatial Web framework which publish information about geographic features uses unique identifiers (within this framework at least) and might be seen as a source of geoidentifiers. Therefore, any gazetteer (Hill, 2006) or OGC Web Feature Service from a SDI might be such a source of geoidentifiers. Currently, there are several instances of WFS services in the SDI of Spain, which frequently contain instances of the same geographic feature. The services model the geographic feature in different manner and use different identifier usually derived keys from relational databases, therefore, there exist problem of individual identification among different contents and consequently common problems of data integration.

One of the earlier proposals from the OGC community to apply common geographic identifiers for linking purposes is a framework based on *Geolinked Data Access Service* and *Geolinking Service.* This approach is dedicated to publish geographically linked information (e.g., statistical data) separately from spatial representation (spatial objects). Since this proposal is based on merging datasets from different sources by using a linkage field found in the sources, it fixes geographic data to only one spatial representation source and geoidentifiers are used only as syntactic link. In practice, it can be seen as technological facilitation for data publishers.

A proposal for providing the integrated view across distributed services is the EuroGeoNames project (Jakobsson, & Zaccheddu, 2009), a prototype of an integrated gazetteer for Europe from the INSPIRE directive. Apart from defining the data model to be followed by all community members, it provides rules for identifier definition. The named place identifier has to be composed of (1) its name, (2) two-letter ISO 3166 code (e.g., ES, NL) and (3) a code generated according to the BASE36-system (e.g. "2YC67000B"). However, these identifiers still remain unique only in this distributed gazetteer.

Interlinking the corresponding instances of the same feature across different providers might be interesting for the Geospatial Web. It might be the way of adding logic relations among features and avoid the necessity of providing a new separate platform. The next section shows the application of an administrative geography of Spain published as Linked Data for the improvement of a SDI services catalog.

3 OGC services catalog application

The services catalog is one of the elements of a SDI. Since it is responsible for service discovery, its functionality determines the reusability of the offered services in the SDI and might be improved by applying best practices from the Semantic Web. This paper proposes a framework, where the administrative geography published as Linked Data is one of the core elements. A services catalog dedicated to support the visualization applications based on on-the-fly data integration is presented as a use case. The principal advantage of this approach is reflected by the improvement of the functionality of the end application.

3.1 Coverage issue

In INSPIRE, service discovery requests allow restricting the geographic extent of searched datasets and services to a required MMB. Frequently, the geographic extent of published data and service corresponds with the coverage of an individual from a geographic ontology (e.g., Europe from a geographic region ontology, Europe Union from a political organization ontology). Therefore, using MMB to describe available resources usually introduces false positives in the collection of results. For example, in the SDI of Spain the coverage of published service frequently corresponds with an administrative unit area of provider (e.g., council of Zaragoza). The administrative boundaries are far away from being rectangular and their MMBs overlaps significantly. The figure 2 shows the issue of overlapping MMBs of administrative areas. The shadowed rectangle represents the required MMB from the service discovery request. In this scenario the application would return services which geographic extend MMB corresponds with BBOX1, BBOX2 or BBOX3 when in reality it should provide only those services whose MMB corresponds with BBOX3.

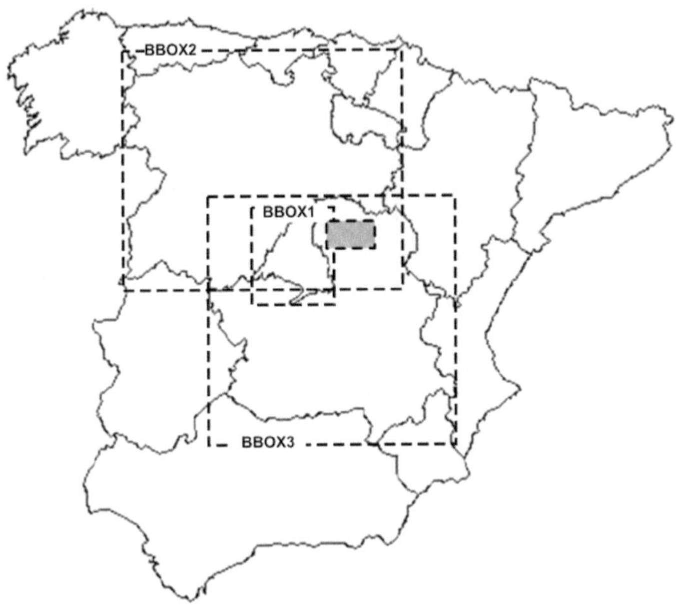

Fig. 2. Overlapping MMBs of administrative areas issue (Spain case)

The functionality of the catalog might be improved by operating on more precise spatial objects. A more promising approach could be application of the identifiers from an administrative geography which not only provides footprints but also permits reasoning on logic relations among concepts.

3.2 Administrative geography of Spain

Currently, there is no administrative geography published as Linked Data for Spain. Within the Spanish SDI there are various OGC WFS services which publish information about administrative unit entities. These services are provided by central and local authorities. Some of them offer administrative boundaries with different resolution (e.g., the *Infraestructura de datos Espaciales de España-WFS* service, IDEE-WFS[7]) separately for each level of administrative division, and others such as the gazetteers focused on gathering named places (e.g., IDEE-WFS-Nomenclator-NGC[8]), contain administrative units among published features. However, neither of these models permits to express the full administrative model that exists

[7]http://www.idee.es/IDEE-WFS/ogcwebservice?
[8]http://www.idee.es/IDEE-WFS-Nomenclator-NGC/services?

in Spain. The existing domain ontologies for modelling political organizations of the territory are usually based on the 'part-of' relation (parental relation). Such model is not enough flexible to scale the complexity of the territorial organization of countries, which apart from main division units (e.g., municipality, province and autonomous community in Spain) has to involve the units of different status (e.g., autonomous cities of Ceuta and Melilla, or associations of administrative units in Spain). Therefore, it is required usually a dedicated administrative ontology (e.g., Ordnance Survey) to model the political organization of territory of a country. In the case of Spain the Administrative Unit Ontology has been proposed as the domain ontology (Lopez-Pellicer, et.al., 2008). Apart from "part-of" and "has-part" relations, the "is-member-of" and "has-member" relations were defined to distinguish the association of the administrative units whose spatial representation might overlay the boundaries of direct parental units. An example of such association might be "comarca" of Aragon Autonomous Community which groups municipalities. Each municipality might lay in boundaries of only one province, however, one comarca might aggregate municipalities from different provinces.

The D2R Server has been used to publish the administrative geography of Spain, and to generate a data dump. The national Gazetteer, *IDEE-WFS-Nomenclator-NGC*, has been used as the reference source to extract the part of administrative geography. Although this model does consider logical relations among features (e.g., "parent-child"), the location of features in the administrative structure is defined indirectly by their names offered via *LocationEntity* element, whose structure contains concepts from the territorial organisations of Spain (e.g., *autonomous community*, *province*, *municipality* or *island*). Since the published data are not complete (e.g., there is no assignation of comarca names to municipalities), the INE online catalog (*Instituto Nacional de Estadística*, the National Statistics Institute of Spain) was used to complement the data. Then, the result data has been linked to their corresponding instances from different WFS services via the *skos:relatedMatch* relation. This approach produced the *administrative geography of Spain* published as Linked Data (see figure 3), and one of its advantages might be the maintenance of the references to different instances across the SDI of Spain.

```
@prefix au: <http://purl.org/iaaa/sw/gsw/ont/au-spain.owl#>.
@prefix agont: <http://purl.org/iaaa/sw/gsw/ont/app/agont.owl#>.
@prefix ag: <http://purl.org/iaaa/sw/gsw/georef/ag/>.
@prefix rdf: <http://www.w3.org/1999/02/22-rdf-syntax-ns#>.
@prefix rdfs: <http://www.w3.org/2000/01/rdf-schema#>.
@prefix skos: <http://www.w3.org/2009/08/skos-reference/skos.rdf#>.

ag:ZaragozaMun rdf:type au:Municipality;
      rdfs:label "Zaragoza"@es,"Zaragoza"@en;
      au:part-of ag:Spain, ag:AragonComunidad;
      au:member-of ag:ZragozaComarca;
      skos:relatedMatch <http://www.idee.es/IDEE-WFS/
ogcwebservice?SERVICE=WFS&VERSION=1.1.0&REQUEST=GetFeature&MAXFEATUR
ES=1&NAMESPACE=xmlns(ideewfs=http://www.idee.es/
wfs)&TYPENAME=ideewfs:BDLL1000Municipio&FILTER=%3CFilter%20xmlns:ide
ewfs=%22http://www.idee.es/
wfs%22%3E%3CPropertyIsEqualTo%3E%3CPropertyName%3Eideewfs:nombre%3C/
PropertyName%3E%3CLiteral%3EZARAGOZA%3C/Literal%3E%3C/
PropertyIsEqualTo%3E%3C/Filter%3E>;
(. . .)
      agont:bond-100 <http://www.idee.es/IDEE-WFS/
ogcwebservice?SERVICE=WFS&VERSION=1.1.0&REQUEST=GetFeature&MAXFEATUR
ES=1&NAMESPACE=xmlns(ideewfs=http://www.idee.es/
wfs)&TYPENAME=ideewfs:BDLL1000Municipio&FILTER=%3CFilter%20xmlns:ide
ewfs=%22http://www.idee.es/
wfs%22%3E%3CPropertyIsEqualTo%3E%3CPropertyName%3Eideewfs:nombre%3C/
PropertyName%3E%3CLiteral%3EZARAGOZA%3C/Literal%3E%3C/
PropertyIsEqualTo%3E%3C/Filter%3E>.
```

Fig. 3. An example of the RDF description that represents the Zaragoza municipality

The section 3.3 describes an architecture and implementation of a services catalog component which applies the administrative geography of Spain during the service selection process.

3.3 Architecture and Implementation

The usage of geographic feature identifiers requires the annotation of registered resources in a catalog with corresponding geoidentifiers. Creation of metadata of registered services is one of the characteristics of the services catalog deployed in the SDI of Spain (Nogueras-Iso, et.al., 2009). Its architecture has been extended with the *Knowledge Content (KC)* that is responsible for the service search process (see Nogueras-Iso, et.al., 2009 for the description of the services catalog architecture). Figure 4 presents the main elements of the KC component. The KC has access to two RDF dataset sources: the *Administrative Geography* (AG), i.e. the administrative geography of Spain, and *Service Description Register* (SDR) which contains RDF serialization of the registered service description. The reference ontologies, i.e. *Administrative Geography Ontology* (*agont*) and the *Service Description Ontology* (*svont*), are applied during the reasoning process. The concepts from the administrative geography (1) are linked to the entities from the reference WFS service, the source of boundary spatial definitions, and (2) the URI of the administrative units are used in the service description as indication of the geographic extent (*dc:coverage* property). Figure 5 represents an example of the link between the administrative geography and the service description.

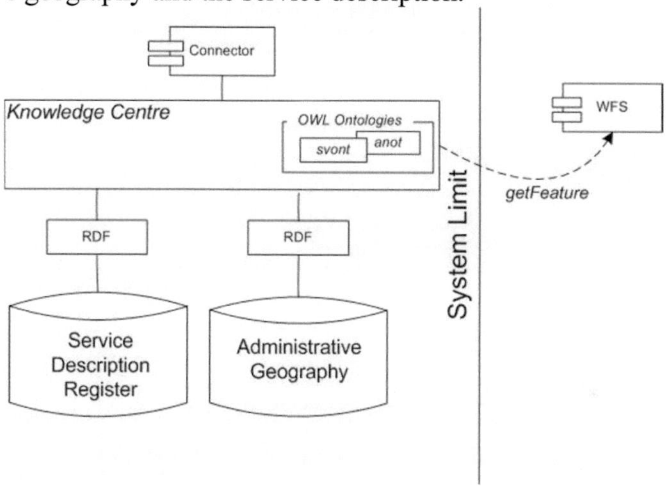

Fig. 4. Administrative geography as support for services catalog

During the registration of a new service, the catalog uses the *getCapabilities* response to create proper description of the service. One of the elements is the service geographic extent expressed via MMB. The metadata model of the description has been extended with the *geoidentifier* metadata to contain an URI from the administrative geography of Spain. This element is not an ISO19119 element; therefore, it is neither visible to users nor published by the OGC CSW. The geoidentifier value is obtained from the analysis of the service MMB offered by provider. This MMB is used to request the KC that identifies the most extended administrative unit within the MMB (i.e., the *prime administrative unit*). Validation of the result consists in checking if the service provides any data from the disjoint area of the MMB and the prime administrative unit coverage (a set of retrieval tests). If it is impossible to identify the prime administrative unit (e.g. in the case of the hydrography service of the Ebro river basin, some data lies on France as well) or the validation fails, the URI of the *Non* concept (i.e., the disjoint concept with *administrativeUnit*) is returned. The service registration ends by deploying the registered service description in the RDF container.

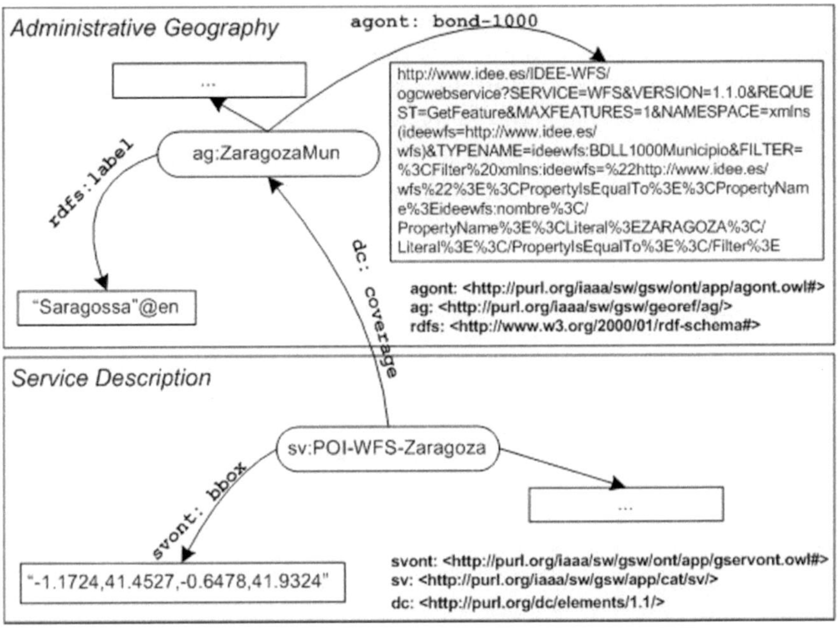

Fig. 5. Relations between the service description and the administrative geography

The searching process for services with a MMB restriction exploits the semantic links between the semantic description of services and the features from the AG. The simplified example of a request (only the spatial part) consumed by KC is shown in figure 6. The request pattern uses as the input the searched MMB (*$BBOX*) expressed as literal (e.g., "1.16311, 41.0937, 1.7132, 41.6686"). The first part extracts those services which spatial reference of coverage (*dc:coverage*) refers to *administrativeUnits* which boundaries are in an interaction with the searched MMB (within it or intersects it). The *administrativeUnit* boundary is defined via the *bond-1000* property containing reference to the correspondent WFS entity, which is converted automatically into the spatial object by applying the profile instructions. The second part of the request extracts those services which geoidentifier is defined as *Non* and the collection is filtered similarly as in the previous version of services catalog, i.e. comparing the service MMB and the requested one.

```
PREFIX svont: <http://purl.org/iaaa/sw/gsw/ont/app/gservont.owl#>
PREFIX agont: <http://purl.org/iaaa/sw/gsw/ont/app/agont.owl#>
PREFIX rdf: <http://www.w3.org/1999/02/22-rdf-syntax-ns#>
PREFIX dc: <http://purl.org/dc/elements/1.1/>
PREFIX sf: <http://purl.org/iaaa/sw/gsw/app/gfun>
SELECT ?s
WHERE {
  ?s rdf:type svont:service;
     dc:coverage ?x.
?x rdf:type agont:au;
     agont:bond-1000 ?g.
FILTER(sf:INTERACT(?g, $BBOX))
}UNION
SELECT ?s
WHERE {
?s rdf:type svont:service;
     dc:coverage agont:Non;
     svont:bbox ?g.
FILTER(sf:INTERACT(?g, $BBOX))
}
```

Fig. 6. SPARQL request pattern for service selection via a MMB ($BBOX is the MMB literal, eg., "-1.1724,41.4527,-0.6478,41.9324")

The KC component has required implementation of the spatial functions, such as *intersect,* or *within*, known form spatial data bases. Therefore, the Jena framework[9] has been chosen to deploy RDF datasets as its proprietary extension to SPARQL RDF query language (Jena ARQ[10]) permits to implement such additional functionality.

3.4 The use case

The catalogs that use precise spatial representation instead of MMB approximation might offer better functionality for the applications based on on-the-fly data integration. An example might be an application which allows displaying geographic information from different OGC services found in the services catalog (see figure 7). For improving the user experience, the list of selectable layers should depend on current displayed area.

[9] http://openjena.org/
[10] http://jena.sourceforge.net/ARQ/

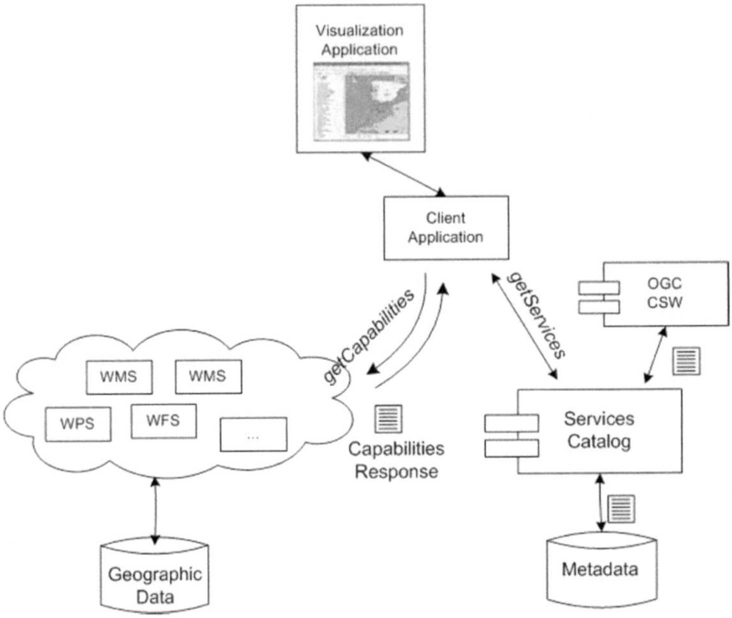

Fig. 7. The services catalog as the support component for application based on on-the-fly data integration.

The prototype of the improved services catalog has been used as the core component of Web application[11] which displays spatial data provided from different OGC services. The main disadvantage of this proposal is the response time of the WFS service. To solve this problem we have created a local repository of spatial objects retrieved previously from reference service. The cache techniques have also improved the catalog response time and the behaviour of the end application.

4 Results and future works

This paper shows the current approaches in referencing and identification of geographic features in the Semantic Web and the Geospatial Web. Applying best practices from the Semantic Web might be useful for the Geospatial Web. An administrative geography published in accordance with Linked Data principles might be useful for data integration as it permits referencing the geographic concept to the corresponding instances from other sources. Such ontology might be used as source of geoidentifiers in

[11] http://www.idee.es/IDEE-ServicesSearch/ServicesSearch.html

geospatial solutions and its main advantage lies in using more precise spatial representation and spatial reasoning on semantic level.

Additionally, the usage of geoidentifiers along with minimum bounding boxes to represent the service geographic extents might improve the recall of OGC services catalog. For instance, we have demonstrated that this improvement has enabled the development of web-based applications that facilitate on-the-fly data integration.

The principal advantage of using Linked Data technology in geospatial solutions is the possibility of explicit identification of features and abstraction of their spatial definition from footprint and computational representation. The different spatial representation might be accessible via linked instances and chosen according to the application requirements.

One of the future tasks will be applying the administrative geography of Spain as guideline to map instances of the same geo-concept individuals between two different gazetteers for merging purpose. Next, there will be investigated an adaptable framework to support complex spatial requests applying different spatial representations of features.

Acknowledgments

This work has been partially supported by Spanish Government (projects "España Virtual" ref. CENIT 2008-1030, TIN2007-65341 and PET2008_0026), the Aragón Government (project PI075/08), the National Geographic Institute (IGN) of Spain, GeoSpatiumLab S.L. The work of Aneta Jadwiga Florczyk has been partially supported by a grant (ref. AP2007-03275) from the Spanish government.

References

Auer, S., Dietzold, S., Lehmann, J., Hellmann, S. and Aumueller, D. (2009) Triplify – Light-Weight Linked Data Publication from Relational Databases, Proceedings of the 18th International World Wide Web Conference, April 20—24, 2009, Madrid, Spain, pp. 621—630.

Auer S., Lehmann J. and Hellmann S. (2009b) LinkedGeoData - Adding a Spatial Dimension to the Web of Data, In: 8th International Semantic Web Conference (ISWC2009), October 25—26, 2009, Washington, DC, USA.

Auer, S., Bizer, C., Kobilarov, G., Lehmann, J., Cyganiak, R. and Ives, Z. (2008) DBpedia: A Nucleus for a Web of Open Data, The Semantic Web, Proceedings of the 6th International Semantic Web Conference, the 2nd Asian Semantic Web Conference (ISWC+ASWC 2007), November 23—30, 2008, Karlsruhe, Germany, pp. 722—735.

Becker, C. and Bizer, C. (2008) DBpedia Mobile: A Location-Enabled Linked Data Browser, Proceedings of the Linked Data on the Web Workshop, April 22, 2008, Beijing, China.

Bizer, C. and Seaborne A. (2004) D2RQ -Treating Non-RDF Databases as Virtual RDF Graphs, Poster at the 3rd International Semantic Web Conference, November 7—11, 2004, Hiroshima, Japan, http://www4.wiwiss.fu-berlin.de/bizer/pub/Bizer-D2RQ-ISWC2004.pdf, Last date accessed 10.2009.

Bizer, C., Heat, T. and Berners-Lee, T. (2009) Linked Data – The Story So Far, International Journal on Semantic Web and Information Systems.

Brickley, D. and Guha, R.V. (2004) RDF Vocabulary Description Language 1.0: RDF Schema, W3C Recommendation, 2004, http://www.w3.org/TR/rdf-schema/. Last date accessed 10.2009.

Craglia M. (Eds.) (2009) INSPIRE Metadata Implementing Rules: Technical Guidelines based on EN ISO 19115 and EN ISO 19119 (Revised edition), Drafting Team Metadata and European Commission Joint Research Centre, 2009.

Dolbear, C. and Hart, G. (2008) Ontological Bridge Building - using ontologies to merge spatial datasets, Proceedings of the AAAI Spring Symposium on Semantic Scientific Knowledge Integration, AAAI/SSS Workshop, March 26—28, 2008, Stanford U, CA.

Egenhofer, M.J. (2002) Toward the Semantic Geospatial Web, Proceedings of the 10th ACM International Symposium on Advances in Geographic Information Systems, pp.1—4.

Erling, O. and Mikhailov, I. (2007) RDF Support in the Virtuoso DBMS, Conference on Social Semantic Web, Vol. 113, 2007, pp. 59—68.

Goodwin, J.; Dolbear, C. and Hart, G. (2009) Geographical Linked Data: The Administrative Geography of Great Britain on the Semantic Web Transactions in GIS, 2009

Gruber, T.R. (1993) A translation approach to portable ontology specifications, Knowledge Acquisition, 5(2), pp. 199—220.

Haklay, M. and Weber, P. (2008) OpenStreetMap: User-Generated Street Maps, IEEE Pervasive Computing, 7(4), pp. 12—18.

Heath T. (2009) The Linked Data community homepage, http://linkeddata.org/, Last data accessed 10.2009.

Hill, J. (2006) Georeferencing, The Geographic Associations of Information, Digital Libraries and Electronic Publishing, MIT Press.

Inspire (2007) Directive 2007/2/EC of the European Parliament and of the Council. Official Journal of the European Union, 2007.

Jakobsson, A. and Zaccheddu P.G. (2009) EuroGeoNames (EGN) - A Prototype Implementation for an INSPIRE Service, GSDI 11 World Conference and the 3rd INSPIRE Conference, Rotterdam 15—19 June 2009.

Klyne, G. and Carroll, J.J. (2004) Resource Description Framework (RDF): Concepts and Abstract Syntax W3C Recommendation, 2004, http://www.w3.org/TR/rdf-concepts/, Last date accessed 10.2009.

Langegger, A., Wöß, W. and Blöchl, M. (2008) A Semantic Web Middleware for Virtual Data Integration on the Web, The Semantic Web: Research and Applications, pp. 493—507.

López-Pellicer, F.J.; Florczyk, A.J.; Lacasta, J.; Zarazaga-Soria, F.J. and Muro-Medrano, P.R. (2008) Administrative Units, an Ontological Perspective. ER Workshops, 5232, Springer, pp. 354—363.

Maue, P., Schade, S. and Duchesne, P. (2009) OGC Discussion Paper 08-167r1: Semantic annotations in OGC standards, Technical report, Open Geospatial Consortium.

McGuinness, D.L. and van Harmelen F. (2004) OWL Web Ontology Language Overview, W3C Recommendation, 2004, http://www.w3.org/TR/owl-features/. Last date accessed 10.2009.

McKee, L. (2004) The Spatial Web, An Open GIS Consortium (OGC) White Paper, http://portal.opengeospatial.org/files/?artifact_id=3859. Last date accessed 10.2009.

Nebert, D., Whiteside, A. and Vretanos, P. (Eds.) (2007) OpenGIS - Catalogue Services Specification (version: 2.0.2). OGC 07-006r1, Open Geospatial Consortium Inc., http://portal.opengeospatial.org/files/?artifact_id=20555, Last date accessed 10.2009.

Nogueras-Iso, J., Barrera, J., Rodrígez, A.F., Recio, R., Laborda, C. and Zaragaza-Soria, F.J. (2009) Development and deployment of a services catalog in compliance with the INSPIRE metadata implementing rules, SDI Convergence: Research, Emerging Trends, and Critical Assessment. The Netherlands Geodetic Commission (NGC), 2009, pp. 21—34.

A Machine Learning Approach for Resolving Place References in Text

Bruno Martins, Ivo Anastácio, Pável Calado

Instituto Superior Técnico / INESC-ID Lisboa, PT

Abstract. This paper presents a machine learning method for resolving place references in text, i.e. linking character strings in documents to locations on the surface of the Earth. This is a fundamental task in the area of Geographic Information Retrieval, supporting access through geography to large document collections. The proposed method is an instance of stacked learning, in which a first learner based on a Hidden Markov Model is used to annotate place references, and then a second learner implementing a regression through a Support Vector Machine is used to rank the possible disabiguations for the references that were initially annotated. The proposed method was evaluated through gold-standard document collections in three different languages, having place references annotated by humans. Results show that the proposed method compares favorably against commercial state-of-the-art systems such as the Metacarta geo-tagger and Yahoo! Placemaker.

1 Introduction

Recently, Geographical Information Retrieval (GIR) has captured the attention of many researchers that work in fields related to text mining and data retrieval. The general motivation is that geographical information is pervasive over textual documents, since most of them contain references to place names over the text. However, a fundamental task that needs to be addressed in GIR relates to resolving these place references, i.e. linking the character strings in the documents that correspond to place references to their true locations on the surface of the Earth. Place reference resolution

M. Painho et al., (eds.), *Geospatial Thinking*, Lecture Notes in Geoinformation and Cartography,
DOI 10.1007/978-3-642-12326-9_12,

presents non-trivial problems due to the inherent ambiguity of natural language (e.g., place names often have other non geographic meanings, different places are often referred to by the same name, and the same places are often referred to by different names).

Over the last few years, the place reference resolution problem has been addressed by many different researchers. The problem is generally divided into two separate sub-tasks, namely (i) place reference identification, and (ii) place reference disambiguation by matching the identified references against a gazetteer, i.e. a database associating place names to the corresponding geospatial footprints. The first sub-task is deeply related to the Named Entity Recognition problem, which has been thoroughly studied in the natural language processing (NLP) community (Sang and Meulder, 2003). The second sub-task has deserved less attention, although previous studies have, for instance, attempted heuristics inspired in the techniques proposed for the NLP problem of word sense disambiguation (Leidner, 2007; Buscaldi and Rosso, 2008). The majority of the current approaches are based on rule-based methods and hand-tuned heuristics, which are not particularly robust or generalizable. A particularly interesting challenge relates to the application of data-driven methods in the place reference resolution task, using principled approaches for combining the different sources of evidence that are available from training data.

This paper proposes a machine learning method for resolving place references in text. The proposed method is an instance of stacked learning (Wolpert, 1992), in which a first learner based on a Hidden Markov Model (HMM) is used to tag place references, and then a second learner implementing a regression through a Support Vector Machine (SVM) is used to rank the possible disambiguations for the place references that were initially tagged. In terms of the implementation, the first learner is based on the HMM tagger available in the LingPipe package (Carpenter, 2007), which uses character language models for each state in the HMM, together with a maximum likelihood *n*-gram transition model. The second learner is based on the implementation of SVM regression available in the Weka machine learning toolkit (Witten and Frank, 2000). The proposed method was evaluated through gold-standard document collections in three different languages (i.e., English, Spanish and Portuguese), having place references annotated by humans. The results show that the proposed method compares favorably against commercial state-of-the-art systems like the Metacarta geo-tagger and Yahoo! Placemaker.

The rest of this paper is organized as follows: Section 2 presents related work, covering both named entity recognition and place reference resolution. Section 3 presents the proposed machine learning method, detailing the individual learners that were used for place reference identification and for place reference resolution. Section 4 presents the experimental validation of the proposed method. Finally, Section 5 presents our conclusions and points directions for future work.

2 Related Work

Place reference resolution can be divided into two separate sub-tasks, namely place reference identification and place reference disambiguation. Place reference identification concerns delimiting, in a document, the character strings that refer to places. This is a particular instance of the more general problem of Named Entity Recognition, which has been extensively studied in the Natural Language Processing (NLP) community. Place reference disambiguation refers to associating the recognized references into the corresponding entries in a gazetteer. The sub-task has been addressed by the Geographic Information Retrieval (GIR) community. This section surveys relevant previous works.

2.1 Named entity recognition

The named entity recognition (NER) task, as proposed in the NLP community, refers to locating and classifying atomic elements in text into predefined categories such as the names of persons, organizations, locations, expressions of times, quantities, monetary values, percentages, etc. Current state-of-the-art systems can achieve near-human performance, effectively handling the annotation of ambiguous cases (e.g. the same proper name can refer to distinct real world entities) and achieving F_1 scores around the mark of 90% (Sang and Meulder, 2003).

Initial approaches, which are nonetheless still commonly used, were based on manually constructed finite state patterns and/or dictionary lists of entity names. Despite being robust and capable of achieving state-of-the-art performances, these initial methods lack the ability of coping with the problems of robustness and portability. Domain experts are needed to build the patterns and each new source of text requires significant tweaking of patterns and/or dictionaries to maintain optimal results.

The current trend in NER is to use machine-learning approaches, relying on features extracted from training data that reflect properties of (i) individual named entities (e.g., capitalization, entity type and frequency) and (ii) general statistical measures either at the document scale or at the corpus scale. Machine learning approaches are more attractive in that they are trainable and adaptable, at the same time maintaining a state-of-the-art performance. Several different classification methods have been successfully applied on this task (Sang and Meulder, 2003). For instance Mayfield et al. (2003) applied Support Vector Machines (SVM) to classify each individual name entity. Chieu and Ng (2003) and Bender et al. (2003) applied Maximum Entropy (ME) approaches, using local features occurring near individual tokens and global features from the whole document. Zhou and Su experimented with Hidden Markov Models (HMMs), also using a large variety of features (Zhou and Su, 2002). Conditional Random Fields (CRFs), a generalization of ME and HMMs, were explored by McCallum and Li (2003). Florian et al. (2003) combined Maximum Entropy and Hidden Markov Models (HMM) under different conditions. Recent works have shown that CRFs have advantages over traditional HMMs in the face of many features, although CRFs need fairly extensive feature engineering to outperform a good HMM baseline (Carpenter, 2007). The method reported in this paper is based on the HMM tagger implementation from LingPipe, a suite of Java libraries for text mining. Section 3.1 details the HMM method used for place reference identification.

2.2 Place reference identification and disambiguation

For most Geographic Information Retrieval (GIR) applications, the handling of place references requires recognizing the mentions to places given over text (i.e., delimiting their occurrences), as well as disambiguating these occurrences into the corresponding locations on the surface of the Earth (i.e., assigning geospatial footprints to the place references). Having the geospatial footprints is, for instance, crucial for supporting geospatial analysis and map-based visualization. While NER approaches can be made to rely entirely on features internal to the documents, place reference disambiguation requires always external knowledge in the form of a dictionary (i.e., a gazetteer) for translating place names into geospatial footprints. Geonames[1] is an example of a modern wide-coverage gazetteer, describing over 6.5 million unique places from all around the world and having been used in many GIR experiments (Wick and Becker, 2007).

Similarly to the general case of named entity recognition, the main challenges in resolving place references are related to ambiguity in natural language. Amitay et al. characterized ambiguity problems according to two types, namely geo/non-geo and geo/geo (Amitay et al., 2004). Geo/non-geo ambiguity refers to the case of place names having other, non geographic meanings (e.g., Reading in England). Some of the most common

[1] http://www.geonames.org

words are for instance also place names (e.g., Turkey). On the other hand, geo/geo ambiguity arises when two distinct places have the same name. For instance almost every major city in Europe has a sister city of the same name in the New World. The geo/non-geo ambiguity is addressed when identifying mentions to places, while the geo/geo ambiguity is latter addressed while disambiguating the recognized place references.

Several different approaches for place reference identification and disambiguation have been described in the past. In the context of his PhD thesis, Leidner (2007) surveyed a variety of approaches for handling place references on textual documents. He concluded that most methods usually rely on gazetteer matching for performing the identification, together with natural language processing heuristics such as default senses (i.e., disambiguation should be made to the most important referent, estimated with basis on population counts) or geographic heuristics such as the spatial minimality (i.e., disambiguation should minimize the bounding polygon that contains all candidate referents) for performing the disambiguation. Leidner (2004) also concluded that a major requirement for achieving significant progress in the area is the availability of standard test collections. These would allow not only the comparative evaluation of different approaches, but also the exploration of data-driven methods. The experiments reported in this paper are based on three such collections, developed by extending the annotations made in the context of previous NER experiments.

Despite the technology being very recent, with the GIR area still at its infancy, commercial services offering place reference resolution functionalities are starting to appear. Metacarta is an example of a commercial company that sells state-of-the-art GIR technology. The company also provides a freely-available Web service that can be used to recognize and disambiguate place references over text. An early version of the Metacarta geo-tagger[2] has been described by Rauch et al. (2003). The Yahoo! Placemaker[3] Web service also provides a functionality for geo-tagging text. Both these commercial services were used in experiments where we compared their results against those obtained with the proposed machine learning approach for the recognition and disambiguation of place references.

3 Proposed approach

The proposed method for resolving place references is an instance of stacked learning, a machine learning paradigm that suggests constructing a

[2] http://ondemand.metacarta.com/?method=GeoTagger

[3] http://developer.yahoo.com/geo/placemaker

combined model from several simpler learners (Wolpert, 1992). The basic concept behind stacking is to train two or more learners sequentially, with each successive learner incorporating the results of the previous ones in some fashion. Figure 1 provides an illustration of the processing pipeline used in our work.

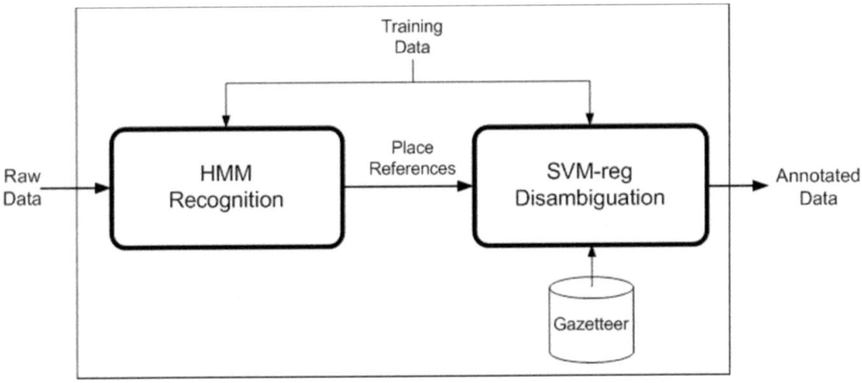

Fig. 1. Proposed approach for resolving place references

In the approach proposed here, a first level learner corresponds to a Hidden Markov Model (HMM) tagger for identifying place references in text, based on the implementation available on the LingPipe[4] package. The second learner takes as input the place references identified in the first level, and uses a Support Vector Machine (SVM) regression to rank a set of possible disambiguations (obtained by querying the geonames gazetteer) and then select the best candidate. The rest of this section details the learning approaches used in both levels.

3.1 Place reference identification

LingPipe's machine learning approach for named entity recognition employs first-order HMMs, where the hidden states correspond to the tags to be assigned. The general approach is described by Carpenter (2007) and here we present a brief overview.

The recognition process starts with a tokenization scheme that deterministically breaks an input text into a sequence of tokens. Recognizing place references in a given text is transformed into the problem of tagging each token as belonging to a place reference or not. When modeling the problem through HMMs, the joint probability of observing a token sequence

[4] http://alias-i.com/lingpipe

$\delta_1, \ldots, \delta_n$ associated with a tag sequence $tag_1, \ldots, tag_n$ is defined by the formula $p(\delta_1, \ldots, \delta_n, tag_1, \ldots, tag_n) = p(tag_1, \ldots, tag_n) \cdot p(\delta_1, \ldots, \delta_n | tag_1, \ldots, tag_n)$. In a first-order HMM, for a given sequence we have that the probability $p(tag_1, \ldots, tag_n) = p_{start}(tag_1) \cdot \prod_{i>1} p(tag_i | tag_{i-1}) \cdot p_{end}(tag_n)$.

In general, applying HMMs to the recognition task involves two main problems, namely learning and decoding. Learning refers to generating the most probable HMM model from a sequence of observed tokens, known to represent a set of hidden tags, i.e., learning from a tagged corpus. A HMM model is defined as the triple (A,B,π), where π is a vector with the initial state (i.e., tag) probabilities, A is the matrix containing the transition probabilities of moving from a state to a new state, and B is the matrix containing the emission probabilities associated with having a state generating a given token. In LingPipe, the forward-backward algorithm is used to compute both the A and B matrices.

The second problem, decoding, refers to finding the most probable sequence of tags given some observations, i.e., determining the probability $p(tag_1, \ldots, tag_n | \delta_1, \ldots, \delta_n)$. The default choice for solving this problem, and also the one used here, is to apply the Viterbi dynamic programming algorithm (Viterbi, 1967). Lingpipe also provides more advanced decoders.

In typical HMMs, emissions are estimated as multinomials, with some smoothing technique for handling previously unseen tokens. LingPipe's use of HMMs is unusual, in that it estimates the emission probabilities using bounded character *n*-gram language models, one for each tag. Traditionally, when an unseen word occurs, an HMM model outputs a default probability, therefore increasing the number of assignment errors. By working with *n*-grams at the character level, the probabilistic models used in LingPipe have the advantage of considering sub-words, which are more general and robust features in this task. LingPipe also interpolates all orders of maximum likelihood estimates using Witten-Bell smoothing (Maning and Schutze, 1999). In this work, we use the default values for the maximum order of *n*-grams and for the interpolation parameter for the language models, namely the value of 8.0 for both parameters.

The tagging problem addressed in this paper is modeled through an encoding that is sensitive to position (i.e., the BMEWO+ encoding), considering that tokens (i.e., *n*-grams) can either belong to a place reference, to a named entity that is not a place, or to neither of the previous cases. LingPipe's BMEWO+ encoding distinguishes begin-of-entity (B) tokens, mid-entity tokens (M), end-of-entity (E) tokens, single-token entities (W) and non-entity tokens (O). The non-entity tokens are subdivided into non-entity tokens with the previous token being an entity (BB_O), non-entity tokens with the following token being an entity (EE_O), non-entity tokens with the previous and following tokens being an entity (WW_O), and non-entity tokens with the previous and following tokens not being an entity (MM_O). The tag transition constraints (e.g., B must be followed by M or E, etc.) encode a kind of long-distance information about preceding or following words, for instance, appropriately modeling the fact that strong location-preceding words trigger a location. The HMM model is used to annotate *n*-gram sequences according to the above tags (i.e., BMEWO+ tags for both locations and non-locations), from which the system then generates the final results for place reference identification.

3.2 Place reference disambiguation

Our approach for place reference disambiguation involves (i) generating disambiguation candidates by querying the geonames gazetteer, (ii) scoring possible candidates using SVM regression, and (iii) selecting the highest scoring candidate. The scoring criteria used in the second step are based on estimating the distance between the true geospatial footprints and the geospatial footprints of the candidates.

The token sequences identified as place references by the first level HMM learner are used to query the geonames gazetteer. The top ten geographic concepts returned for the query are considered as candidates. For each pair <*place-reference, candidate*>, the following set of features is computed:

- The Levenshtein distance between the place reference, as recognized in the text, and the most similar place name associated with the candidate. Geonames describes multiple alternative names for each geographic concept and, more exactly, this feature corresponds to the minimum Levenshtein distance computed from the set of alternative names.
- The population count for the candidate geographic concept, as described in the geonames gazetteer.
- The number of alternative names described in the geonames gazetteer for the candidate geographic concept.
- The geospatial distance between the candidate geographic concept and the closest (in terms of metric distance) candidate geographic concept corresponding to a place reference recognized in the same document.
- The area of the convex hull computed with the centroid coordinates of the candidate concept and the centroid coordinates of all candidates corresponding to a place reference recognized in the same document.
- The area of the concave hull computed with the centroid coordinates of the candidate concept and the centroid coordinates of all candidates corresponding to a place reference recognized in the same document.

For training and evaluating the regression model, we compute the geospatial distance between each candidate and the true geospatial footprint, as described in the golden collection. The regression model attempts to estimate this distance, with basis on the considered features. More formally, let X be the training set consisting of m instance pairs $xi = \{f_{i,1}, \ldots, f_{i,n}, d_i\}$ where $f_{i,j}$ is the j-th input feature of a given example i and d_i is the corresponding geospatial distance. The regression goal is to estimate a function $reg(x)$ with basis on the training set X that takes as input a set of instance pairs $x'_i = f_{i,1}, \ldots, f_{i,n}$, is as close as possible to the target values d_i for every x'_i, and at the same time is as flat as possible for good generalization.

Support Vector Machines (SVMs) are a set of machine learning approaches used for classification and regression, developed in the mid 90's by Vapnik and his co-workers at AT&T Bell Labs (Vapnik, 1995). In the case of SVM regression, the basic idea is to map the features into a high-dimensional feature space via a kernel function (which may be linear or not), and then do a linear regression in this new space. Many different algorithms have been proposed for training SVM regression models. The Weka machine learning toolkit, which we used in this work, implements the approach proposed by Shevade et al. (1999), which in turn is an improvement over Smola and Scholkopf's sequential minimal optimization (SMO) algorithm for training a support vector regression (Smola and Scholkopf, 1998). Here we used Weka's implementation with a Radial Basis Function (RBF) kernel.

The final step of the proposed approach involves selecting the candidate with the least estimated distance as the result for the disambiguation. Although this was not considered in the experiments reported in this paper, some applications can benefit from keeping the several possible disambiguations together with the estimated distance as a disambiguation score. In the experiments reported here, we only tested disambiguation using the top scoring candidate.

4 Experimental validation

In this section, we describe the details of our empirical evaluation. This includes the details of our experimental design (i.e., the datasets and the evaluation metrics that were considered), as well as results for the experiments that evaluated the effectiveness of the methods under study.

4.1 Gold-standard collections and evaluation metrics

Our validation methodology is based on the standard NER evaluation setup, which involves partitioning a collection of labeled data into training and test segments. Models are developed with the training data, and afterwards evaluated against the test data. Given our two-stage approach, we take separate measures for the identification and disambiguation sub-tasks.

In terms of evaluation metrics, we measured results with the commonly used F_1 rate, which is equal to the geometric average between precision and recall. Precision is the percentage of correct place references identified/disambiguated by the system. Recall is the percentage of place references present in the test collection that are identified/disambiguated by the system. A place reference is correctly identified only of it is an exact match with the corresponding reference in the test collection. Correct disambiguation corresponds to the case where the assigned geospatial footprint respected at least one of the following conditions:

1. The angular distance between the geospatial coordinates for the candidade and the true location is less than 1 degree.
2. The candidate has a name that exactly matches the place reference in the text, and is at an angular distance from the true location of less than 5 degrees.

Since different gazetteers (e.g., Geonames and Yahoo! Placemaker) use different names and geospatial footprints for the same geographic concept, we cannot hope to have a disambiguation scheme that exactly matches the coordinates in the gold-standard collection. Instead of an exact match, we use the thresholds above, setting them to the values of 1 and 5 through empirical observation over a subset of the identified place references.

Notice that correct disambiguation only occurs when a correct recognition has first taken place, since it does not make sense to disambiguate textual expressions that are not even place references.

In terms of the labeled data, we started with the datasets made available in the context of the two NER evaluation experiments made at the Conference on Computational Natural Language Learning (CoNLL) (Sang and Meulder, 2003), as well as from the contest for named entity recognizers in Portuguese (HAREM) (Mota and Santos, 2008). These original datasets represent three different languages (i.e., Spanish, English and Portuguese) and contained general named entity annotations, identifying persons, organizations, locations, etc.

The source of the data was journalistic texts in the case of CoNLL and documents from multiple sources (e.g., newswire texts, Web pages or emails) in the case of HAREM. We kept the original annotations referring to locations, extending them with associations to geospatial coordinates and identifiers in the gazetteer of the Yahoo! GeoPlanet platform. Other classes of named entities were generalized into a single *non-location* class. The places annotated in the text served as queries to the Yahoo! gazetteer, from which we retrieved the corresponding geospatial footprints. We latter manually validated and corrected these automated annotations with the help of fellow researchers from our group. Table 1 presents a statistical characterization for the labeled datasets that were used in our experiments, also showing the state-of-the-art results that have been reported for the task of identifying place references in these collections.

For both editions of the CoNLL NER evaluation, the organizers provided three data files for each language (Spanish in the 2002 edition and English in 2003): a training file, a file for testing systems during the development stage (i.e., the Test 2 row in Table 1) and a file for final tests (i.e., the Test 1 row in Table 1). The HAREM evaluation campaign had three editions and we used the gold-standard collection from the first HAREM as training data, the gold-standard collection from mini-HAREM as the first test set (i.e., the Test 1 row in Table 1), and the gold-standard collection from the second HAREM as the second test set (i.e., the Test 2 row in Table 1).

Table 1. The labeled datasets used in the validation experiments

		Words	Paragraphs	Locations	Non-Loc.	F1 Score
Test 1	CoNLL-02 (Esp.)	47.937	1.916	978	3.346	**0.795**
	CoNLL-03 (Eng.)	46.759	3.245	1.835	4.099	**0.961**
	HAREM (Por.)	5.729	320	73	413	**0.647**
Test 2	CoNLL-02 (Esp.)	47.076	1.517	1.095	2.454	**0.825**
	CoNLL-03 (Eng.)	42.677	3.445	1.649	3.965	**0.912**
	HAREM (Por.)	2.385	116	43	157	**0.599**
Train	CoNLL-02 (Esp.)	241.435	8.323	4.911	13.811	-
	CoNLL-03 (Eng.)	186.373	14.023	7.115	16.299	-
	HAREM (Por.)	7.952	606	113	316	**0.709**

4.2 Obtained results

We compared the proposed stacked learning approach against two state-of-the-art commercial geo-tagging systems, namely Yahoo! Placemaker and Metacarta geo-tagger. We also compared the proposed approach against two simpler baselines in terms of the disambiguation sub-task, which respectively corresponded to (i) using a default sense heuristic in which the disambiguation is made to the most important referent as estimated with basis on population counts, and (ii) using a spatial minimality heuristic in which disambiguation is made by minimizing the distance between the referent and all possible referents to references in the same sentence.

We used the training datasets for separately training recognition and disambiguation models for each of the three languages. We then evaluated the trained models, as well as the commercial geo-tagging systems, using the two separate test collections for each language. Table 2 presents the obtained results, separating the two collections for each language.

Table 2. The obtained results when comparing different approaches

		Place Reference Recognition			Place Reference Disambiguation		
		Precision	Recall	F_1	Precision	Recall	F_1
Esp. T1	HMM+SVMReg				0.398	0.617	**0.484**
	HMM+Population	0.505	0.783	**0.614**	0.400	0.610	0.481
	HMM+Distance				0.360	0.560	0.436
	Yahoo! Placemaker	0.477	0.076	0.131	0.477	0.076	0.131
	Metacarta GeoTag	0.453	0.814	0.582	0.216	0.388	0.278
Esp. T2	HMM+SVMReg				0.472	0.549	**0.508**
	HMM+Population	0.662	0.769	**0.712**	0.470	0.540	0.503
	HMM+Distance				0.430	0.500	0.461
	Yahoo! Placemaker	0.657	0.061	0.111	0.627	0.058	0.106
	Metacarta GeoTag	0.577	0.751	0.277	0.302	0.393	0.342
Eng. T1	HMM+SVMReg				0.684	0.666	**0.675**
	HMM+Population	0.837	0.815	**0.826**	0.680	0.660	0.670
	HMM+Distance				0.590	0.580	0.584
	Yahoo! Placemaker	0.546	0.638	0.588	0.540	0.631	0.581
	Metacarta GeoTag	0.651	0.891	0.752	0.396	0.542	0.457
Eng. T2	HMM+SVMReg				0.627	0.665	**0.645**
	HMM+Population	0.733	0.777	**0.755**	0.620	0.660	0.638
	HMM+Distance				0.540	0.570	0.557
	Yahoo! Placemaker	0.523	0.658	0.582	0.508	0.640	0.566
	Metacarta GeoTag	0.547	0.868	0.671	0.307	0.488	0.377
Por. T1	HMM+SVMReg				0.150	0.123	0.135
	HMM+Population	0.267	0.219	0.241	0.250	0.210	0.226
	HMM+Distance				0.230	0.190	0.211
	Yahoo! Placemaker	0.250	0.014	0.027	0.250	0.014	0.027
	Metacarta GeoTag	0.242	0.589	**0.343**	0.174	0.425	**0.247**
Por. T2	HMM+SVMReg				0.389	0.326	0.354
	HMM+Population	0.500	0.419	0.456	0.440	0.370	**0.405**
	HMM+Distance				0.390	0.330	0.354
	Yahoo! Placemaker	0.438	0.062	0.109	0.438	0.062	0.109
	Metacarta GeoTag	0.380	0.628	**0.458**	0.168	0.292	0.213

The best results were achieved for the English language (an average score of 0.791/0.660 for the two test collections, in terms of the F_1 metric in recognition/disambiguation), while for Portuguese we could only achieve the modest result of 0.349/0.245 in terms of F_1. It should nonetheless be noticed that the dataset used for training the Portuguese models was much smaller, making it difficult to draw conclusions from the Portuguese experiments. In terms of the recognition performance, the obtained results are comparable against those obtained in the previous NER evaluation campaigns, although somewhat inferior - see Table 1. However, the main focus of this paper is on place reference resolution, which was not addressed in previous NER evaluations.

Leidner (2007) evaluated different heuristics for place reference resolution, using a gazetteer specifically developed for his experiments and a subset of the CoNLL-03 English collection with hand-made disambiguations to geospatial coordinates. The best-performing heuristic (i.e., spatial minimality) achieved an F_1 score of 0.365, although his results are not directly comparable to ours.

A manual inspection on the produced annotations revealed that some of the frequent errors in the produced annotations corresponded to problems such as those exemplified in the list bellow:

- The context in which the references are given is often difficult to interpret. For instance, in the test collections, names like *Barcelona* were often given as names of football clubs instead of locations.
- Metacarta often interprets complete sentences like *10Km south of Lisbon*, instead of just recognizing the name of the city as it is annotated in the test collections. The geospatial coordinates returned by Metacarta are also adapted accordingly.
- Placemaker often interprets expressions like *Lisbon, Portugal* as a single place reference to a capital city, instead of two separate geographic references as they are annotated in the test collections.
- The geonames service often fails in returning the correct disambiguation, in the top ten results, for places corresponding to abbreviations in the text (e.g., U.S. for the United States of America) or for points of interest (e.g., names for monuments or football stadiums) anotated as locations in the test collections.
- In the case of places corresponding to large geographical areas, the geospatial coordinates returned by the different services (i.e., geonames, Metacarta and PlaceMaker) often show considerable variation, among themselves and against the coordinates available in the test collections.

Notice that some of the above issues relate to problems in the evaluation methodology, and not in the systems themselves.

Despite the above problems, it is our belief that the achieved results are good enough to be used in subsequent processing stages of GIR applications (e.g., assigning documents to geographic scopes or retrieving documents geographically). Our currently ongoing work is pursuing this path, aiming at measuring the effect of the place reference resolution performance in other GIR tasks such as geographic scope assignment (Anastácio et al., 2009) or document retrieval (Martins and Calado, 2010).

5 Conclusions and future work

This paper presented a machine learning method for resolving place references in text. The proposed method is an instance of stacked learning (Wolpert, 1992), in which a first learner based on a Hidden Markov Model (HMM) is used to tag place references, and then a second learner implementing a regression through a Support Vector Machine (SVM) is used to rank the possible disambiguations for the place references that were initially tagged. Experiments with labeled datasets in three different languages attested for the adequacy of the proposed approach, showing that it outperforms two commercial state-of-the-art systems, namely Yahoo! Placemaker and the Metacarta geo-tagger.

Despite the interesting results, there are also many challenges for future work. For instance LingPipe's HMM method uses a relatively simple set of features (Carpenter, 2007). It would be interesting to explore richer feature sets or even more complex probabilistic models (e.g., Conditional Random Fields) for the identification sub-task. Similarly, in the disambiguation sub-task, one can also consider other sources of evidence as disambiguation features (e.g. semantic similarity in a taxonomy of places). Resolving temporal references over text also has important applications and equally challenging problems (Ahn et al., 2007). In the future, we plan to study integrated machine learning methods for resolving spatio-temporal references in text. Finally, we also plan to address the usage of spatio-temporal reference disambiguation in the context of spatial data infrastructures, for instance following the ideas of Ladra et al. (2008) related to exploring the recent OGC Web Processing Service (WPS) specification for deploying toponym resolution services.

Acknowledgements

This work was supported by the Fundação para a Ciência e a Tecnologia (FCT), through project grant PTDC/EIA/73614/2006 (GREASE-II).

References

Ahn, D., Rantwijk, J., and Rijke, M. (2007) A Cascaded Machine Learning Approach to Interpreting Temporal Expressions. Proceedings of the 2007 Annual Conference of the North American Chapter of the Association for Computational Linguistics.

Amitay, E., Har'El, N., Sivan, R., and Soffer, A. (2004) Web-a-where: Geotagging web content. Proceedings of the 27th Annual international ACM SIGIR Conference on Research and Development in information Retrieval.

Anastácio, I., Martins, B., Calado, P. (2009) A Comparison of Different Approaches for Assigning Geographic Scopes to Documents. Proceedings of IN-Forum: 1st Simpósio de Informática.

Bender, O., Och, F., and Ney, H. (2003) Maximum Entropy Models for Named Entity Recognition. Proceedings of the 7th Conference on Natural Language Learning.

Buscaldi, D., and Rosso, P. (2008) A conceptual density-based approach for the disambiguation of toponyms. International Journal of Geographic Information Science, 22(3).

Carpenter, B. (2007) LingPipe for 99.99% Recall of Gene Mentions. Proceedings of the 2nd BioCreative Workshop.

Chieu, H., and Ng, H. (2003) Named Entity Recognition with a Maximum Entropy Approach. Proceedings of the 7th Conference on Natural Language Learning.

Florian, R., Ittycheriah, A., Jing, H., and Zhang, T. (2003) Named Entity Recognition through Classifier Combination. Proceedings of the 7th Conference on Natural Language Learning.

Garbin, E., and Mani, I. (2005) Disambiguating toponyms in news. Proceedings of the 2005 Conference on Human Language Technology and Empirical Methods in Natural Language Processing.

Ladra, S., Luaces, M., Pedreira, O., and Seco, D. (2008) A Toponym Resolution Service Following the OGC WPS Standard. Proceedings of the 8th international Symposium on Web and Wireless Geographical information Systems.

Leidner, J. (2004) Towards a Reference Corpus for Automatic Toponym Resolution Evaluation. Proceedings of the 1st Workshop on Geographic Information Retrieval.

Leidner, J. (2007) Toponym Resolution in Text. PhD thesis, University of Edinburgh.

Manning, C., and Schutze, H. (1999) Foundations of Statistical Natural Language Processing. MIT Press.

Martins, B. and Calado, P. (2010) Learning to Rank for Geographic Information Retrieval. Proceedings of the 6th ACM Workshop on Geographic Information Retrieval

Mayfield, J., McNamee, P., and Piatko, C. (2003) Named Entity Recognition using Hundreds of Thousands of Features. Proceedings of the 7th Conference on Natural Language Learning.

McCallum, A., and Li, W. (2003) Early results for Named Entity Recognition with Conditional Random Fields, Feature Induction and Web-Enhanced Lexicons. Proceedings of the 7th Conference on Natural Language Learning.

Mota, C., and Santos, D. (eds., 2008) Desafios na avaliação conjunta do reconhecimento de entidades mencionadas: O Segundo HAREM. Linguateca.

Rauch, E., Bukatin, M., and Baker, K. (2003) A confidence-based framework for disambiguating geographic terms. Proceedings of the HLT-NAACL 2003 Workshop on Analysis of Geographic References.

Sang, E., and Meulder, F. (2003) Introduction to the CoNLL-2003 shared task: Language-Independent Named Entity Recognition. Proceedings of the 7th Conference on Natural Language Learning.

Smith, D. and Crane, G. (2001) Disambiguating geographic names in a historical digital library. Proceedings of the 5th European Conference on Research and Advanced Technology for Digital Libraries.

Smith, D. and Mann, G. (2002) Bootstrapping toponym classifiers. Proceedings of the HLT-NAACL 2003 Workshop on Analysis of Geographic References.

Shevade, S., Keerthi, S., Bhattacharyya, C., Murthy K. (1999) Improvements to the SMO Algorithm for SVM Regression. IEEE Transactions on Neural Networks, 11(5).

Smola, A., Scholkopf, B. (1998) A Tutorial on Support Vector Regression. NeuroCOLT2 Technical Report Series - NC2-TR-1998-030.

Vapnik, V. (1995) The Nature of Statistical Learning Theory. Springer-Verlag.

Viterbi, A. (1967) Error bounds for convolutional codes and an asymptotically optimum decoding algorithm. IEEE Transactions on Information Theory, 13(2).

Wick, M., and Becker, T. (2007) Enhancing RSS Feeds with Extracted Geospatial Information for Further Processing and Visualization. In A. Scharl and K. Tochtermann (eds.) The Geospatial Web - How Geobrowsers, Social Software and the Web 2.0 are Shaping the Network Society, Springer.

Witten, I., and Frank, E. (2000) Data Mining: Practical Machine Learning Tools and Techniques with Java Implementations. Morgan Kaufmann.

Wolpert, D. (1992) Stacked Generalization, Neural Networks, 5(2).

Zhou, G., and Su, J. (2002) Named Entity Recognition using an HMM-based Chunk Tagger. Proceedings of the 40th Annual Meeting of the Association for Computational Linguistics.

Towards a Spatial Semantics to Analyze the Visual Dynamics of the Pedestrian Mobility in the Urban Fabric

Thomas Leduc[1], Francis Miguet[1], Vincent Tourre[1,2], Philippe Woloszyn[3]

[1] CERMA laboratory UMR CNRS 1563, 6 quai François Mitterrand, BP 16202, 44262 Nantes cedex 2, France
{thomas.leduc, francis.miguet, vincent.tourre}@cerma.archi.fr
[2] Ecole Centrale de Nantes, rue de la Noë, BP 92101, 44321 Nantes cedex 3, France
[3] RESO laboratory, ESO, Maison de la Recherche en Sciences Sociales, Université de Haute Bretagne - Rennes II, place du recteur Henri Le Moal, CS 24307, 35043 Rennes cedex, France
philippe.woloszyn@univ-rennes2.fr

Abstract. The aim of this paper is to evaluate the impact of visual ambiences (visualscape) onto an urban pedestrian pathway. This work takes place within the interdisciplinary research project called Ambioflux, dealing with the sustainable mobility issues in the city. We add to the spatial semantic layer GDMS (Generic Datasource Management System) an innovative method based on partial isovists fields in order to compute the visibility of the pedestrian all along his pathway. This method allows "concatenating" several partial visibility polygons in order to represents the visual perception of the pedestrian.
After a brief overview of the visibility analysis context, we justify the need of a specific semantic tool to develop the type of dynamics visual analysis we focus on. The remainder of this paper is dedicated to the methodology of the mobile pedestrian pathway's visual fingerprint characterization using the spatial formalism already described. At last, we present a use case based on a real city tour so as to identify the best rotation's direction from the visual perception point of view.

M. Painho et al., (eds.), *Geospatial Thinking*, Lecture Notes in Geoinformation and Cartography,
DOI 10.1007/978-3-642-12326-9_13,

1 Introduction

The Ambioflux's interdisciplinary project deals with the urban ambiences in the context of the city re-qualification. The aim of this project is to focus on both environmental and eco-systemic aspects of the study but also on its psychological and anthropological characteristics.
In the context of this research project, the objective of the current task is to evaluate, so as to characterize, the impact of visual ambiences (visua-lscape) onto an urban pedestrian pathway using and extending the spatial semantic layer called GDMS (Generic Datasource Management System).
(Zacharias, 2001) wrote: the sensory experience of the environment may dominate the information field and hold sway over many of our actions. We personally adhere to (Piombini, 2006)'s simple hypothesis: visualscape plays an important role in the pedestrian's perception of the surrounding urban environment. Thus, the optimization of the visualscape's assessment is a good solution to enhance walking practices.

After a brief overview of the visibility analysis context, we justify the need of a specific semantic tool to develop the type of dynamics visual analysis we focus on. The remainder of this paper is dedicated to the methodology of the mobile pedestrian pathway's visual fingerprint characterization using the spatial formalism already described. At last, we present a use case based on a real city tour so as to identify the best rotation's direction from the visual perception point of view.

2 Overview of visibility analysis

In (Gibson, 1979), an ecological theory of visual perception, based on optic flow patterns, has been defined. In this theory, real movement plays a vital part in the perception process. The three key ideas of it are: optic array, textured gradients, and affordance. The optic array corresponds to the set of all visual information (about the layout of objects in space) reaching the eye. It is a set of nested solid angles corresponding to surface elements in the environment. The textured gradient provides information about the characteristics (distance, speed, etc.) of the perceived environment. This mechanism requires almost a small signal processing by the cognitive system. At last, the affordance implies that a particular meaning is enclosed in each visual information: the potential use of an object is directly perceivable (e.g. a chair 'affords' sitting).

In the 1970s, two main approaches emerge in the visibility analysis context: the concept of viewshed in terrain and landscape analysis and the concept of isovist in architecture and urban space.

The viewshed analysis is a traditional way of analyzing a visibility field. It is defined as the part of terrain visible from a viewpoint, and is basically applied to the landscape with terrain and topographic differentiation (Lynch, 1976). As noticed in (Yang, 2007), viewshed analysis in GIS is rarely applied to urban settings because the operation is based on raster data or TIN (triangular irregular network) data structure, which have problems of accuracy in representing complex geometry of urban form.

(Benedikt, 1979) defines an isovist as "the set of all points visible from a given vantage point in space and with respect to an environment". This concept has been previously introduced in (Tandy, 1967). As mentioned in one of (Benedikt, 1979) footnotes, "readers familiar with Gibson's work will see the affinity of isovists with [ambient] optic arrays". The isovist is simply this optic array, with the wavelength and intensity information omitted (Benedikt and Burnham, 1985). To sum up, an isovist is usually a 2D bounded polygon which is a useful tool to define the open space concept. From a morphological point of view, open spaces are usually defined as the empty space, the void, between the surrounding buildings. However, although these open spaces are not material components of the physical world, they can be conceived as part and parcel of our urban heritage (Teller, 2003). (Batty, 2001) puts the emphasis on the fundamental motivations of conducting visibility analysis research. He noticed that the key questions "how far can we see", "how much can we see", and "how much space is enclosed" are relevant to develop good urban design.

The Gibsonian panoramic visual world can be seen as a set of temporal series of excitations (sort of visual snapshots) one might experience moving about inside the urban fabric. The pedestrian faces urban morphology in the context of its natural movements. As written in (Yang, 2007), human perception is influenced and to a certain extent can be manipulated by reconfiguring physical urban morphology, under particular ambients constraints (Woloszyn, 2002).

Several different spatial analysis methods have been defined to study the plenum conception (Couclelis, 1992) of urban space: isovist fields (Benedikt, 1979; Batty, 2001), space syntax (Hillier and Hanson, 1984), visibility graphs (Turner et al., 2001), agent-based approach (Batty and Jiang, 1999), convex partitions (Peponis et al., 1997), sky opening maps (Teller, 2003; Sarradin et al., 2007) and ambianscape formalization (Woloszyn et al., 2000). Some of them – agent-based and visibility ones - are based on a regular grid medium on which all subsequent analysis is developed, that's the reason why we have dismissed them. Some other assumes that urban open spaces can be uniquely partitioned into convex sets that meet some criteria of minimality in number. The scientific community is divided over this issue, that's why we did not focus on convex partitions. At last, according to (Yang, 2007), the Teller's sky opening indicator, which requires double projections of its spherical surface for its visual indicator, makes the original 3D spherical surface distorted and deformed, and easily induces incorrect calculation.

For all these reasons, we have decided to focus on the well-known 2D isovist fields (vector based approach) method. Indeed, (Hillier and Hanson, 1984) noticed "human space is in fact full of strategies [...] to reduce the three-dimensional structures to the two dimensions in which human beings move and order space". Essentially, isovists describe local geometrical properties of spaces with respect to individual observation points and weight all the possible view directions equally. An isovist is a 2D horizontal slice of pedestrian's surrounding space.

For the specific analysis of motion trajectories, (Meilinger, 2009) processes view-specific partial isovists. Partial isovists consider only a restricted part of the theoretically available visual field (for example, 90° instead of 360°). They correspond better to the restrictions of the human visual apparatus. To reproduce the dynamic qualities of the Gibsonian visual world that takes into account not only the bifocal vision but also the relatively free movement of the head and shoulders (Teller, 2003), one may consider the partial isovist angle as an input parameter of the parametric model so as to produce range of results.

3 Why do we need to merge a GIS&T[1] semantic tool and visibility analysis?

What seems obvious with all previous models is the lack of wide scale simulations on motion approach. Indeed, the mathematical combination of a set of isovists results in a multitude of heterogeneous geometric shapes that needs to be analyzed. It seems important to address it regarding the most appropriate scientific tool. Thus we have decided to focus on one of the Association of American Geographers (AAG) reference studies. The Geographic Information Science and Technology Body of Knowledge[2] (BoK) is a community-developed inventory of the knowledge and skills that define the GIS&T field. In order to achieve the analysis of the visibility all

[1] GIS&T: Geographic Information Science and Technology.

[2] http://aag.org/bok/ (valid on Jan. 6th 2010).

along a pedestrian pathway, we have clearly identified several involved sub-domains of this BoK such as the following:

- in term of analytical methods:
  - query operations and query languages (spatial queries),
  - geometric measures (distances and lengths, areas, etc.),
  - basic analytical operations (buffers),
  - basic analytical methods (spatial interaction),
- in term of cartography and visualization (data considerations, principles of map design),
- in term of design aspects
  - analysis design (coupling scientific model with GIS),
  - application design (workflow analysis and design),
- in term of conceptual foundations (fields in space and time, metrical relationship),
- in term of data modeling (classic vector data model),
- in term of geocomputation (simulation modeling).

This observation has already been implicitly made even if it does not seem to have yet been translated into such an inventory. So, GIS-based visibility analyses are nowadays quite a common approach. Indeed, an urban pedestrian pathway and a visualscape are both data that include a spatial component. What is required here is some tool able to process these spatial data using: on one hand the table-oriented programming paradigm (for its table-friendly syntax, its fundamental and consistent collection operations, and its easiness of understanding) and, on the other hand, batch processing with parametric ability and procedural extension. We pretend that the use of a spatial SQL with semantics ability is essential to perform such an objective. That is the reason why we need to take benefits from the GDMS specific layer, aside the features of robustness, scalability, and easy to use main characteristics.

GDMS (Generic Datasource Management System) is related with a similar layer, called GDBMS (Anguix and Carrion, 2005), used in the gvSIG project to manage the alphanumeric data access. The main limitation of the original GDBMS layer is that it can only be used for alphanumeric purposes. The GDMS contribution is a general refactoring that mainly adds spatial functionalities and a more powerful SQL processor. It has been defined and implemented in the context of the OrbisGIS project, and extended in its recent GearScape customization (with a sort of procedural ability). The relevance of this layer for spatial processing has already been exhibited by (Leduc et al., 2009a, 2009b).

The main objectives of GDMS are to provide to the user not only a simple and powerful API but also a spatial SQL derived language. Moreover, as an intermediate layer between the user and the information source, GDMS intends to reduce the coupling between the processes and the specificities of each underlying format. As a consequence, former work may easily be reused in a much larger set of scenarios. The learning curve is consequently even simpler.

With GDMS, the idea is to move all problems of interoperability across data repositories, but also about the SQL semantics by developing a highly flexible, portable and standards compliant tool to build SQL queries. GDMS provides a SQL processor that lets the execution of the common Data Manipulation Language statements against any source mapped by a driver. To avoid introducing a new grammar, GDMS fully preserves the SQL-92 grammar and adds to this standard geometric concepts and spatial functions as in OGC simple features SQL specification. As an analogy to spatial SQL for R-DBMS, GDMS provides an extended SQL query language on heterogeneous data types.

It is the main purpose of GDMS to improve data creation and sharing. As in a Spatial Data Infrastructure (SDI) the consumption of data is as important as the production and sharing of data, the SQL processor in GDMS allows data feedback as if they were new data sources (materialized views). This means that the result of the SQL queries can easily be integrated into the SDI as a new data source. Those data will be ready to be used by further SQL statements as any other existing data source.

GDMS allows an extension to the semantics of the SQL language in terms of functions and custom queries. The functions and custom queries are artifacts that contain the implementation of some operations on the data and can be reused just by referencing their name into a SQL statement. In this way, some user can implement a buffer operation and other can reuse it just by calling *ST_Buffer()* in a SQL query:

SELECT ST_Buffer(the_geom, 20) FROM mydata;

4 Methodology

4.1 Prerequisites

We want our model to rely on "classical" data of vector type so as to let it be easily and widely repeatable. That is why, in the following use case, we present a simulation that is only based on both a polygonal buildings layer and a linear pathway layer. Though, it is obvious that it could be enriched with several other layers such as urban vegetation or street furniture. In the same manner than the buildings layer, they would be treated as visual masks (the simpler is probably to union all the mask layers in a preprocessing phase).

What needs to be remembered here is that a pedestrian walk in the urban fabric is fundamentally a time-line process. To simplify the model, we have decided to adopt a discrete time approach of the perception mechanism: a slice-time has to be set so as to sequence each pathway. The mean speed of the pedestrian is stable, at each time tick, he reaches a new following position. The continuous linear pathway is thus transformed into a set of ordered punctual equidistant positions. In each of these positions, an isovist is computed and the corresponding polygon is then added to the global isovists field.

As written in the visibility analysis overview section, for the specific analysis of motion trajectories, (Meilinger, 2009) processes view-specific partial isovists. To reproduce the dynamic qualities of the Gibsonian visual world, we have decided to consider the partial isovist angle as an input parameter of our parametric model so as to produce range of results. To illustrate the topic in the following use case, five distinct aperture angles are mentioned: 30°, 60°, 90°, 140° and 200°. Even if none of them are clearly identified in both the visual perception theory and the cognitive psychology science, we may admit that a visual object is less avoidable and much stronger in term of continual awareness, if it is located in front of us (small aperture angle) than far left- or right-sided.

We are aware that such preconditions are extremely simplistic and thus unadapted to model the visual perception of a pedestrian immersed in a composite visualscape. The aim here is to simplify so as to be able to simulate. In the conclusion we will suggest several possible enhancements.

As mentioned in the visibility analysis overview, we clearly need a model able to produce a directional visibility analysis that take also into account visual dynamics all along the pathway (motion approach).

4.2 Sequence of spatial processing

The simplified sequence diagram presented in Fig. 1, sums up the main phases of the visual dynamics analysis method we have developed. The global process is composed of the following different phases and sub phases:

1. sample the continuous pathway in both directions so as to obtain a set of equidistant punctual positions and, for each of them, the corresponding local speed vector,
2. perform the isovists field computation. This step is achieved using a dedicate development that makes a strong use of the GDMS efficient spatial index implementation. It is a vector based process much more accurate than a raster one;
3. loop on the aperture angle values (parametric study) and, for each of them:

- evaluate the visual cones in each punctual position (according to the speed vector direction) and calculate the intersection with the corresponding isovist so as to produce a partial isovist,
- compute the weighted coverage of the resulting partial isovists field. The aim is here to produce, using a dedicated polygonal coverage algorithm, a partition of the input (partial) isovists fields. That is, a division of it into non-overlapping and non-empty subsets that cover all of it. These parts are collectively exhaustive and mutually exclusive.

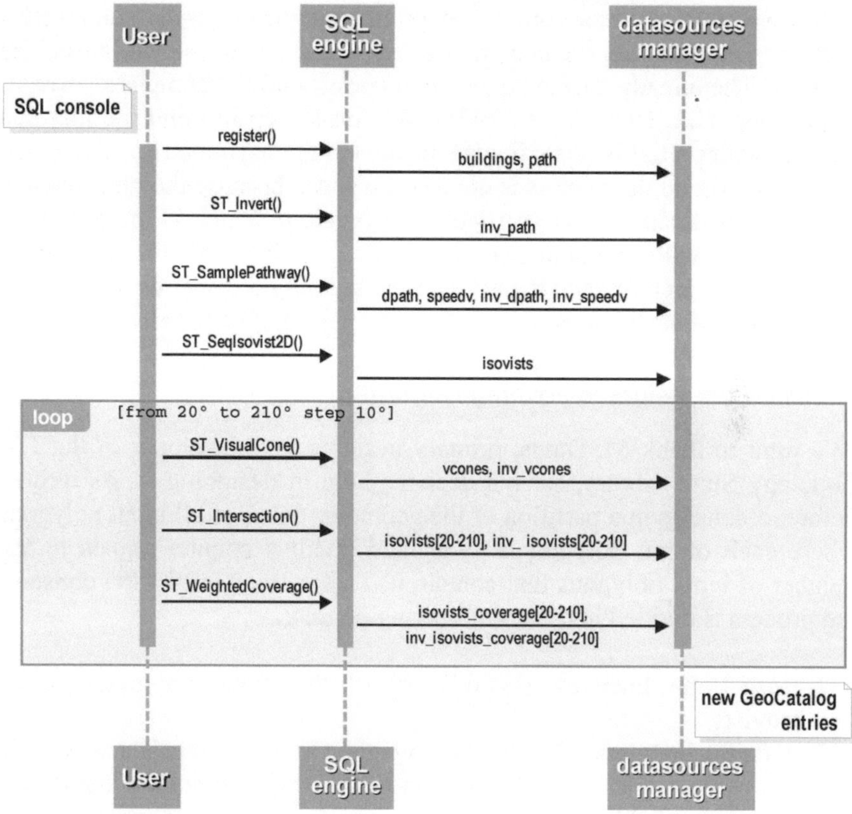

Fig. 1. Simplified sequence diagram of the spatial processes involved in the visual analysis. The aim is here first to produce a set of partial isovists fields and then to produce a weighted coverage of the global visible area

4.3 Post-processing

The objectives of the processing phases presented above are to produce both a set of partial isovists fields (for different values of aperture angles) and a set of weighted coverage of all the corresponding global visible areas. The weighted coverage process is illustrated in Fig. 2. It consists in a partition of the geometry union of all input polygons. In this output partition, each item is "weighted" with a counter equals to the number of input polygons that contain it. Producing such a weighted coverage map is useful to analyze the urban areas that are promoted from a temporal point of view (that is, all those the pedestrian is able to look upon for a long time) and all those that are neglected (almost outside his field of vision).

Another post-process consists in producing indicators to characterize each isovists field. So, we need to identify a few sets of relevant visual indicators. The already defined ones often correspond to "complete" isovists fields (Benedikt, 1979; Batty, 2001). We must keep in mind the fact that we deal with partial isovists fields (fundamentally asymmetric). The translation into formal descriptors is not a so easy task because the meaning and relevance of descriptors are difficult to estimate *a priori*. In the following use case, we will comment two simple scalar indicators. The first one is the polygonal area of each isovist, and the second corresponds to the number of edges of each isovist.

Focus on the weighted coverage algorithm

We want to thank M. Davis, primary designer and developer of the JTS Topology Suite[3] library, for his decisive help in designing it. As written before, it consists in a partition of the geometry union of all input polygons where each output polygon is "weighted" with a counter equals to the number of input polygons that contain it. This memory and CPU consuming process is divided into the 4 following phases:

1. extract the linework (boundaries) of the input polygonal partial isovists,
2. node the linework. The JTS UnaryUnionOp is a convenient way to do this (Union on a set of LineStrings has the effect of noding them);
3. pass the noded linework into JTS Polygonizer and polygonize it,
4. for each resultant polygon, determine an interior point and then find the source polygons which contain this point so as to weight it (using spatial index and *PreparedGeometry* speed the process up).

[3] http://tsusiatsoftware.net/jts/main.html (valid on Jan. 6th 2010).

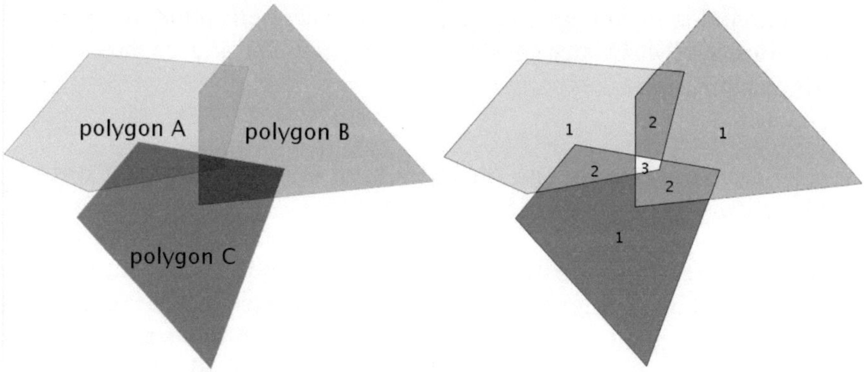

Fig. 2. Illustration of the weighted coverage process on a set of three overlapping polygons. The result is a partition of the geometry union of all input polygons. In this output partition, each item is "weighted" with a counter equals to the number of input polygons that contain it

5 Use case

In order to exhibit the relevance of our model and its potentialities, we have decided to apply it to a pedestrian pathway called "*parcours confort*" proposed[4] by the *Nantes Métropole* Tourist Office (NMTO). Nantes is a west-coast located city in France. This tour, accessible to all (even for people in a wheelchair), proposes to those who have some time to spend wandering in Nantes a discovery of the historical city centre from medieval to 19th century morphologies (see Figures 3 and 4). This 1590 meters long tour starts from the tramway central station *Commerce* and run through:

- the medieval district called *quartier du Bouffay*,
- the *Cours des 50 otages* main street, named in memory of the fifty hostages who were executed during the Second World War. This boulevard has been redesigned in 1990 by the Italian architect Italo Rota;
- the *Place Royale*, which embeds a 19th century fountain representing the city of Nantes with the Loire River sitting on her feet,
- the *Passage Pommeray*, the only multilevel covered street in Europe, containing a staircase bordered by statues.

[4] This tour is available on-line on the website: http://www.nantes-tourisme.com/ (valid on Jan. 6th 2010).

Our goal is to compare, in term of the visual perception of a pedestrian, the former NMTO tour in both clockwise (CW) and counter-clockwise (CCW) directions.

5.1 Comparison of the global visual coverage rate of both directions on the NMTO tour

We assume that this 1590 meters long tour has been divided into 983 punctual equidistant positions or time steps. The distance between each of these positions is equal to 1.6 meter (time-slice between two successive "snapshots" is approximately equals to 1.5 second for a walking speed of about 4 km/h). In each position, several isovists have been computed according to the variation of the aperture angle.

Fig. 3 presents a weighted coverage of each partial isovists field (they both share the same aperture angle equals to 60°). The left one corresponds to the counter-clockwise oriented pathway, while the right one corresponds to the clockwise oriented pathway. One may notice that the visual aspect of those both maps are rather different: visual-walk is a one-way process.

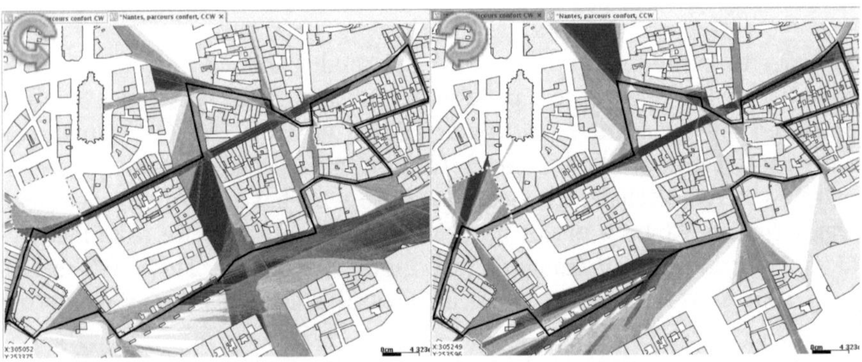

Fig. 3. Comparison of two partial isovists fields (with an aperture angle equals to 60° in both cases) of the same NMTO tour in the historical city centre (black line). On the left hand side, pathway is counter-clockwise (CCW) oriented while, on the right hand side, it is clockwise (CW) oriented. Shapes underlined correspond to some specific locations that will be developed later in this use case

The three different shades of grey presented in the Fig. 3 correspond to an interval classification:

- the light grey corresponds to an interval from 0% to 1%,
- the medium grey corresponds to an interval from 1% to 5%,
- and the dark grey corresponds to an interval from 5% to 22%.

These percent values correspond, for a given polygonal area, to the time it has been seen (with an aperture angle equals to 60°) by a given pedestrian all along the 983 punctual positions.

In the Fig. 4, we identify the indices of the punctual positions so as to be able to analyze the results we present in both the Fig. 5 and the Fig. 6.

Fig. 4. Screenshot of both buildings and pedestrian pathway layers. As may be noticed, the continuous pathway is sampled into 983 punctual positions so as to let us perform the isovists fields computations

The Fig. 5 and Fig. 6 illustrate the ability of our tool to produce sort of signatures of the partial isovists field (with an aperture angle always equals to 60°) according to the evolution of a given scalar indicator all along each pathway. Aim is here to compare, for a given pedestrian punctual position, the area of the partial isovist (in both directions) and its number of edges (in both directions also).

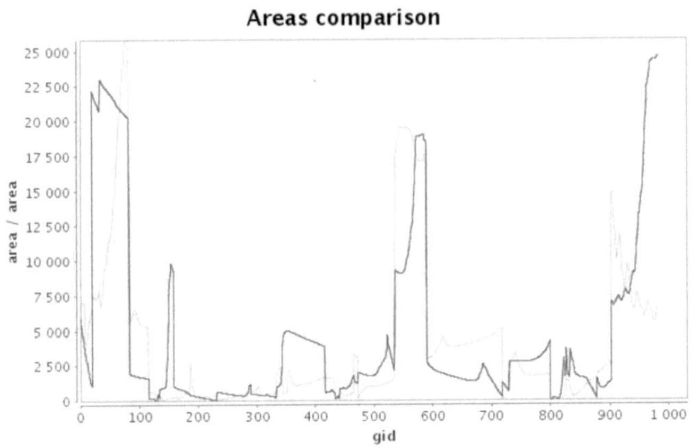

Fig. 5. This line chart represents the evolutions of the area indicator all along each pathway (CCW one is represented by the dark grey line and CW one is represented by the light grey line)

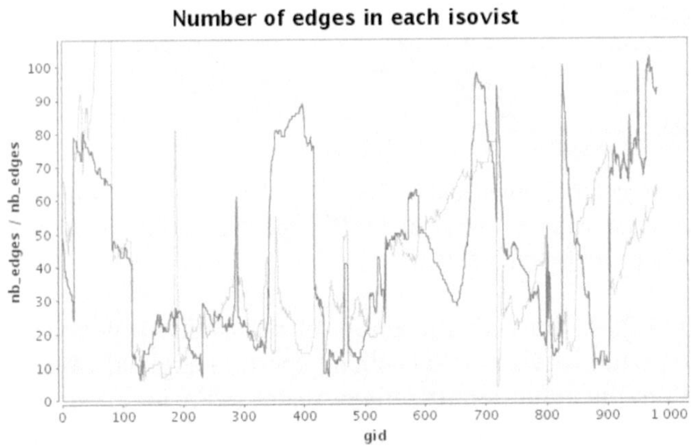

Fig. 6. This line chart represents the evolutions of the number of edges indicator all along each pathway (CCW one is represented by the dark grey line and CW one is represented by the light grey line)

One may notice that around position #425 (*Sainte Croix* church), the CCW number of edges (in dark grey) is 3 times greater than the corresponding CW number of edges (in light grey). As a consequence, the pedestrian isovist is quite fuzzy in the CCW direction.

Concerning the punctual position around #720 the visual area of the CW oriented pathway (in light grey) is two times greater than the visual area of the CCW oriented pathway (in dark grey). Indeed, *Saint Nicolas* church is only visible, at this position, from the CW oriented tour!

5.2 Zoom in some specific areas so as to show different comparison methods

In this section, we will compare different methods to analyze the perceived environment. Each of them will be applied to a specific urban object in an adapted context.

Visible surfaces ratio of an urban square: the example of the Place Royale

The surface ratio is the visible surface in the studied tour over the total surface of the *Place Royale* square (see Fig. 7). In this figure, the light grey polygons correspond to partial isovists of aperture angle equals to 30°. The medium grey polygons (that contain also the light grey one) correspond to partial isovists of aperture angle equals to 60° and, at last, the dark grey polygons correspond to partial isovists of aperture angle equals to 90°. One may notice than this square and especially its nice central fountain is best seen with the CW tour.

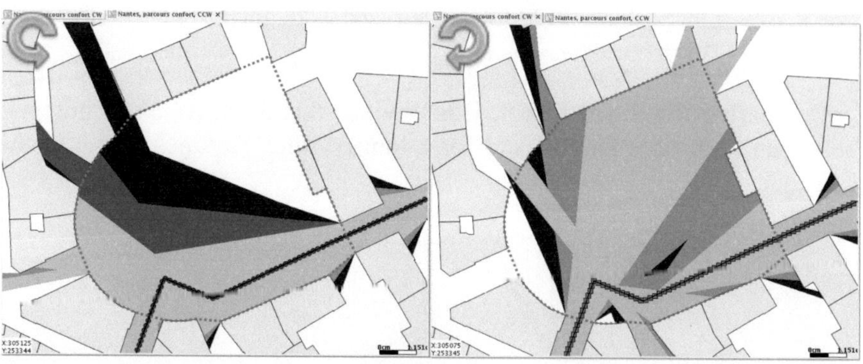

Fig. 7. Amount of visible areas for each pathway. The CW tour (on the right) covers obviously much more the square than the CCW one (on the left). In those both screenshots, light grey polygons correspond to a partial isovist field of 30°, medium grey ones to 60° partial isovists fields and dark grey ones to 90° partial isovists fields

As an illustration, we have computed coverage ratios for each of the aperture angles:

- for the CW tour: 64.7% is visible with an isovist of 30°, 88.7% with an isovist of 60°, and 93.7% with an isovist 90°;
- for CCW tour: 36.6% is visible with an isovist of 30°, 52.4% with an isovist of 60°, and 67.7% with an isovist 90°.

Visibility of a specific valuable urban object: the example of the Saint Nicolas church

In the Fig. 8, we demonstrate that the front of *Saint Nicolas* church is only seen during 2 time steps (from positions #727 to #728) over 983 with an aperture angle greater of equal than 200°. Those both isovists give the pedestrian the possibility to see only 6.89% of the total façade of this church.

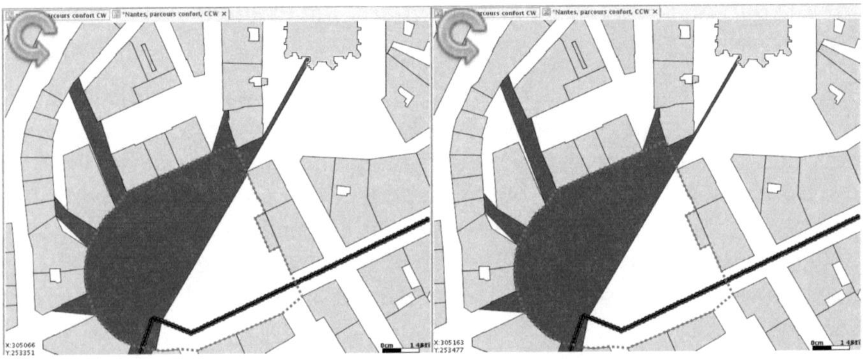

Fig. 8. Isovist for positions #727 (on the left) and #728 (on the right) for the CCW tour

Comparing Fig.8 and Fig. 9, it is obvious that *Saint Nicolas* church is best seen (2.5 more length) and for a longer time (12 times more) in the CW tour.

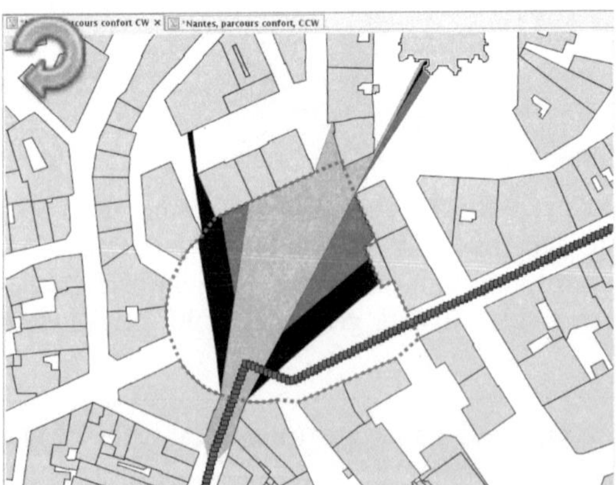

Fig. 9. Partial isovists fields (aperture angles from 30° - light grey polygon, to 90° - dark grey polygon) from position #230 to position #253 in the CW oriented tour

Visibility of a specific valuable urban front: the example of the Sainte Croix church

In the specific case of *Sainte Croix* church, in the CCW oriented tour (on the left) the pedestrian is once again badly positioned to see the front of the monument (see Fig. 10).

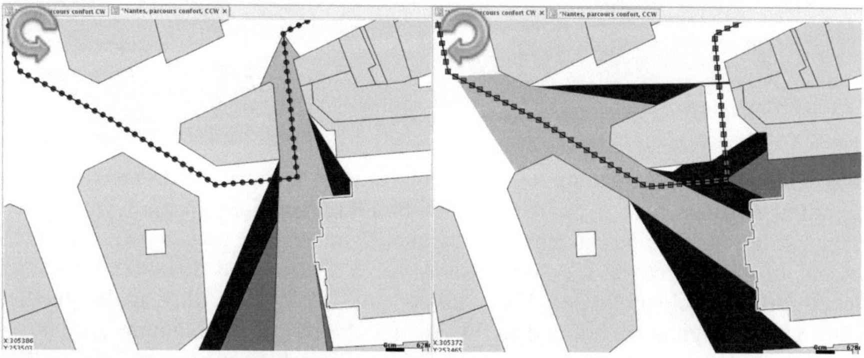

Fig. 10. The CCW oriented tour (on the left) gives the pedestrian the possibility to see at most 50% of the front of this nice building. On the contrary, with the same aperture angle, in the CW oriented tour, the front of Sainte Croix church is much more seen. Moreover, in the CW tour, the incidence angle is far better (40° according to the normal to the façade). In those both screenshots, the color code corresponds to some different notions. Indeed, the light grey polygon is the first isovist of the pedestrian walk that contains part of the façade (positions #417 in the CCW tour and #518 in the CW one). The same way, the medium grey polygon is the last isovist that contains part of the façade (positions #432 in the CCW tour and #550 in the CW one). By contrast, the dark grey polygons are partial isovists fields from position #417 to #432 in the CCW tour and from position #518 to #550 in the CW one

Visibility and distance: the example of the place de Sarajevo square

In this last example, it is once again obvious that the front of the old stock exchange is much more visible in the CW oriented tour than in the CCW one (see Fig. 11).

In conclusion of the use case, these results clearly show that some of the most important buildings are not well seen in the CCW tour and that the CW tour allows a better discovery of the city. A good feedback of the person in charge of the tourist tours, encourage us to apply this method to analyze some other city tours presented on the *Nantes Métropole* Tourist Office website.

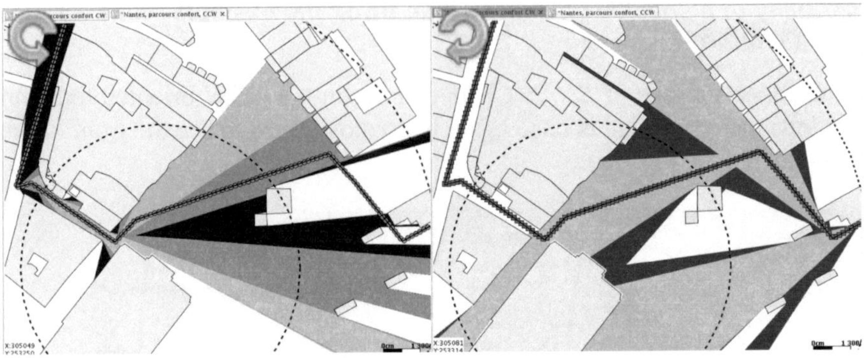

Fig. 11. In the CCW oriented tour (on the left), even with an aperture angle equals to 140°, less than half of the front of this building is visible. In the CW oriented tour (on the right), a partial isovist with an aperture angle equals to 60° is enough to see it fully. Moreover, it is fully visible with a distance less than 90 meters (and a comfortable incident angle). The color code of this figure is once again specific: light grey polygons correspond to 30° partial isovists fields, medium grey polygons correspond to 60° partial isovists fields and dark grey ones to 140° partial isovists fields

6 Conclusion and outlook

The use case that has been presented in this paper is relative to Nantes historical center. It is however obvious that the methodology we described is transposable to any other pedestrian pathway in an urban fabric and, more generally but with some limits, to any journey through a built environment independently from the means of transportation. As an example, in a public transportation such as a bus, a trolley bus or a tramway, passenger's partial isovist main direction is often perpendicular to the motion one. What differs the most is probably the distance between two consecutive snapshots; indeed it depends on the displacement's speed.

Another practical application case is relative to the visual impact of a new structure, such as a building, on its surrounding open space. What kind of visual masks does it involves? Does it hide significantly some interesting view? It is also probably a useful tool in the decision making process for the positioning of urban road signs, etc.

Concerning the decision we took to couple GIS&T and visibility analysis, this study is another proof of the relevance of the GDMS layer in term of spatial knowledge enhancement. Much more than a syntax with spatial abilities, this layer provides an extensible engine and a set of tools that let us combine the physical features perceived so as to infer the pedestrian feeling and therefore behavior. Empowered with efficient semantics ability one may admit that a new GIS dedicated to the characterization of a visual-walk in the urban fabric has thus been developed.

Regarding next steps of our study, we aim to focus on some specific and relevant indicators (such as jaggedness, radial variance and radial skew, convex deficiency… (Stamps, 2005)) so as to characterize partial isovists fields. The visualscape indicators we plan to implement have mainly to integrate the psychophysical dimension of the perception process (Woloszyn, 2002). At last, we have decided to focus on a 2.5D implementation of this model (coupling it with the ones presented in (Morello and Rati, 2009) and (Yang et al., 2007)) so as to take surrounding buildings, with their elevation component, into account for enhanced mask effects computation.

Acknowledgements

The AMBIOFLUX project was funded by CNRS and the French MEEDM Ministry under PIRVE's (*Programme Interdisciplinaire de Recherche Ville et Environnement*) contract #1004. Part of the GearScape developments is funded by the AMBIOFLUX project.

Special thank to Fernando GONZÁLEZ CORTÉS (Spain) for all the developments performed on the GearScape platform.

References

Anguix, A. and Carrion, G. (2005). gvSIG: Open Source Solutions in spatial technologies. In GIS Planet, Estoril, Portugal.

Batty M (2001). Exploring isovist fields: space and shape in architectural and urban morphology. Planning and design: Environment and planning B, 28(1):123–150.

Batty M and Jiang B (1999). Multi-agent simulation : new approaches to exploring space-time dynamics in GIS. CASA Working Papers Series, London, UK, (10):25.

Benedikt ML (1979). To take hold of space: isovists and isovist fields. Environment and Planning B: Planning and Design, 6(1):47–65.

Benedikt ML and Burnham CA (1985). Perceiving architectural space: From optic arrays to isovists. In Persistence and Change, pages 103–114. edited by WH Warren and RE Shaw. Hillsdale, NJ: Lawrence Erlbaum Associates.

Couclelis H (1992). People Manipulate Objects (but Cultivate Fields): Beyond the Raster-Vector Debate in GIS. In Frank AU, Campari I, and Formentini U, editors, Theories and Methods of Spatio-Temporal Reasoning in Geographic

Space, International Conference GIS - From Space to Territory: Theories and Methods of Spatio-Temporal Reasoning, Lecture Notes in Computer Science, pages 65–77, Pisa, Italy. Springer.

Gibson JJ (1979). The Ecological Approach to Visual Perception. Boston: Houghton Mifflin.

Hillier B and Hanson J (1984). The Social Logic of Space. Cambridge University press.

Leduc, T., Bocher, E., González Cortés, F., and Moreau, G. (2009a). GDMS-R: A mixed SQL to manage raster and vector data. In GIS Ostrava 2009 - Symposium on Seamless Geoinformation Technologies, Ostrava, Czech Republic.

Leduc T, Woloszyn P, and Joanne P (2009b). GDMS: A spatial semantics to evaluate soundmarks effects on an urban pedestrian pathway. In 12th AGILE International Conference on Geographic Information Science — AGILE'2009, Hanover, Germany.

Lynch KA (1976). Managing the sense of a region. Cambridge: MIT Press.

Meilinger T, Franz G, and Bülthoff HH (2009). From isovists via mental representations to behaviour: first steps toward closing the causal chain. Environment and Planning B: Planning and Design. Advance online publication.

Morello E and Ratti C (2009). A digital image of the city: 3D isovists in Lynch's urban analysis. Environment and Planning B: Planning and Design, 36(5):837–853.

Peponis J, Wineman J, Rashid M, Hong Kim S, and Bafna S (1997). On the description of shape and spatial configuration inside buildings: convex partitions and their local properties. Environment and Planning B: Planning and Design, 24(5):761–781.

Piombini A (2006). Modélisation des choix d'itinéraires pédestres en milieu urbain approche géographique et paysagère. PhD thesis, Université de Franche-Comté. Ecole Doctorale "langages, espaces, temps, sociétés".

Sarradin F, Siret D, Couprie M, and Teller J (2007). Comparing sky shape skeletons for the analysis of visual dynamics along routes. Planning and design: Environment and planning B, 34(5):840–857.

Stamps AE (2005). Isovists, enclosure, and permeability theory. Environment and Planning B: Planning and Design, 32(5):735–762.

Tandy CRV (1967). The isovist method of landscape survey. Methods of Landscape Analysis. Ed. H C Murray, Landscape Research Group, PO Box 53, Horspath, Oxford, OX33 1WX.

Teller J (2003). A spherical metric for the field-oriented analysis of complex urban open spaces. Planning and design: Environment and planning B, 30(3):339–356.

Turner A, Doxa M, O'Sullivan D, and Penn A (2001). From isovists to visibility graphs: a methodology for the analysis of architectural space. Planning and design: Environment and planning B, 28(1):103– 121.

Woloszyn, P. and Follut, D. (2000). The visualization of the urban "ambients" parameters. In 14th International symposium "Computer Science for environmental protection" of Gesellschaft für Informatik (GI), pages 173-186, Bonn, Germany. METROPOLIS Verlag.

Woloszyn, P. (2002). From fractal techniques to subjective quantification: towards an urban ambient metric? In Landscape and Architectural Modeling Symposium, pages 1-6, Sousse, Tunisia.

Yang PPJ, Putra SY, and Li W (2007). Viewsphere: a GIS-based 3D visibility analysis for urban design evaluation. Planning and design: Environment and planning B, 34(6):971–992.

Zacharias J (2001). Pedestrian behaviour and perception in urban walking environments. Journal of Planning Literature, 16(1):18.

Managing Collapsed Surfaces in Spatial Constraints Validation

Alberto Belussi[1], Sara Migliorini[1], Mauro Negri[2], Giuseppe Pelagatti[2]

[1]Dipartimento di Informatica, Università degli Studi di Verona, Verona, Italy, {alberto.belussi|sara.migliorini}@univr.it
[2]Dipartimento di Elettronica e Informazione, Politecnico di Milano, Milan, Italy, {mauro.negri|giuseppe.pelagatti}@polimi.it

Abstract. In recent years many Spatial Data Infrastructures (SDI) have been built or are under construction in several European countries and geographical data have been collected in many different ways, sometimes following traditional approaches (like, remote sensing techniques of photogrammetry and topography) or more "up to date" methods (like, GPS or others). However, regardless of the used methods, the geometries resulting from the surveying process describe the geographic object according to the considered accuracy level; therefore, in some cases such geometries, usually modeled as surfaces, are collapsed to set of curves and/or points.

In this paper we propose a methodology which aims to handle this collapsed behavior of surfaces by preserving the conceptual schema of the database, which means that we preserve the geometric types (i.e., the surface type) of the spatial attributes even if their values can collapse. Moreover, we extend the semantics of integrity constraints included in the conceptual model in order to handle during data validation the collapsed surfaces as much as possible as real surfaces.

1 Introduction

The conceptual schema of a spatial database is an abstract model of real world entities, called *geographic objects*, and their relationships. In particular, a conceptual schema defines the objects of interest and how they are

M. Painho et al., (eds.), *Geospatial Thinking*, Lecture Notes in Geoinformation and Cartography, DOI 10.1007/978-3-642-12326-9_14, © Springer-Verlag Berlin Heidelberg 2010

represented by means of attributes. In more detail, a geographic object is characterized by one or more spatial attributes, besides the several thematic ones. A conceptual schema is composed of some *classes*, each of which collects objects with the same characteristics. Therefore in the following, a geographic object can also be referred to as a *class instance* or *feature*.

Geographic objects can be captured using different kind of instruments and technologies, each one characterized by a particular level of accuracy and a threshold of acquisition, i.e. a measure under which the used instruments are not able to define the shape of the objects. It follows that in certain cases, the spatial component of a geographic object actually acquired can be of a geometric type which is different from the one declared at conceptual level or may not be captured at all (Mustière and Devogele 2008).

The term *collapsed surface* is used in this paper for denoting a geographic object whose spatial extent is defined at conceptual level of surface type, but due to the chosen acquisition threshold is captured as a curve or point feature. The collapse of a surface can be categorized into two types: complete collapse and partial collapse (Su et al. 1998). The *complete collapse* regards the transformation of a surface into a curve, when it is a long but thin feature, or else into a point, when it is smaller than the given threshold. On the other hand, a *partial collapse* occurs when a part or some parts of the surface are collapsed, while the others are acquired as surfaces.

The amount of geographic objects in a class whose spatial extent is collapsed, or the number of collapsed parts in case of a partial collapse, varies depending on the threshold chosen for the data acquisition process. Therefore, the collapse of a surface is a phenomenon which mainly dependents on the acquisition process and is independent from the conceptual design of the reality. Given these considerations, this paper treats the problem of collapsed surfaces in a transparent and automatic manner with respect to the conceptual schema. In particular, no 'ad hoc' spatial attributes are added to the classes, in order to represent geometries. Each class maintains the spatial attribute originally defined at conceptual level and the possible collapsed geometries are treated into the implementation level. This choice increases the interoperability in an open and distributed environment, such as in a Spatial Data Infrastructure (SDI), where the various involved organizations share the same conceptual schema, but can acquire the geographic objects of interest using different techniques and threshold acquisition parameters.

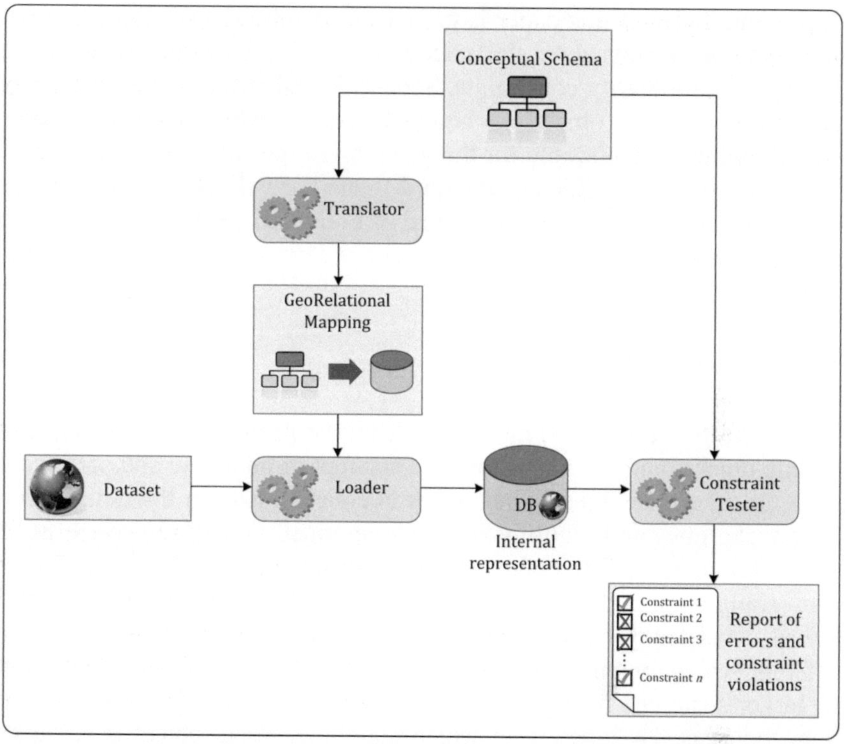

Fig. 1. Architecture of the methodology presented in (Belussi et al. 2009) for the validation of spatial integrity constraints defined at conceptual level

This paper proposes a methodology for automatically treating collapsed surfaces at implementation level, without changing the conceptual specification of a spatial database to handle the collapsed geometries; in particular as regards to the validation of spatial integrity constraints. For treating the validation problem we consider the architecture depicted in Fig. 1, which has been originally proposed by Belussi et al. in (Belussi et al. 2009). In more detail, given a dataset encoded into one allowable format, it is automatically loaded into a geo-relational internal representation, on the basis of the conceptual specification. Then spatial constraints defined at conceptual level are checked on this representation via some SQL queries that return the set of violating objects. Our work extends this methodology abstracting from the use of the GeoUML language and adopting the proposed architecture for treating collapsed surfaces. In this context our approach requires that the implementation level must be able to automatically treat the geometries representing collapsed surfaces as normal surfaces, in particular as regards to the constraint validation activity performed by the Tester module.

The remainder of this paper is organized as follows: Section 2 presents some previous results about the treatment of collapsed surfaces and validation of spatial integrity constraints. Section 3 analyzes how the surface extent of a geographic object can be collapsed and which kinds of collapse transformations are possible for the geometric types defined by the Simple Feature Model (OGC 2006). Section 4 treats the validation of topological and part-whole constraints in presence of collapsed surfaces. Finally, Section 5 summarizes the results of this paper and discusses future work.

2 Related Work

Su et al. in (Su et al. 1998) discuss in detail the problem of collapsing geographic objects and they introduce the concept of complete and partial collapse for a surface object. Their work is concentrated on the definition of some techniques and algorithms for automatically collapse surface in the raster format, in order to obtain a multi-scale database.

In (Kang et al. 2004) the authors treat the problem of verifying the topological consistency between two datasets that represent the same area at different scales, in particular where the small-scale dataset is derived from the larger-scale one by a generalization method. More specifically, the authors focus on the situation in which a 2-dimensional object is collapsed into a 1-dimensional one. They consider the 9 Intersection Model (9IM) (Egenhofer et al 1991) and define some strategies to convert the 8 possible topological relations between two polygons (called P-P relations) into the 19 possible topological relations between a polygon and a curve (called P-L relations). These strategies are based on the comparison between the possible 9IM matrices for the P-P relations and the P-L relations, and the definition of a measure of distance between them. The primary goal of the authors is to define some rules for guiding the collapse operation in order to preserve the topological relations that are valid in the source database. On the contrary, this paper analyzes the problem of validating spatial integrity constraints (not only topological but also part-whole ones) when a collapse operation has already been performed in the acquisition process.

In (Belussi et al. 2009) the authors discuss the importance of automatically validate spatial integrity constraints defined at conceptual level. They propose a methodology for automatically load a dataset, given in any desired format, into a common internal geo-relational representation on the basis of the conceptual schema. Given that representation, a set of SQL queries are automatically generated on the basis of the spatial constraints defined into the conceptual schema and some predefined SQL templates.

Those queries are executed on the internal representation in order to discover the violating objects.

3 Possible Surface Collapse Transformations

This section analyzes the possible collapse transformations that can be applied to a geometric object of surface type, with reference to the geometric types defined in the Simple Feature Model Standard (OGC 2006). Let us notice that our treatment refers to geometric objects in the 2D space.

The possible surface types defined in the Standard are: Surface and MultiSurface. A *Surface* is a 2-dimensional geometric object consisting of a single "patch" that is associated with one exterior boundary and zero or more interior boundaries. The boundary of a surface is the set of closed curves corresponding to its exterior and interior boundaries. A *MultiSurface* is a 2-dimensional geometry collection whose elements are surfaces, moreover it undergoes to the following constraints: (1) all elements in the collection use the same coordinate reference system and (2) the geometric interiors of any two surfaces in the collection do not intersect in the full coordinate system. The boundaries of any two surfaces in the collection may intersect, at most, at a finite number of points.

The Standard defines two types for linear objects: Curve and MultiCurve. A *Curve* is a 1-dimensional geometric object usually stored as a sequence of points. The boundary of a curve is given by its two endpoints. A *MultiCurve* is a 1-dimensional geometry collection whose elements are curves. The boundary of a MultiCurve is obtained by applying the *mod 2* union rule: a point is in the boundary of a MultiCurve if it is in the boundaries of an odd number of elements of the MultiCurve. In particular, this paper considers only MultiCurve without self overlapping, namely the curves in a MultiCurve can overlap only at a finite number of points.

Finally, there are two types of point objects: Point and MultiPoint. A *Point* is a 0-dimensional geometric object and represents a single location in the coordinate space. The boundary of a point is the empty set. A *MultiPoint* is a 0-dimensional geometry collection whose elements are points. The boundary of a MultiPoint is the empty set.

This paper deals with the most general case of collapse: the partial one, where some parts of the surface are collapsed into curves or points, while the others are represented as surfaces. Therefore, the collapse of a surface g produces a MultiSurface, denoted as *surface,* a MultiCurve, denoted as *curve* and a MultiPoint, denoted as *point*. In particular, the *surface* component represents the non-collapsed parts, while the *curve* and the *point* are composed of the parts that have been collapsed into curves and points respectively. Some admitted collapse situations are depicted in Fig.2. In more detail, the components generated during a collapse operation have to satisfy some properties, as formalized in the following.

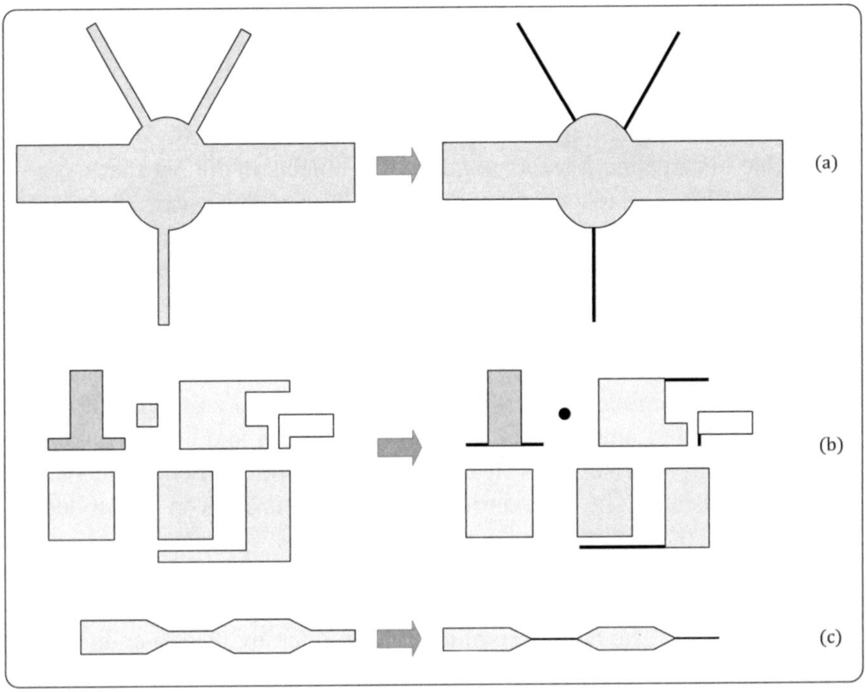

Fig. 2. Some examples of allowable collapses. In situation (a) there are some roads that converge into a roundabout, while the width of the main road is above the chosen threshold, the width of the secondary roads is smaller than the threshold and they are collapsed into curves. Situation (b) presents a set of buildings in which some components have a width smaller than the threshold and therefore are acquired as curves, moreover a building is too small and is represented as a point. Finally, in situation (c) there is a unique road whose width changes along the path, where this width is smaller than the threshold the corresponding road portion is collapsed into a curve

Property 1 (*Collapse of a Surface*)
The collapse of a Surface g has to satisfy the following conditions:

1. The components cannot be all null at the same time:
   g.point is not null or g.curve is not null or *g.surface is not null*
2. The geometries of the *g.surface* and *g.curve* components have to be weakly connected, that is they have to be connected and the connection is satisfied even if it happens on the boundary.
3. Between the *g.surface* and *g.curve* components have to exist a touch relation, that is these two components have not to overlap themselves:
   g.surface touches g.curve

In Fig. 3 some cases of collapsed surface that violate the above constraints are presented. In particular, in (a) only the constraint 2) is violated, while in (b) only the constraint 3 is.

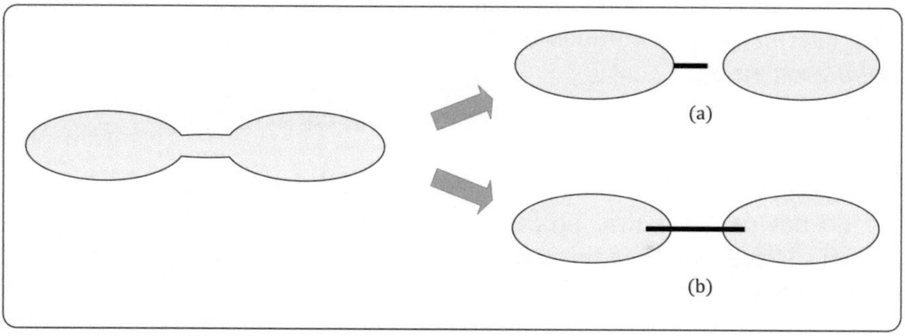

Fig. 3. Some examples of not admitted collapses

Property 2 (*Collapse of a MultiSurface*)
A MultiSurface *g* can partially degenerate to a set of points, i.e. some of its components can degenerate to points while others can degenerate to curves or not degenerate. Moreover, the connection between the components is not required. The only condition that a collapsed MultiSurface *g* has to avoid is the overlapping among the components:

- Between the *g.surface*, *g.curve* and *g.point* components of a collapsed MultiSurface *g* have to exists the following relations:
  g.surface (touches or disjoint) g.curve
  g.surface disjoint g.point
  g.curve disjoint g.point

The notion of boundary and interior for a collapsed surface is given below. These definitions highlight the fact that the collapse operation has produced an overlap between the boundary and the interior of a collapsed surface *g*, due to the loose of accuracy. In particular, the geometries of the *curve* and *point* components are included both in the interior and boundary of *g*. Moreover, the boundary and interior of a collapsed surface *g* is a geometry collection.

Definition 1 (*Boundary of a collapsed surface*)
Given a Surface (or MultiSurface) *g* whose extent is collapsed into a MultiSurface *surface*, a MultiCurve *curve* and a MultiPoint *point* components. The boundary of *g* is composed of the *surface* boundary and the *curve* and *point* geometries:

$$g.boundary = g.surface.boundary \cup g.curve \cup g.point$$

Definition 2 (*Interior of a collapsed surface*)
Given a Surface (or MultiSurface) *g* whose extent is collapsed into a MultiSurface *surface*, a MultiCurve *curve* and a MultiPoint *point* components. The interior of *g* is composed of the *surface* interior and the *curve* and *point* geometries:

$$g.interior = g.surface.interior \cup g.curve \cup g.point$$

We do not introduce other properties of a collapsed surface, for example we do not define metric properties (like the area or perimeter of a collapsed surface), since in this paper we focus on topological and part-whole constraints validation.

4 Validation of Spatial Integrity Constraints in Presence of Collapsed Surfaces.

The validation of spatial integrity constraints defined at conceptual level is an important activity, both for checking the quality of a given dataset and for monitoring the consistency of information stored in a spatial database. The presence of collapsed surfaces makes more difficult the validation activity, because constraints defined on surface objects have to be valid even if the geometry actually stored is of another type. Therefore, the defined constraints have to be relaxed, reflecting the loose of accuracy occurred during the acquisition process.

The set of spatial integrity constraints considered in this paper can be classified into two categories: topological and part-whole constraints, as proposed also in (Price et al. 2001) and (Belussi et al. 2009). In particular, a topological constraint requires that a certain topological relation exists among some given objects, while a part-whole constraint requires that the extent of a certain object is composed of the extents of some other given objects.

Section 4.1 formally defines the set of topological relations considered in this paper and analyzes how these relations have to be relaxed in presence of collapsed surfaces. Section 4.2 deals with part-whole constraints by formally defining their meaning and discussing the possible relaxations.

4.1 Topological Integrity Constraints

The topological relationships analyzed in this paper are those defined in (Clementini et al. 1993) and also reported in (OGC 2006), that is: *Equals*, *Disjoint*, *Touches*, *Within*, *Overlaps*, *Crosses*, *Intersects* and *Contains*.

The formal meaning of the considered topological relations is given in Table 1, where `f` denotes the global extends of the object *f* (interior plus boundary) in the common point set representation, `f°` denotes the interior of the object *f*, while `dim(f)` denotes the dimension of the object *f*: 0 for points, 1 for curves and 2 for polygons. Finally, $\cup$ ($\cap$) is the common union (intersection) operator defined on the point set representation of *f* and *g*.

The rewriting rules presented in the following section extends the definition of these relations to collapsed surfaces. In particular, they consider the relations presented in Table 1 and apply them to the relaxed notion of boundary and interior of a collapsed surface. However, two cases have to be distinguished: the case in which only one of the involved objects is collapsed, treated in the first subsection, and the case in which both involved objects are collapsed, treated in second subsection. In both cases, the rewriting process has to deal with the situation in which some of the components of a collapsed surface *g* do not exist (i.e. are *null*). A missing component has not to influence the evaluation of the topological relation, unless it appears in a condition like "*g.component=null*". In particular, any atomic condition *c* that involves a missing component is replaced with *false*. Finally, if the missing component appears into a union or intersection operation, then it is replaced with an empty set. We also need to introduce a specialization of the *Crosses*(*Overlaps*) relation, called $Crosses_{through}$($Overlaps_{through}$) that can be applied only to two curves c_1 and c_2, where we require that at least one intersection point between c_1 and c_2 is a pass through point and not a pure tangency point. Formally, this is obtained by considering the sign of the cross products calculated between the two pairs of vectors that describe how the curves (first vectors pair) meet and how they leave each other (second vectors pair).

Rewriting of Topological Relation with One Collapsed Object

Given the topological relations presented above, this section analyzes how a relation between two objects *f* and *g* has to be rewritten when *g* is a collapsed surface, while *f* can be an object of any geometric type. A subscript *S*, *C* or *P* is added to *f* in order to specify its geometric type: surface, curve or point respectively. If more than one subscript is added to *f*, then it can be of any of these types. Moreover, a *c* subscript is added to the relation name *r*, in order to highlight that it is the relaxed form of the relation *r*.

Table 1. Formal meaning of topological relations presented in (OGC 2006)

Topological relation	Semantics	Abbreviation
f Disjoint g	$f \cap g = \emptyset$	DJ
f Touches g	$(f^\circ \cap g^\circ = \emptyset) \wedge (f \cap g \neq \emptyset)$	TC
f Within g	$(f^\circ \cap g^\circ \neq \emptyset) \wedge (f \cap g = f) \wedge (f \cap g \neq g)$	IN
f Contains g	g Within f	CT
f Overlap g	$(f^\circ \cap g^\circ \neq \emptyset) \wedge (f \cap g \neq f) \wedge (f \cap g \neq g) \wedge \dim(f^\circ \cap g^\circ) = \max(\dim(f), \dim(g))$	OV
f Crosses g	$(f^\circ \cap g^\circ \neq \emptyset) \wedge (f \cap g \neq f) \wedge (f \cap g \neq g) \wedge \dim(f^\circ \cap g^\circ) < \max(\dim(f), \dim(g))$	CR
f Intersects g	not f Disjoint g	INT
f Equals g	$f = g$	EQ

Disjoint

The disjoint relation requires that the intersection between two objects *f* and *g* is the empty set. Therefore, a collapsed surface *g* is disjoint from an object *f*, if each of its components has an empty intersection with *f*, regardless the type of *f*.

```
f_SCP DJ_C g ⇔
  (f_SCP DJ g.surface or g.surface=null) and
  (f_SCP DJ g.curve or g.curve=null) and
  (f_SCP DJ g.point or g.point=null)
```

Touch

The touch relation requires that two object *f* and *g* intersect but not on their interiors. Therefore, if *f* is a curve or a point, it touches *g* only if *f* touches the surface component of *g* or it is disjoint from *g.surface* and intersects one of the other components. We also exclude the case in which *f* passes through *g.curve*.

```
f_C TC_C g ⇔
  not(f_C CR_through|OV_through g.curve) and
  (f_C TC g.surface or
   ((f_C DJ g.surface or g.surface=null) and
    (f_C INT g.curve or f_C INT g.point)))

f_P TC_C g ⇔
  f_P TC g.surface or
  ((f_P DJ g.surface or g.surface=null) and
   (f_P INT g.curve or f_P INT g.point))
```

On the other hand, if *f* is a surface, it touches *g* only if it touches one or more components of *g* and is disjoint from the others.

```
f_S TC_C g ⇔
  (f_S TC g.surface and
    (f_S TC|DJ g.curve or g.curve=null) and
    (f_S TC|DJ g.point or g.point=null)) or
  ((f_S DJ g.surface or g.surface=null) and
    (f_S TC g.curve or f_S TC g.point))
```

Within

Three cases can be distinguished, depending on the type of *f*. If *f* is a surface, then it has to be contained into *g.surface.*

```
f_S IN_C g ⇔
  not(g.surface=null) AND f_S IN g.surface
```

If *f* is a curve then it can be contained into the surface or curve component of *g*, or into the union of them.

```
fC INC g ⇔
  (not(g.surface=null) and
   ((fC∩g.surface)∪(fC∩g.curve)) EQ fC and
    not(fC IN|EQ g.surface.boundary) or
  (g.surface=null and not(g.curve=null) and
   fC IN|EQ g.curve
```

Finally, if *f* is a point then it can be contained into one of the components of *g*, or into the union of them if *f* is a point aggregate.

```
fP INC g ⇔
  (not(g.surface=null) and not(g.curve=null) and
   ((fP∩g.surface)∪(fP∩g.curve)∪(fP∩g.point)) EQ fP
   and not(fP IN g.surface.boundary)) or
  (g.surface=null and g.curve=null and
   fP IN|EQ g.point)
```

Contains

A surface *f* contains a collapsed surface *g*, if each component of *g* (*g.surface*, *g.curve* and *g.point*) is contained in *f*. The conditions on *fS.boundary* are necessary to exclude some cases that already satisfy the touch relation. If *f* is a curve or a point then *g* cannot be contained in *f* since *g* is a *surface*, thus the relation is false.

```
fS CTC g ⇔
  (g.curve=null or fS∩g.curve EQ g.curve) and
  (g.point=null or fS∩g.point EQ g.point) and
  (fS CT g.surface or
   (g.surface=null and
    ((g.curve≠null and not(g.curve IN|EQ fS.boundary))
     or
     (g.point≠null and not(g.point IN fS.boundary)))))

fCP CTC g ⇔ false
```

Overlap

The overlap relation requires that the interior of the two objects *f* and *g* intersect and that the dimension of the intersection is equals to the maximum dimension between the involved objects. Therefore, *f* can be only a surface and have to overlap at least one component of *g*, otherwise it can contain one or two components of *g* but it has to be disjoint or contain only partially at least one of the others.

```
fS OVC g ⇔
  fS OV g.surface or
  (fS CT g.surface and
   (not(fS∩g.curve EQ g.curve) and g.curve≠null) or
    not(fS∩g.point EQ g.point) and g.point≠null)) or
  (fS DJ|TC g.surface and
   (g.curve IN|OV fS or g.point IN|OV fS)) or
  (g.surface=null and
   (g.curve OV fS or g.point OV fS or
     (g.curve DJ|TC fS and g.point IN fS) or
     (g.point DJ|TC fS and g.curve IN fS)))

fCP OVC g ⇔ false
```

Crosses

The cross relation is defined only between curves (or points) and surfaces and it is equivalent to overlap with the only difference that the dimension of the intersection has to be smaller than the maximum dimension of the involved objects. Therefore, it can be relaxed as follows.

```
fC CRC g ⇔
  (((fC∩g.surface)∪(fC∩g.curve)) IN fC or
   fC INT g.point) and
  (fC TC g.surface ⇒ (fC INT g.curve or
                      fC INT g.point))

fP CRC g ⇔
  (((fP∩g.surface)∪(fP∩g.curve)∪(fP∩g.point)) IN fP)
   and
  (fP TC g.surface ⇒ (fP INT g.curve or
                      fP INT g.point))

fS CRC g ⇔ false
```

Intersects

By definition two objects intersect if they are not disjoint, it follows that this relation is satisfied only if the negated of the disjoint condition is satisfied.

Equals

An equal condition can be satisfied only if *g* is compared with another surface object, thus if *f* is not a surface equal is false.

```
fS EQC g ⇔ g.curve is null and g.point is null and
             fS EQ g.surface

fCP EQC g ⇔ false
```

Rewriting of Topological Relation with Both Collapsed Objects

Let us now consider the case in which both *f* and *g* are collapsed surfaces. In this case, first *f* is transformed into its collapsed components: *f.surface*, *f.curve* and *f.point*, then the definition of the relation is given using the above defined relaxed topological relation r_c evaluated between the components of *f* and *g*. A *cc* subscript is added to the relation name *r*, in order to highlight that it is the relaxed form of *r*, when both *f* and *g* are collapsed surfaces.

Disjoint

In presence of a disjoint relation, each component of the collapsed surface *f* has to be disjoint from the components of *g*. Therefore, the disjoint relation $\mathtt{DJ}_C$ is checked between each component of *f* and *g*.

```
f DJCC g ⇔
    (f.surface=null or f.surface DJC g) and
    (f.curve=null or f.curve DJC g) and
    (f.point=null or f.point DJC g)
```

Touches

A collapsed surface *f* touches a collapsed surface *g* if at least one of *f* components touches *g* with respect to the relation $\mathtt{TC}_C$.

```
f TCCC g ⇔
   (f.surface TCC g and
    (f.curve=null or f.curve TCC|DJC g) and
    (f.point=null or f.point TCC|DJC g)) or
   ((f.surface=null or f.surface DJC g) and
    (f.curve TCC g or f.point TCC g))
```

Within

The within relation is satisfied if each components of *f* is contained into the collapsed surface *g*, with respect to the relation $\mathtt{IN}_C$.

```
f INCC g ⇔
    (f.surface=null or f.surface INC g) and
    (f.curve=null or f.curve INC g) and
    (f.point=null or f.point INC g)
```

Contains

A collapsed surface *f* contains a collapsed surface *g* if *g* is within *f*, therefore it is enough to swap *f* with *g* and apply the $\texttt{IN}_{CC}$ relation.

```
f CTcc g ⇔ g INcc f
```

Overlap

Two collapsed surfaces *f* and *g* are in overlap if at least one component of f overlaps (or crosses) g with respect to the $\texttt{OV}_{C}$ relation defined above or vice versa.

```
f OVcc g ⇔
  f.surface OVc g or f.curve CRc g or f.point CRc g or
  g.surface OVc f or g.curve CRc f or g.point CRc f
```

Crosses

Since the crosses relation is defined only between curves (or points) and surfaces it is always false when evaluated on a pair of collapsed surfaces.

```
f CRcc g ⇔ false
```

Intersects

By definition two objects intersect if they are not disjoint, therefore this relation is valid only if the negated of the disjoint condition is valid.

```
f INTcc g ⇔ not(f DJcc g)
```

Equals

Two collapsed surfaces are equal if each of their corresponding components coincides or are both null.

```
f EQcc g ⇔
  ((f.surface=null and g.surface=null) or
      f.surface EQ g.surface) and
  ((f.curve=null and g.curve=null) or
      f.curve EQ g.curve) and
  ((f.point=null and g.point=null) or
      f.point EQ g.point)
```

Validation of Topological Integrity Constraints

A topological integrity constraint between two geographic objects *a* and *b* specifies the topological relation that must exist between *a* and *b*. More specifically, a topological constraint is defined by two aspects: the desired topological relation and the logical structure of the constraint. For example, an existential topological constraint requires that for each instance *a* of the constrained class *A*, there exists an instance *b* of the constraining class *B* such that between the spatial attribute *g* of *a* and the spatial attribute *f* of *b* a particular topological relation *r* (or disjunction of topological relations) is satisfied.

$$\forall\ a \in A\ .\ \exists\ b \in B\ .\ check(g, f, r)$$

The function *check*(*g*,*f*,*r*) returns true if between *g* and *f* the topological relation (or disjunction of topological relations) *r* is valid, false otherwise.

By replacing the existential quantifier with the universal one, a universal topological constraint is obtained, which requires that the topological relation *r* (or disjunction of topological relations) exists between the spatial attribute *g* of the constrained object and the spatial attribute *f* of all the instances of the constraining class *B*.

$$\forall\ a \in A\ .\ \forall b \in B\ .\ check(g, f, r)$$

Given a topological integrity constraint between *f* and *g* and the rewriting rules defined above, the relaxed version of the constraint which considers the presence of collapsed surfaces has to take care of the four possible cases:

1. Neither *g* nor *f* have been collapsed.
2. Only *g* has been collapsed.
3. Only *f* has been collapsed.
4. Both *g* and *f* have been collapsed.

<u>Example 1</u>

Let us consider the situation depicted in Fig. 4, containing some roads and buildings. Suppose now to represents at conceptual level the roads objects with a class *Road* characterized by a MultiSurface spatial attribute, and the buildings with a class *Building* containing a MultiSurface spatial attribute. Then a universal topological constraint is defined between the *Road* and the *Building* class, for imposing that each instance of *Road* is disjoint from each instance of the *Building* class:

$$\forall\ a \in Road\ .\ \forall b \in Building\ .\ check\ (g, f, DJ)$$

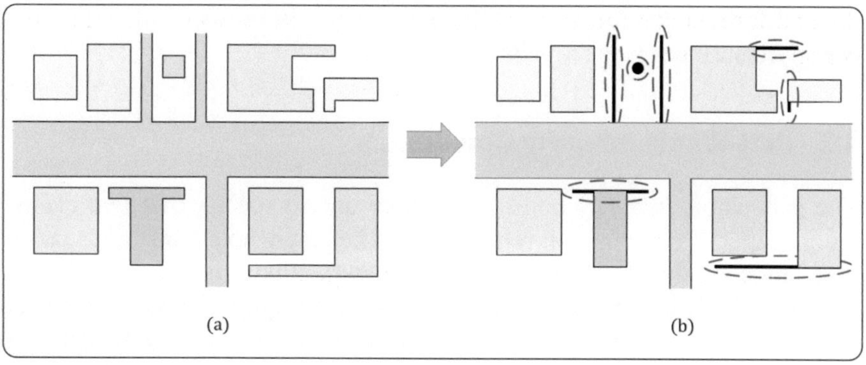

Fig. 4. A situation composed of a main road with some branches and a set of buildings placed on the main road sides. Figure (b) is a representation of figure (a) obtained using a different threshold, such that some roads or building elements are collapsed into points or curves. The collapsed elements are the ones highlighted with a dashed circle around it

However since both classes may have some collapsed instances, the spatial integrity constraint has to be rewritten in the following manner:

```
∀ a ∈ Road . ∀b ∈ Building .
 (g.curve=null and g.point=null and
  f.curve=null and f.point=null and
  check(g.surface, f.surface, DJ))
```

Neither *g* nor *f* have been collapsed

```
 OR
 ((g.curve≠null or g.point≠null) and
  f.curve=null and f.point=null and
  check(g, f.surface, DJc))
```

Only *g* has been collapsed.

```
 OR
 (g.curve=null and g.point=null and
  (f.curve≠null or f.point≠null) and
  check(g.surface, f, DJc))
```

Only *f* has been collapsed.

```
 OR
 ((g.curve≠null or g.point≠null) and
  (f.curve≠null or f.point≠null) and
  check(f, g, DJcc))
```

Both *g* and *f* have been collapsed

Let us notice that when a geometry is not collapsed, then only the corresponding *surface* component is not null, in all the other cases a partial or complete collapse occurred. The term DJ stands for the classical disjoint relation between two surfaces, while the terms DJ_c and DJ_{cc} denotes the relaxed form of the disjoint topological relation in presence of one or two collapsed surfaces respectively.

4.2 Part-Whole Integrity Constraints

The part-whole integrity constraints allow one to specify that two classes are in a spatial composition relationship. The part-whole constraints considered here are the one defined in (Belussi et al. 2006), that is: *composition*, *membership* and *partition*. All these constraints require that given a spatial attribute f of an instance of the composite class B, there exist some instances of the component class A whose spatial attributes g together compose f.

More specifically, the *composition* constraint requires that for each instance of the constrained class B, its spatial attribute f must be equal to the union of the spatial attribute g of one or more instances of the constraining class A. The *membership* constraint is similar to an *IN* topological constraint between the spatial attribute g of the constrained class A and the spatial attribute f of the constraining class B, but a disjunction or touch relation among the components of the same whole is also required. Finally, the *partition* relation combines the strong composition constraint with the membership one, since it requires that the union of the spatial attribute g of the A instances is equal to the spatial attribute f of the B instance and that the spatial attribute g of the A instances that compose the same f do not overlap.

The presence of collapsed surfaces into a part-whole constraint can be treated in different way, depending on how many components have been collapsed and if the whole object is also collapsed or not, four cases can be distinguished:

1. All components have been collapsed and the whole has been collapsed: the constraint can be checked on the collapsed geometries, without any particular extension.
2. All components have been collapsed but the whole has not been collapsed: the composition constraint cannot be satisfied; the validator tool can return a warning message or not consider the constraint at all. The validator is able to distinguish this case from the one in which there is no component for the whole.
3. Only some components have been collapsed and the whole has not been collapsed: the constraint can be checked considering only the non-collapsed components. In particular, the constraint is valid only if the non-collapsed components spread out to fill the extent originally occupied by the collapsed components.

4. Only some components have been collapsed and also the whole has been collapsed: this situation is equivalent to the second case.

<u>Example 2</u>

Let us consider the area roads which are composed of some vehicular areas and sidewalks. In some cases the width of the sidewalks is smaller than the chosen acquisition threshold, therefore they are not acquired at all or they are acquired as curves, while the extent of the vehicular areas is spread out to occupy all the extent of the area road, as illustrated in Fig. 5. This example corresponds to the case 3 presented above and the constraint can be checked considering only the non-collapsed components.

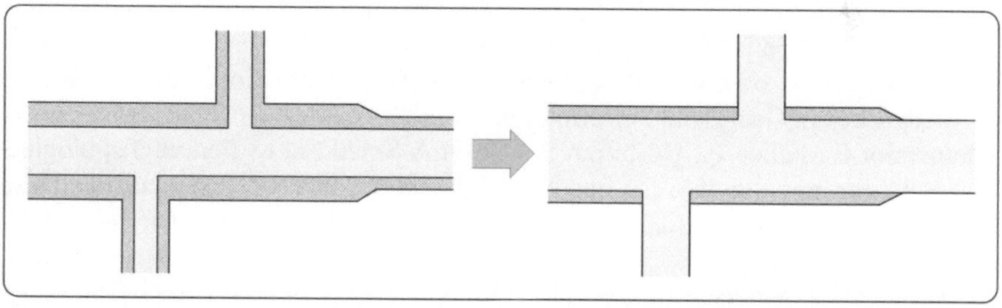

Fig. 5. An area road composed of a vehicular area, represented by the light grey surfaces, with a sidewalk on each side, represented in dark grey. When the width of the sidewalk is smaller than the chosen acquisition threshold, it is captured as a curve and the vehicular area spread out to occupy the extent originally occupied by the collapsed sidewalk

5 Conclusion and Future Work

This paper proposes a methodology for automatically treating collapsed surfaces into the implementation level, without modifying the conceptual specification of a spatial database. The designer has not to worry about what and when geographic objects can be collapsed, the underlying architecture is able to automatically treat collapsed surfaces as normal surfaces, in particular as regards the validation of spatial integrity constraints. In particular, we analyze how topological and part-whole constraints have to be rewritten in presence of collapsed surfaces. Moreover, the presented validation tool is able to recognize when a surface has been collapsed and to apply a relaxed form of the constraint, when it is necessary.

As future work we are studying the integration of the proposed approach for managing collapsed surfaces in a multi accuracy spatial database. The aim is to build a system which is able to integrate spatial data coming from different sources with different accuracy levels and with or without collapsed geometries.

References

Belussi A., Negri M., Pelagatti G. (2006) Modelling Spatial Whole-Part Relationships Using an ISO-TC211 Conformant Approach. Information and Software Technology, vol. 48, 2006, pp. 1095-1103.

Belussi A., Migliorini S., Pelagatti G. and Negri M. (2009) From the Conceptual Design of Spatial Constraints to their Implementation in Real Systems. In Prooceedings of the 17th ACM SIGSPATIAL International Conference on Advances in Geographic Information.

Clementini E., Felice P., Oosterom P. (1993) A Small Set of Formal Topological Relationships Suitable for End-User Interaction. In Proceedings of the Third International Symposium on Advances in Spatial Databases (SSD 1993), pp.277-295

Egenhofer M.J. and Herring J.R. (1991) Categorizing Binary Topological Relationships Between Regions, Curves, and Points in Geographic Databases. Tech. Rep., Dep. of Surveying Engineering, University of Maine, Orono.

Kang H.K., Kim T.W., Li K.J. (2004) Topological Consistency for Collapse Operation in Multi-scale Databases. In Proceedings of the First International Workshop on Conceptual Modeling for Geographic Information Systems (CoMoGIS - ER Workshops 2004), pp.91-102

Mustière S. and Devogele T. (2008) Matching Networks with Different Levels of Detail. Geoinformatica (2008) 12, pp. 435–453.

OGC (2006) OpenGIS® Implementation Specification for Geographic Information - Simple Feature Access - Part 1: Common Architecture. Open GIS Consortium, Inc. Project Document OGC 05-126.

OGC (2006) OpenGIS® Implementation Specification for Geographic Information - Simple Feature Access - Part 2: SQL option. Open GIS Consortium, Inc. Project Document OGC 06-104r3.

Price R., Tryfona N., Jensen C.S (2001) Modeling Topological Constraints in Spatial Part-Whole Relationships. In Proceedings of 20th International Conference on Conceptual Modeling (ER 2001) pp. 27-40

Standard ISO 19109:2005, Geographic information: Rules for application schema, http://www.iso.org/iso/catalogue_detail.htm?csnumber=39891

Su B., Li Z. and Lodwick G. (1998) Morphological Models for the Collapse of Area Features in Digital Map Generalization. GeoInformatica (1998) 2:4, pp. 359-382.

A Spatio-Temporal Model Towards Ad-Hoc Collaborative Decision-Making

Martin Raubal[1], Stephan Winter[2]

[1] Department of Geography, University of California, Santa Barbara, USA
raubal@geog.ucsb.edu
[2] Department of Geomatics, The University of Melbourne, Australia
winter@unimelb.edu.au

Abstract. For an autonomous agent, performing a task in a spatio-temporal environment often requires interaction with other agents. Such interaction can be initiated by ad-hoc collaborative planning and decision-making, which then leads to physical support on site. On-site collaboration is important for a variety of operations, such as search-and-rescue or pick-up-and-delivery. Tasks are performed through sequences of actions, and agents perceive possibilities for these actions in terms of affordances from the environment. Agent collaboration therefore requires the communication of affordances between agents with different capabilities. This paper introduces a spatio-temporal model for the decentralized decision-making of autonomous agents regarding on-site collaboration. Based on Janelle's time-geographic perspective on communication modes, we demonstrate that different task situations lead to different spatio-temporal constraints on communication, involving both physical presence and telepresence. The application of such constraints leads to an optimized message distribution strategy and therefore efficient affordance communication with regard to maximizing support in performing a given task.

1 Introduction

Performing a task in a spatio-temporal environment often involves collaboration between various agents. Examples include rescue teams in emergency response, intelligent robots supporting humans in elderly care, or ride-sharing with clients seeking hosts for transportation. In order to identify potential hel-

M. Painho et al., (eds.), *Geospatial Thinking*, Lecture Notes in Geoinformation and Cartography, DOI 10.1007/978-3-642-12326-9_15, © Springer-Verlag Berlin Heidelberg 2010

pers, agents need to communicate their tasks and needs to others. Such communication and subsequent decision-making involves at least three different perspectives: The *spatio-temporal* view includes issues such as the distribution of helping agents in space, the urgency of solving a task, and the task location. The *social* point-of-view deals with the willingness of other agents to cooperate, and also involves institutional and legal constraints regarding agent cooperation. The *technical* level tackles issues related to the communication infrastructure. Within the framework of peer-to-peer communication, these perspectives can be viewed as different network levels. Previous research in the area of agent collaboration has focused on the social and technical levels. In this paper we propose a *spatio-temporal model towards collaborative decision-making*, which integrates all three network levels. We hypothesize that the integration of spatio-temporal constraints within a model of agent collaboration makes an optimized message distribution possible and therefore results in efficient communication leading to optimal support in performing a task.

The presented model is based on the communication of affordances between agents in a network of peers. The use of affordances allows us to consider action possibilities, which can be formally represented in a functional framework. The model accounts for the fact that different task situations—with respect to urgency, risk, and location—result in different time-geographic communication constraints. Negotiation between client agents and potential helping agents is represented within a *request-offer-choice* process, which includes affordance-based similarity measurement that takes agent capabilities into account.

Section 2 presents related work in the area of agent planning, peer-to-peer communication, and agent collaboration. Section 3 introduces the individual components and theories, on which the model is founded. In Section 4, we develop the collaborative agent process model, detailing the negotiation process. Section 5 applies this process model to a hypothetical emergency scenario. Section 6 discusses the application. The final section presents conclusions and directions for future research.

2 Related work

2.1 Agent planning

According to the heterogeneity of the involved fields there is no common agreement about a definition of the term *agent* [1]. An agent can be anything, such as a robot that perceives its environment through sensors and acts upon it through effectors [2]. More specifically, agents are considered computer systems that are situated in some environment and can act autonomously [3]. *Multi-agent systems* (MAS) depict systems as a combination of multiple autonomous and independent agents and are therefore well suited to simulate collaboration of different actors.

Agents can be represented as functions that map percepts to actions. Abstract models of agents distinguish between purely reactive agents, agents with subsystems for perception and action, and agents with state. These abstract models can be implemented in different ways, depending on how the decision-making of the agent is realized. Here, we consider *utility-based agents* [2], which have valuation functions that allow them to compare between different action sequences to achieve a goal. Such functions map world states to real numbers, which describe associated degrees of happiness.

Agent-based modeling and simulation has gained much popularity in the field of Geographic Information Science due to the disaggregate nature of agents and their ability to move across different spatial scales and representations [4]. Application scenarios include the modeling of urban phenomena [5], pedestrian movement [6], and shared-ride trip planning [7].

Planning is the development of a strategy for solving a task. For an agent, a plan is an action sequence, where each action to be performed depends on some pre-conditions. Every action causes effects or post-conditions that affect or trigger subsequent actions in the chain. The plan terminates when the goal is reached. A planner in Artificial Intelligence takes three input variables: a representation of the initial state of the world, a representation of the intended outcome (goal), and a set of possible actions to be performed to reach the goal. Formally, a plan is a triple $<O, I, AC(p, q)>$ [8] where O is the intended outcome, I the initial state of the world, and AC a set of actions—each defined via pre- and post-conditions p, q. After executing actions the state of the world is changed, which impacts the future plan, therefore planning is a non-linear process. One of the main challenges within dynamic environments is that one can neither assume complete knowledge nor the availability of objects and other agents supporting certain actions.

2.2 Peer-to-peer communication

Peer-to-peer (P2P) communication is ad-hoc communication between distributed agents, without involvement of a dedicated server providing communication services to its clients, or any other hierarchic communication infrastructure. It enables mobile agents to collaborate in an ad-hoc manner provided that they agree on a communication protocol. In a P2P communication network each node is of equal importance. Nodes can take the role of a communication client, receiving services from other nodes, but they are also service providers for the other nodes. P2P communication networks are transient in nature, with nodes entering and leaving the network freely and frequently.

A special class of P2P communication is characterized by mobile nodes. For this class the communication is realized wirelessly by radio, which is short-range, due to the typically limited on-board energy resources. This means connectivity in mobile networks depends on the physical distance of nodes, which is constantly changing. Communication over larger distances relies on message forwarding and routing [9, 10].

2.3 Agent collaboration

In a MAS, autonomous agents may cooperate with others to achieve their goals. How to achieve meaningful coordination is a difficult issue and requires interdisciplinary work within the recently established field of *computational cognitive social science* [11]. Agent collaboration is based on cognitive architectures, representing and explaining cognitive processes of individual agents during the performance of tasks. The most prominent architectures are ACT-R [12] and Soar [13]. The CLARION cognitive architecture [14] extends cognitive modeling to social simulation. It accounts for agents' socially oriented goals and bases agent cooperation on the fact that social interaction between agents is made possible through their understanding of each other's motivations.

Previous research has focused on technical aspects of P2P collaboration [15] and hierarchical multi-agent models integrating knowledge-based communication, such as for the RoboCupRescue[1] simulation [16]. Luo and Bölöni [17] presented a game-theoretic model for the canonical problem of spatio-temporal collaboration with the goal of optimizing individual benefits. A typical application of intelligent agent collaboration is elderly care. For example, the PEIS (Physically Embedded Intelligent Systems) ecology is a network of heterogeneous smart devices that ranges from simple gadgets, such as refrigerators with sensors, to sophisticated mobile robots or even humans [18]. These intelligent entities communicate and collaborate with each other by providing information, and combining physical and virtual functionalities to perform complex tasks, such as supporting human inhabitants in their flats. Shared-ride trip planning is an urban transportation application of ad-hoc agent collaboration: client agents representing customers, e.g., pedestrians, seek transportation by host agents representing vehicles, e.g., public transportation vehicles, in an ad-hoc manner [7, 19]. The goal of collaboration is to bring clients to their destinations, for some host benefits. In typical scenarios multiple clients compete for free capacity and multiple hosts compete for clients.

[1] http://www.robocup.org/

Recently, it was demonstrated how decentralized time geography can be applied to ad-hoc collaborative agent planning [20]. Agents performed a spatio-temporal analysis based on local knowledge in a distributed environment, thereby evaluating whether they can independently contribute to physical support at a specific site by a specified time. Experiments in a multi-agent simulation framework investigated the impact of different combinations of agent density and communication radius, as well as behavioral strategies on task performance.

This paper takes previous work a step further by explicitly considering the interaction of mobile agents in terms of a combination of *spatio-temporal*, *social*, and *communication network* levels. The goal is to develop a general and comprehensive model of spatio-temporal decision-making for ad-hoc agent collaboration. We specifically investigate the cooperation process, which includes communication of affordances between agents. Such communication enables clients and service providers to form better decisions by taking the individual capabilities of agents into account. The model also provides a way to determine the influence of task dimensions and spatio-temporal constraints on the efficiency of communication.

3 Spatio-temporal framework of collaborative decision-making

This section specifies different task dimensions, whose values lead to spatio-temporal constraints for communication. We start by introducing affordances, which are communicated by the collaborating agents, and a recently proposed similarity measure for them. Time geography serves as the basis for modeling the spatio-temporal decision-making process.

3.1 Affordance representation and similarity

The theory of affordances [21] is based on the tenet that agent and environment form an inseparable pair. Affordances have to be described relative to the agent. For example, a chair's affordance 'to sit' results from a bundle of attributes, such as 'flat and hard surface' and 'height', many of which are relative to the size of an individual agent. Norman [22] recasts affordances as the results from the mental interpretation of things, based on people's past knowledge and experiences, which are applied to the perception of these things.

In order to supplement Gibson's theory of perception with elements of cognition, situational aspects, and social constraints, Raubal [23] presented an extended theory of affordances suggesting that affordances belong to three different realms: *physical*, *social-institutional*, and *mental*. This distinction was formally specified in a functional model [24]. The agent is represented through its physical structure, spatial and cognitive capabilities, and a goal. *Physical affordances (Paff)* for the agent result from invariant compounds—unique combinations of physical, chemical, and geometrical properties—and the physical structure of the agent. *Social-institutional affordances (SIaff)* are created through the imposition of social and institutional constraints on physical affordances. *Mental affordances (Maff)* arise for the agent when perceiving a set of *Paffs* and *SIaffs* in an environment at a specific location and time. Affordances offer possibilities for action as well as possibilities for the agent to reason about them and decide whether to utilize them or not, i.e., mental affordances.

We specify an affordance A as a triple $<O, E, \{AC\}>$ [25]. The outcome O is the change of world state after executing the actions AC with respect to manipulated entities of type E. Each action is represented by physical (ph) and social-institutional (si) constraints or pre-conditions, AC therefore being defined as a set of actions $\{ac_1(ph_1,si_1),..., ac_n(ph_n,si_n)\}$. Constraints are tied to a certain action with respect to an entity, while the outcome is equal for all actions defined for the affordance A. An affordance can be utilized through several actions, e.g., the move-ability affordance of a desk may include the actions *carry* and *push*.

When communicating affordances between agents and evaluating whether an offered affordance is good enough compared to the requested affordance to help in solving a task, it is necessary to determine their similarity. Here, we apply a similarity measure for affordances [26], which uses the action and outcome specification from $<O, E, \{AC\}>$. Affordances are more similar the more similar their descriptors are. The overall similarity Sim_A between affordances A_s and A_t is defined as the weighted sum for the individual similarities computed for actions (sim_{AC}) and outcomes (sim_O) (Equation 1). The former depend on the similarity values computed for their physical and social-institutional constraints.

$$Sim_A(A_s,A_t) = \omega_{ac} * {}^1/_n \Sigma\, sim_{AC} + \omega_o * sim_O;\ where\ \Sigma\omega = 1 \quad \textbf{(1)}$$

3.2 Task dimensions

Agents can perform tasks by utilizing various affordances. For example, to change a light bulb, an agent can move a chair below the light, step onto it, and change the bulb [27]. When planning how to solve a task, the agent must take several aspects into account. For the purpose of modeling spatio-temporal collaborative decision-making, we consider the following task dimensions: *collaboration*, *urgency*, *risk*, and *location*. Table 1 describes and explains the possible values for each of them.

Table 1. Task dimensions with possible values and explanations.

Task dimension	Value	Explanation
Collaboration	0	Agent can solve task alone.
	1	Another agent needed to either solve task or support requesting agent in (sub)task.
	n	More than 1 agent needed to solve task.
Urgency	immediate	Immediate help required, values depend on context, e.g., within 10 min.
	flexible	Help required within reasonable time frame.
Risk	high	Describes the inverse probability that the agent(s) will solve the task.
	medium	
	low	
Location	(x, y)	Coordinate pair; default location for the task is the current location of the requesting agent.

3.3 Spatio-temporal communication constraints

Agents and resources are available at a limited number of locations for a limited amount of time. Time geography defines the space-time mechanics of locational presence by considering different constraints [28]. The original time geography framework recognized the possibility of telepresence using electronic communication, although it received much less attention than physical presence. Time geography's focus on time as a resource enabling activity participation has received explicit interest by researchers lately [29, 30]. It fits naturally to views of time as the major scarce resource in information economies and accelerated modern lifestyles [31].

Janelle [32] classified communication modes from a time-geographic perspective. Table 2 summarizes classes based on their spatial and temporal constraints. Spatial constraints are either physical presence or telepresence, while temporal constraints are either synchronous or asynchronous. *Synchronous presence* (SP) is the communication mode of face-to-face (F2F) interaction. F2F requires coincidence in both time and space. *Synchronous telepresence* (ST) requires only coincidence in time. *Asynchronous presence* (AP) requires coincidence in space but not in time. *Asynchronous telepresence* (AT) does not require coincidence in space and time.

Table 2. Spatio-temporal communication constraints, based on [32]

	Spatial	
Temporal	Physical presence	Telepresence
Synchronous	**SP** Face to face	**ST** Telephone, instant messaging, radio, teleconferencing
Asynchronous	**AP** Refrigerator notes, hospital charts	**AT** Email, fax, printed media, web pages

The spatio-temporal communication constraints for agent collaboration depend directly on some of the task dimension values. Constraints exist if *collaboration* ≥ 1 and they vary for different *urgency* values. If *urgency* = *immediate*, then only *SP* and *ST* of other agents lead to potential help in solving a task, because if help comes after some time threshold, the utility for the requesting agent is zero. Take, for example, an emergency scenario where someone who cannot get out of a car that fell into a river, needs to be rescued. There is only a small critical time interval for survival. If *urgency* = *flexible*, then *AP* and *AT* may also be viable, depending on how much time the requesting agent has for solving the task. *AP* may lead to a higher risk because it is assumed that some other agent will come by the requesting agent's location within a certain time interval and react to a posted message. In general, flexible task urgencies result in more choices and less constrained communication.

4 Collaborative agent process model

This section develops a high-level framework for ad-hoc negotiation of on-site collaboration between agents. We allow for autonomous agents that follow their individual goals, and only if they cannot reach them on their own, they ask peers for help.

4.1 Communication for collaboration

Agents use P2P communication to negotiate with each other for collaborative action. One defining parameter for the design of a negotiation procedure is the radio range, which is an issue in all P2P communication and depends on the protocol / platform. To expand the search range for help beyond the immediate radio range, message forwarding strategies can be applied, e.g., within a specified search range pre-calculated by spatio-temporal relevance constraints [7]. We assume here that agents have sufficient energy for movement and physical work on board, and may even utilize internal mechanisms for battery recharge. This allows them to broadcast anytime—in contrast to sensor networks, which are rigidly limited by their energy resources and communicate only in synchronized time windows.

Negotiations within a search range require stable communication links in a potentially fragile communication network over the time of the negotiation process. This requirement makes the problem substantially different from simple message dissemination problems [9]. Robust negotiation strategies require negotiating over short distances and by immediate response. Also, in case a message gets lost, agents must be able to continue their work based on their current information. Negotiations can take different forms. One way is a client agent sending a request, interested agents responding with offers, and the client selecting and booking an offer. Another way is providers advertising their services, clients registering, selecting, and booking when needed. Only the prior form allows for synchronous communication, which facilitates robust negotiations.

Figure 1 illustrates the negotiation process. The client initiating the communication reaches four other agents within its radio range (dark gray). Since these agents are located within the search range (light gray), they re-broadcast the request once. Other agents receiving the request will also re-broadcast once if they are located within the search range. Some agents beyond the search range may have received the request, but since they are outside this range they ignore it. For example, for *Agent 8* being within radio range of *Agent 5* but outside the search range means that it receives the client's request, but does not re-broadcast. Therefore, *Agent 9*, although within radio range of *Agent 8*, will not receive the request. In contrast, *Agent 7* is within the search range but outside the radio range of any broadcasting agent and therefore does not receive the request. The request messages traveling through the communication network keep track of their broadcasting agents. This way, the message remembers the shortest route back to the client.

Agent 1 is sufficiently close to be in *synchronous co-presence* (black circle) with the client and can start collaboration without delay. *Agents 2, 3, and 4* are in *synchronous telepresence*: they are able to communicate directly, but have to approach first before an interaction can take place. Agents that can only be reached by the client through mediating agents are in asynchronous telepresence, since communication can be delayed by the requirement to re-broadcast sent messages. In situations with *radio range* ≥ *search range* asynchronous telepresent agents do not exist.

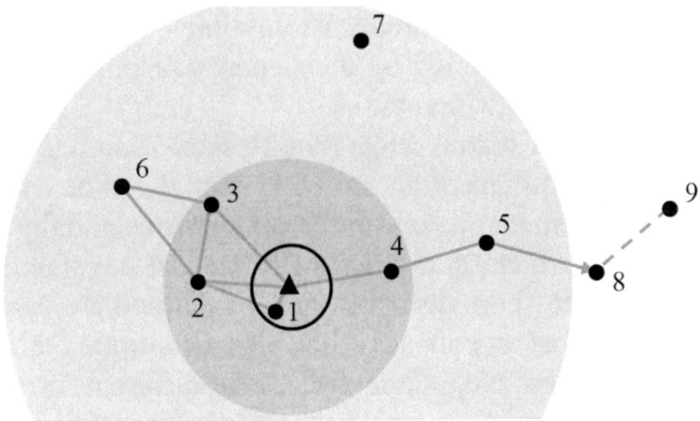

Fig. 1. A client agent (triangle) requesting help via short-range radio and the emerging communication network

4.2 The negotiation process

This section models the individual steps of the negotiation process. We assume that agents can communicate ad-hoc and make decisions based on utility functions. Accordingly, the focus is on *communication and exploitation of spatio-temporal constraints and affordances*.

4.2.1 Client's request

A *client* initializes a negotiation process as soon as it is confronted with a task beyond its capabilities (*collaboration* > *0*). The task focalizes the client's perceived affordances and enables it to create a plan. The client may discover the need for help by learning about the physical and social parameters of the task, or the individual actions involved in the plan. The parameters may be released from the object to be manipulated, or experienced by trial-and-error. At this time the client cannot judge whether a single or multiple collaborative agents are needed to solve the task because it does not know the capabilities of near-by agents. However, the client is able to specify the urgency of performing the task. By setting an upper time limit the client implicitly defines a search range for helping agents. With these parameters a request can be formulated, consisting of:

- *Message type*: 'request' (tells other agents how to treat this message);
- *task location*: client location by default;
- *search range*: a time frame;
- *task*: a set of <*O, E, {AC}*>. In this set of affordances the specific subtask the client needs help for has the highest weight. It is parameterized by the difference of learned properties of the subtask and the client's own capa-

bilities. If the client has experienced the properties by trial-and-error, the parameterization takes the form of an inequality, denoting that the value is beyond its own capabilities;
- list of *broadcasting agents*: agent IDs, initialized by the client's ID.

The request is broadcasted by the client and re-broadcasted by other agents if they are within the search range. Every forwarding agent appends its ID to the list of broadcasting agents.

4.2.2 Service provider's offer

Recipients of the request take the requested affordances and their parameters into account, as well as their own capabilities, duties (urgency of their own current tasks), and utility functions. An agent can formulate two types of offers: (a) contributing to the specific affordances requested by the client, or (b) suggesting different actions to solve the task. For case (a), the simplest situation occurs if the agent can offer to utilize a requested affordance on its own. However, if the request contains a parameter specification in the form of an inequality or a value exceeding its capabilities, the agent can only offer help within the limits of these capabilities. If the agent cannot utilize the requested affordance, but a similar one, it can still offer this similar affordance. For case (b), the agent can apply the affordance similarity measure (Section 3.1) to its own stored affordances and come up with a suggestion. The agent's offer consists of:
- *Message type*: 'offer';
- *travel time distance* to client;
- offered parameterized *affordance* in terms of actions;
- *similarity value* between offered and requested affordance: may be used by client to estimate the risk with booking this agent;
- list of *agents* leading back to client: reverted list of broadcasting agents;
- list of *broadcasting agents*: agent IDs, initialized by offering agent's ID.

Re-broadcasting of *offers* by other agents is conditional to their ID appearing in the list of agents leading back to the client, and their ID not appearing on the list of broadcasting agents (to avoid multiple broadcasting). This strategy assumes that the communication links available for the request are still intact for the offers. While this is realistic in general, in individual cases an offer may not reach the client due to a recently broken link.

4.2.3 Client's choice and booking

The client will compare the travel time distance specified in the offers with its task urgency, determine the similarity between offered and requested affordances, and the amount of support offered. With its own utility function the client is able to rank all incoming offers. The following cases can be distinguished:

1. *No offer arrived.* The client can enlarge the search radius or change its plan.
2. *At least one offer is made matching the highest ranked affordance in the request.* The client can choose either the nearest offering agent (high task urgency) or the agent with the largest capacity for this affordance (risk reduction).
3. *Offers rank other affordances higher than the requested affordance.* The client can choose the offer with the most similar affordance, assuming compatibility in agent capabilities. Alternatively, it can choose the offer with the highest weight for one affordance, assuming that the weight reflects the offering agent's confidence in being helpful. Accepting other than the requested affordance may require revising the plan.

Once the client has made a decision, it will formulate a booking message, which consists of:

- Message *type*: 'booking';
- booked *affordance*;
- list of *agents* leading forward to the offering agent: reverted list of the chosen offer's broadcasting agents;
- list of *broadcasting agents*: agent IDs, initialized by client's ID.

Re-broadcasting of requests by other agents is conditional to their ID appearing in the list of agents in the chain forward to the offering agent and their ID not appearing in the list of broadcasting agents. Booked agents will travel to the client's location and help.

5 Application scenario

This section applies the model to a hypothetical emergency scenario involving a car that got hit by a tree (Figure 2). The client *Agent C* (car driver) tried to move the tree without success and is therefore requesting immediate help from other agents[2] in the communication network. We consider four additional agents (Figure 3): *Agents* H_1 and H_2 are within radio range of *Agent C*; *Agent* H_3 is within the search range; and *Agent* H_4 is outside the search range.

[2] Our model is generic and deals with abstract agents. Here, we focus on the description of the collaboration process between software agents, whether the actors represent humans or not.

Fig. 2. A car driver finding his vehicle blocked by an obstacle contacts other agents to help remove the obstacle

The client's request consists of the following parameters (Section 4.2.1):

```
[request; (34.42, -119.70); 15min;
<O: hasPos (e, Pos(y)) & y≠x; E: tree;
AC: carry (ph: hasPos (e, Pos(x)) & WeightKg (e, >30) &
LengthM (e, >2))>;
(C)]
```

It contains the task location in the form of latitude/longitude coordinates and specifies a search range of 15 minutes. The move-ability affordance is represented through an outcome *O* (entity *e* must have a different position *y* compared to current location *x*), an entity type *E*, and one action specified by physical aspects *ph*. Due to its physical capabilities, *Agent C* can only carry trees with a maximum weight of 30kg and a maximum length of 2m. The requested action (as part of the affordance) therefore exceeds these limits. After the outcome *O*, *C* initializes the list of broadcasting agents being the first sender.

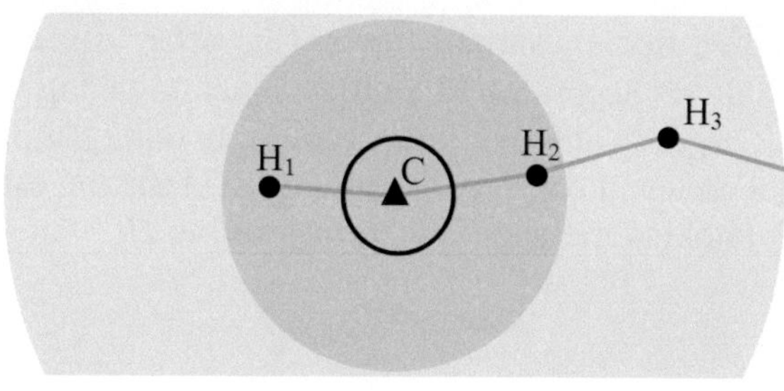

Fig. 3. Client Agent C with four potential service providers in the communication network

All four additional agents receive the client's request. H_1 and H_2 are in synchronous telepresence, and receive the request directly. H_3 and H_4 are in asynchronous telepresence (reached through mediating *Agents* H_2 and H_3). H_2 is occupied with its own task and therefore decides not to make an offer. The other three agents calculate the shortest path [33] to the task location. Only agents that can reach the task location within the specified time will make an offer. This results in H_4 not making an offer. Thus, only H_1 and H_3 are making an offer to the client. Table 3 compares these offers.

Table 3: Offers from *Agents* H_1 and H_3 back to the client

Offer from Agent H_1	**Offer from Agent H_3**

`8min;`
`AC: carry (ph:`
`hasPos (e, Pos(x)) &`
`WeightKg (e, ≤40) &`
`LengthM (e, <1));`
`0.84;`
`(H1, C);`
`(H1);`
	`13min;`
	`AC: lift (ph:`
	`hasPos (e, Pos(x)) &`
	`WeightKg (e, ≤800) &`
	`LengthM (e, [1,15]));`
	`1.00;`
	`(H3, H2, C);`
	`(H3);`

The computed travel times from the locations of H_1 and H_3 to the task location are 8 and 13 minutes. With respect to the parameterized affordance, two different actions—*carry* and *lift*—are offered. *O* and *E* are equal to the client's request (*sim* = *1*) and therefore omitted in the table. The individual components of each action are used to calculate overall affordance similarity according to the measure introduced in Section 3.1. Because the client's request specifies minimum agent capabilities for carrying trees in terms of weight and length, every offer equal to or exceeding this limit results in a similarity value of 1, e.g., `WeightKg (e, >30)` and `WeightKg (e, ≤40)`, and 0 otherwise. Final values are calculated according to Equation 1 with both weights set to 0.5[3].

The offer broadcasted by H_1 is received by *C* directly, and no other agent receiving the message takes an action. In contrast, *C* is not in the radio range of H_3, but H_3 specified that H_2 should forward its offer. *Agent C* then evaluates the incoming offers according to a utility function taking various parameters, such as task urgency and risk, into account. In our example, the client puts a higher weight on agent capacity for solving the task and establishes the ranking (H_3, H_1). A booking message is therefore sent to H_3.

[3] For H_1, sim_{AC} results from the similarities of `hasPos`, `WeightKg`, and `LengthM`, i.e., (1+1+0)/3 = 0.67. Equation 1 then evaluates to Sim_A = 0.5*0.67+0.5*1 = 0.84.

6 Discussion

Compared to an uninformed or brute-force approach such as flooding (every sensor receiving a message broadcasts this message again), the application shows that integrating spatio-temporal constraints within a model of agent collaboration in a P2P network leads to an optimized message distribution among agents and therefore to more efficient support in performing a task. During brute-force search messages are re-broadcasted to every other node within the communication range and this process is repeated consistently. In our case, the *optimization of message distribution* results from the constrained search range based on time-sensitive tasks. Messages are only sent to potential collaborators, resulting in a reduction of overall network traffic and saving bandwidth. Agents (H_4 in the application scenario), whose travel time to the task location exceeds a given limit, do not make offers and therefore further reduce the number of messages. *Efficient task support* results from knowing in advance the helping agents' capabilities. In addition to similarity values between requested and offered affordances, and the client's utility-based decision-making, this is a major step towards spatio-temporal efficiency in agent collaboration.

The client found through trial-and-error that its physical capabilities were insufficient for moving the tree. As a result, some of the affordance parameters could only be specified in terms of lower limits, e.g., `LengthM (e, >2)`. By specifying exact capability values, such as `LengthM (e, 4.50)` an interval rather than a Boolean scale could be used to calculate more precise similarity values, e.g., 4.49 is more similar to 4.50 than 4.35. It is important to note though that the final similarity values are not interpreted on an individual basis, but establish an order from most to least similar. In our scenario, social-institutional (*si*) constraints were only implicitly covered—*Agent* H_2 did not want to make an offer—and did not enter the similarity function. As shown in [26], the similarity function for actions sim_{AC} can easily be extended to represent these aspects, such as lower willingness to help during nighttime versus daytime.

Similar to this application, ad-hoc shared-ride services realize the general model of decision-making demonstrated here. The decision model as presented in [7] also relies on a negotiation process of client requests, host offers, and clients' selection and booking. In light of the present general decentralized decision model, their application-specific negotiation can be interpreted as being based on affordances. Clients in the shared-ride scenario formulate their request by specifying their current and desired locations. They perceive the affordance of moving vehicles with free transportation capacity, and accordingly, specify in their request the task 'move me to a specific location'. Hosts can interpret this task directly by their capabilities to offer rides, depending on their free seats and directions. The other aspects of the general model of decision-making are also present: Urgency is known, a search radius is specified by the client, and the clients' utility function consists of an optimal path algorithm for their own trip. Therefore, protocols and algorithms in shared-ride services can be expressed by the presented model.

7 Conclusions and future work

In this paper we have specified a high-level framework for ad-hoc communication between agents that negotiate for collaboration to perform a task. The underlying model accounts for spatio-temporal constraints, leading to efficient communication and task support. The agents' decision-making is based on affordances, to be able to adapt to any context and task. The framework was demonstrated through an application, which gave insight into how the affordance specification enables clients and service providers to form their decisions.

The presented work suggests several directions for future research:

- The process model needs to be implemented and tested in different real-world application scenarios. Decentralized ride-sharing provides one possible scenario, but there are many others, such as emergency response and various interactions between humans and robots, e.g., in elderly care. Agent-based simulations will provide insights into complexity issues and real-world applicability of spatio-temporal communication constraints.
- The demonstrated application includes only a small number of agents, which leads to the question of scalability. We expect that our approach will scale due to its distributed architecture, local processing, and local evaluations of relevance. Future simulations will address this question.
- The similarity measure for affordances needs to be refined and extended. Strategies for combining offers from different service providers (*collaboration > 1*) must be developed, leading to the classic problem of combinatorial optimization, i.e., determining the set of agents with the largest total attribute value. This becomes even more complex when affordances and their parts cannot simply be added up, e.g., *carry* + *lift*. Determining the similarity between affordances specified through different actions will require the use of action / affordance ontologies.
- There are several ways of specifying a client's utility function. Depending on the context, such function may focus on temporal aspects, risk estimations, and social and institutional issues. In addition, economic models will be needed to balance the costs of the service providers with benefits.

Acknowledgments

The comments from four anonymous reviewers provided useful suggestions to improve the content and clarity of the paper. Dr. Winter acknowledges support from the Australian Research Council, DP0878119.

References

1. Sengupta, R. and R. Sieber, Geospatial Agents, Agents Everywhere ... Transactions in GIS, 2007. 11(4): p. 483-506.
2. Russell, S. and P. Norvig, Artificial Intelligence: A Modern Approach. 2nd ed. Prentice Hall Series in Artificial Intelligence. 2003, London: Prentice Hall.
3. Wooldridge, M., Intelligent Agents, in Multiagent Systems - A Modern Approach to Distributed Artificial Intelligence, G. Weiss, Editor. 1999, MIT Press: Cambridge, MA. p. 27-77.
4. O'Sullivan, D., Geographical information science: agent-based models. Progress in Human Geography, 2008. 32(4): p. 541-550.
5. Benenson, I. and P. Torrens, Geosimulation - Automata-based modeling of urban phenomena. 2004, Chichester, England: Wiley.
6. Batty, M., J. Desyllas, and E. Duxbury, The discrete dynamics of small-scale spatial events: agent-based models of mobility in carnivals and street parades. International Journal of Geographic Information Science, 2003. 17(7): p. 673–97.
7. Raubal, M., S. Winter, S. Teßmann, and C. Gaisbauer, Time geography for ad-hoc shared-ride trip planning in mobile geosensor networks. ISPRS Journal of Photogrammetry and Remote Sensing, 2007. 62(5): p. 366-381.
8. Erol, K., J. Hendler, and D. Nau, Complexity results for HTN planning. Annals of Mathematics and Artificial Intelligence, 1996. 18(1): p. 69-93.
9. Nittel, S., M. Duckham, and L. Kulik, Information dissemination in mobile ad-hoc geosensor networks, in Geographic Information Science - Third International Conference, GIScience 2004, M. Egenhofer, C. Freksa, and H. Miller, Editors. 2004, Springer: Berlin. p. 206-222.
10. Zhao, F. and L. Guibas, Wireless Sensor Networks. 2004, Amsterdam: Elsevier.
11. Sun, R., Prolegomena to Integrating Cognitive Modeling and Social Simulation, in Cognition and Multi-Agent Interaction, R. Sun, Editor. 2006, Cambridge University Press: New York, USA. p. 3-26.
12. Anderson, J., D. Bothell, M. Byrne, S. Douglass, C. Lebiere, and Y. Qin, An integrated theory of mind. Psychological Review, 2004. 111(4): p. 1036-1060.
13. Newell, A., Unified Theories of Cognition. 1990, Cambridge, Massachusetts: Harvard University Press.
14. Sun, R., The CLARION Cognitive Architecture: Extending Cognitive Modeling to Social Simulation, in Cognition and Multi-Agent Interaction, R. Sun, Editor. 2006, Cambridge University Press: New York, USA. p. 79-99.
15. Schoder, D., K. Fischbach, and C. Schmitt, Core Concepts in Peer-to-Peer Networking, in Peer-to-Peer Computing: The Evolution of a Disruptive Technology,

R. Subramanian and B. Goodman, Editors. 2005, Idea Group Inc.: Hershey, PA. p. 1-27.

16. Peng, J., M. Wu, X. Zhang, Y. Xie, F. Jiang, and Y. Liu, A Collaborative Multi-Agent Model with Knowledge-Based Communication for the RoboCupRescue Simulation. in International Symposium on Collaborative Technologies and Systems (CTS'06). 2006.
17. Luo, Y. and L. Bölöni. Children in the Forest: Towards a Canonical Problem of Spatio-Temporal Collaboration. in AAMAS'07, Int. Conference on Autonomous Agents and Multiagent Systems. 2007. Honolulu, Hawai'i, USA: IFAAMAS.
18. Broxvall, M., M. Gritti, A. Saffiotti, B.-S. Seo, and Y.-J. Cho, PEIS Ecology: Integrating Robots into Smart Environments. in IEEE International Conference on Robotics and Automation (ICRA). 2006. Orlando, Florida.
19. Winter, S. and S. Nittel, Ad-hoc shared-ride trip planning by mobile geosensor networks. International Journal of Geographical Information Science, 2006. 20(8): p. 899-916.
20. Raubal, M., S. Winter, and C. Dorr, Decentralized Time Geography for Ad-Hoc Collaborative Planning, in Spatial Information Theory - 9th International Conference, COSIT 2009, Aber Wrac'h, France, September 2009, K. Stewart Hornsby, et al., Editors. 2009, Springer: Berlin. p. 436-452.
21. Gibson, J., The Ecological Approach to Visual Perception. 1979, Boston: Houghton Mifflin Company.
22. Norman, D., The Design of Everyday Things. 1988, New York: Doubleday.
23. Raubal, M., Ontology and epistemology for agent-based wayfinding simulation. International Journal of Geographical Information Science, 2001. 15(7): p. 653-665.
24. Raubal, M. and R. Moratz, A functional model for affordance-based agents, in Towards Affordance-Based Robot Control - International Seminar, Dagstuhl Castle, Germany, June 5-9, 2006. Revised Papers E. Rome, J. Hertzberg, and G. Dorffner, Editors. 2008, Springer: Berlin. p. 91-105.
25. Stoffregen, T., Affordances and events. Ecological Psychology, 2000. 12: p. 1-28.
26. Janowicz, K. and M. Raubal, Affordance-Based Similarity Measurement for Entity Types, in Spatial Information Theory - 8th International Conference, COSIT 2007, Melbourne, Australia, September 2007, S. Winter, et al., Editors. 2007, Springer: Berlin. p. 133-151.
27. Barsalou, L., Ad hoc categories. Memory & Cognition, 1983. 11: p. 211-227.
28. Hägerstrand, T., What about people in regional science? Papers of the Regional Science Association, 1970. 24: p. 7-21.
29. Ren, F. and M.-P. Kwan, Geovisualization of Human Hybrid Activity-Travel Patterns. Transactions in GIS, 2007. 11(5): p. 721-744.
30. Raubal, M., H. Miller, and S. Bridwell, User-Centred Time Geography For Location-Based Services. Geografiska Annaler B, 2004. 86(4): p. 245-265.
31. Miller, H., What about people in geographic information science?, in Re-Presenting Geographical Information Systems, P. Fisher and D. Unwin, Editors. 2005, John Wiley. p. 215-242.
32. Janelle, D., Impact of Information Technologies, in The Geography of Urban Transportation, S. Hanson and G. Giuliano, Editors. 2004, Guilford Press: New York. p. 86-112.

33. Dijkstra, E.W., A note on two problems in connection with graphs. Numerische Mathematik, 1959. 1(1): p. 269-271.

iNav: An Indoor Navigation Model Supporting Length-Dependent Optimal Routing

Wenjie Yuan, Markus Schneider

Department of Computer & Information Science & Engineering,
University of Florida, Gainesville, FL 32611, USA
{wyuan, mschneid}@cise.ufl.edu

Abstract. People may have problems in finding their way to destinations in large buildings. This raises a need of designing and constructing indoor navigation systems. However, none of the available indoor navigation models can automatically calculate shortest paths according to the geometric structure of indoor space. The reasons are that those models which use geometric information produce circuitous routes and that those models which do not consider geometric information only provide very coarse routes. This paper proposes a model to construct a way finding indoor network that is based on the geometry of the indoor space and that supports length dependent optimal routing.

1 Introduction

An important requirement of navigation systems is the ability to find optimal routes for users. Optimal routes can be the least time-consuming routes, the shortest routes, or routes according to user requirements. In outdoor space, they are often calculated on the basis of a road network. However, constructing a network that can support shortest path search in indoor space is more challenging. Although there are some comparable concepts in indoor space and outdoor space like corridors and roads, in in-

---

This work was partially supported by the National Science Foundation under grant number NSF-CAREER-IIS-0347574.

M. Painho et al., (eds.), *Geospatial Thinking*, Lecture Notes in Geoinformation and Cartography,
DOI 10.1007/978-3-642-12326-9_16, © Springer-Verlag Berlin Heidelberg 2010

door space, there are also concepts like rooms and lobbies for which we do not find counterparts in outdoor space. For example, inside a room, there are many implicit paths from one location to another. This makes it difficult to construct path networks for indoor space.

Several efforts have been made in finding routes in indoor space. However, the existing approaches all suffer from at least one of the following problems. First, some models do not take the geometry of indoor space into account so that they can only provide very coarse routes composed of a sequence of object identifiers without detailed directions. As a consequence, these models are not able to determine the length between different locations without manual input. Second, some models ignore the architectural constraints like doors in their models so that the generated routes are not precise enough for practical use. Third, although some models can provide precise routes with detailed information about directions, these routes are not necessarily optimal.

The goal of this paper is to propose a model that can support length-dependent optimal routing based on the geometry of indoor structures. In our model, a network is composed of a set of path segments, and a shortest route represents a finite sequence of path segments that is obtained by applying a shortest path algorithm on the network. The detailed routing information, such as turns and the length of the route, can be obtained from the path segments.

The rest of the paper is organized as follows. Section 2 discusses related work and summarizes the available models for indoor navigation systems. Section 3 discusses the way we explore possible path segments that might be involved in finding optimal routes. In Section 4, we build an indoor network and discuss its benefits to navigation. Finally, Section 5 draws some conclusions and depicts future work.

2 Related Work

Several models have been proposed to support human-oriented indoor navigation. Purely symbolic models [1, 8, 10, 11] are based on a labeling system without considering the geometry of the indoor space. Thus, routes generated by these models are very coarse. Later, the geometry of the indoor space is added to the models in order to determine more detailed routes.

In [7], a time-dependent optimal routing model is proposed for emergency evacuation. The path network is built on the basis of the location of sensors, and the optimal routes are determined after considering environmental information on the positions of evacuees. Although this model can provide a time-dependent optimal route for a quick evacuation, the route it provides is highly dependent on the location of sensors and not on the architectural structure itself. Thus a poorly settled sensor network may lead to improper results.

There are a couple of models that try to convert the architectural structure into path networks. The *node-link* model proposed in [12] and the model in [13] both build path networks based on the reachability of different cells. However, they lack the consideration of constraints, such as doors, windows and walls, in indoor space. Thus, these models cannot lead users to the exact entry of the target cell, and the cost of each link must be provided manually in advance. In addition, these two models can generate circuitous paths from start nodes to end nodes. As shown in Figure 1a, the walls represented by the bolded lines prevent a direct reachability between the rooms. The route generated by [12, 13] is a circuitous one composed of the center points of cell. This problem can be alleviated by the *CoINS* model proposed in [5, 6], which simplify final paths by eliminating some unnecessary nodes from the path and recalculating the segments between two nodes. For example, Figure 1b is the approved path for the situation in Figure 1a. However, the CoINS model still suffers from the problem of a lacking consideration of accessibility constraints.

In [3, 14], a route graph model is proposed to build abstract routes according to exits, walls and some other constraints in indoor environments. However, they do not discuss how to build the route graph for an entire indoor space. In addition, they assume route segments have directions. However, in indoor space, there is no specified direction for places since people can walk in any direction.

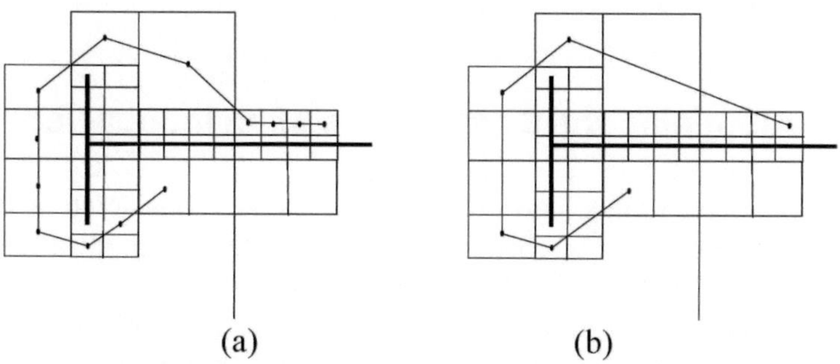

Fig. 1. A circuitous path generated by the node-link model **(a)**, and path simplification by the CoINS model **(b)**

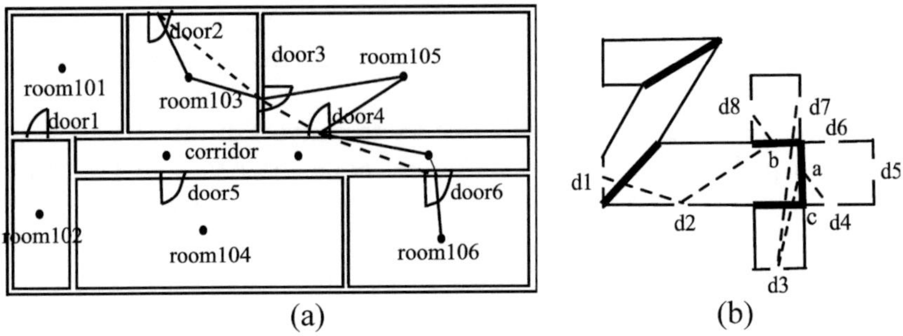

Fig. 2. The solid line in (a) indicates the route calculation in [2], and (b) is an example of the region partitioning in [4]

The model in [2] also takes architectural constraints into account when building the path network. As shown in Figure 2a, the model employs some representative points to represent rooms, corridors and some other objects. Then the calculation of the path is processed among these representative nodes as well as some architectural constraints like doors. In [4], the model is extended by decomposing concave-shaped objects to make sure that users can see the next hop indicated in each instruction during the navigation. However, the routes generated by this model are not the shortest ones. For example, compared to the dashed line in Figure 2a, the generated route indicated by the solid line is not the optimal one. A similar problem exists in the partitioning of concave regions. Figure 2b shows a concave-shaped cell. In this model, point *a* is the intermediate point in the path from *d3* to *d4*. In fact, the optimal way from *d3* to *d4* is from *d3* to *c*, and then to *d4*.

3 Determining the Components of an Indoor Network

In indoor space, rooms, corridors and lobbies, are considered as the basic units; we call them *cells*. However, if we want to provide detailed routing information, it is insufficient to only consider the sequence of the visited cells. In fact, we can notice that there are some implicit routes in indoor space which are commonly used by people. For example, in Figure 2a, if you are in the corridor and you want to go to room105, then you may go straight towards *door4*. The straight line to *door4* is an implicit path as well as the shortest path to room105. In this section, we will explore the shortest path segments in different kinds of cells according to their geometric shapes and architectural constraints, which can support shortest path routing.

3.1 Cells

There are a number of different kinds of architectural cells in indoor space. Some of them have similar shapes but may serve different purposes, and some of them are totally different in shapes but may play the same role during the routing. For example, rooms with multiple doors can be a part of a passage to a certain destination, while rooms with only one door cannot. Thus, in this subsection, we will explore diverse kinds of cells and classify them into different categories according to their geometric and architectural features from the routing perspective.

Simple Cell

A *simple cell* is a cell that is closed by walls and can be accessed by only one *access point*. An *access point* is an architectural constraint controlling the accessibility of the cell. For example, a door is a typical access point in indoor space. Since a simple cell has only one access point, it cannot function as a passage. Thus, it can only play the role of a start object or a target object. Figure 3a shows an example of a simple cell. The solid boundary represents walls, and the black dot represents an access point.

Complex Cell

A *complex cell* is a cell that is closed by walls and can be accessed by multiple access points. A complex room can be considered as either a start object, a target object, or an object that contains paths as passages to destinations. Figure 3b shows an example of a complex cell. Multiple dots represent multiple access points.

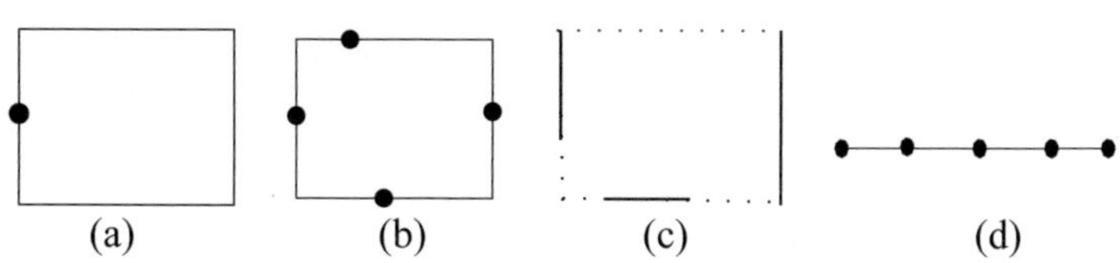

Fig. 3. Simple cell **(a)**, complex cell**(b)**, open cell**(c)**, and connecter**(d)**

Open Cell

An *open cell* is a cell for which at least part of its boundary is not closed by explicit walls or other constraints. For example, concourses in airports, as well as halls and lobbies in buildings, are typical open cells. There are various open boundaries with different widths in different open cells. Even in one cell, multiple open boundaries of different sizes may exist. The variable widths make it difficult to determine the access points in the open boundary. Figure 3c shows an example of an open cell. The dashed line represents the open part of the boundary and the solid line indicates the wall.

Connecter

A *connecter* is an object that connects different floors in a building. A connecter can be a stair, an elevator, or other objects that can be used to reach different floors. The location where a connecter and a floor meet is an access point in this connecter. Figure 3d shows an example of a connecter. The five dots mean that this connecter connects five floors.

3.2 Path Segments

A *route* is a concatenation of *path segments* from a start location to a target location. In indoor space, there are no explicit path segments. However, there are implicit path segments that people often take. For example, people like to go straight to a destination in case they can reach it directly. In the following, we will explore the shortest paths between any pair of access points in different cells.

A simple cell has only one access point; thus, it cannot be used as a passage. We only need to consider the possible path segments in complex cells, open cells and connecters.

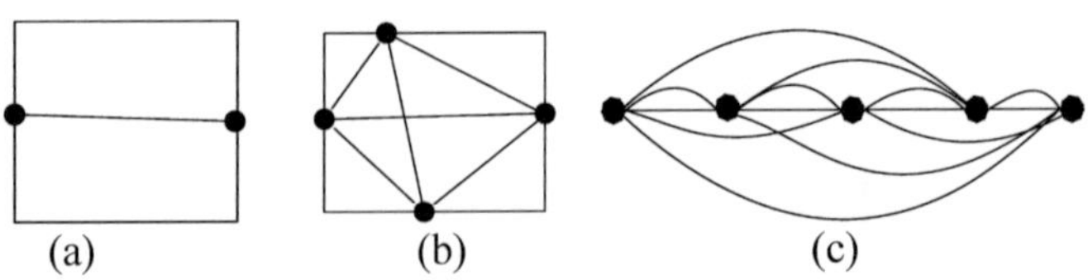

Fig. 4. Path segment in a cell with two access points**(a),** path segments in a cell with multiple access points**(b)**, and path segments in a connecter**(c)**

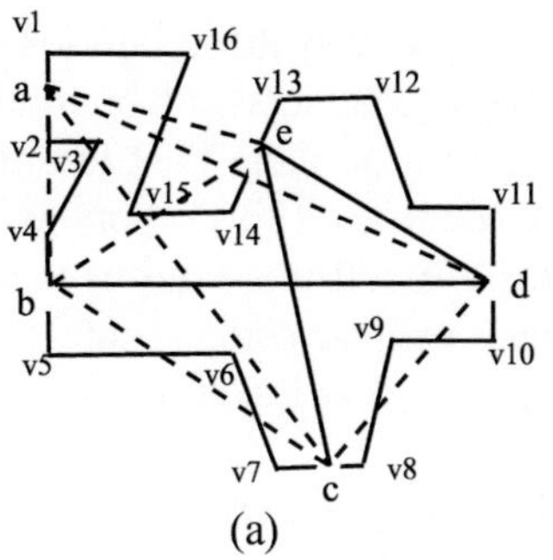

Fig. 5. Two access points in a concave region (like *b* and *c*) cannot reach each other on a straight path segment(a) which leads to a partitioning of straight lines to obtain shortest path segments between two access points(b)

For a complex cell, the approach to determine implicit path segments is based on the shapes of cells and the locations of access points. The simplest case is a cell with only two access points that can be reached through a straight line. Then the shortest path from one access point to the other is the straight line between them (see Figure 4a). If a cell has more than two access points, and all of them can be directly reached from each other, we obtain multiple implicit path segments in the cell. The shortest path segments in this cell are all straight lines connecting any two access points. For example, in Figure 4b, we find six implicit shortest path segments in a cell with four access points.

The above two cases are under the assumption that each pair of access points can be reached through a straight line from each other. This assumption holds if the shape of the cell is a convex region. However, in cells with concave shapes, two access points may not be directly reachable from each other. In Figure 5a, the dashed lines show some cases where a straight line connecting two access points is blocked by the boundary.

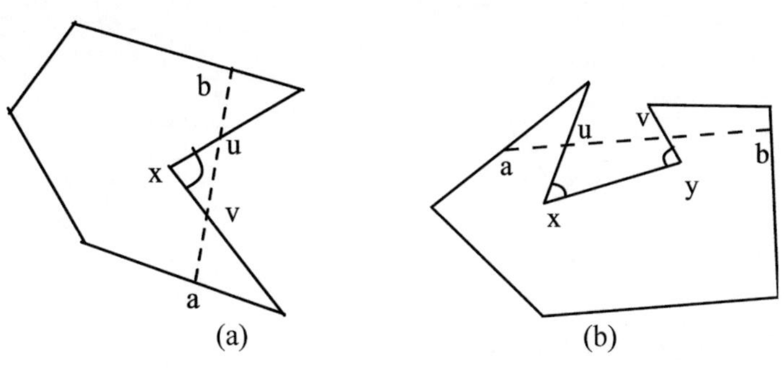

Fig. 6. Examples of intersections with boundaries

From Computational geometry [9], we know that if the interior of a line connecting two boundary points of a polygon intersects the boundary, then this polygon must be a concave polygon. That is, there is at least one vertex whose interior angle is a reflex angle (degree >180°) on one part of the boundary between the two access points. We call this kind of vertex *concave vertex* and the part of the boundary that contains concave vertices *concave boundary*. For example, in Figure 6a, *x* is the concave vertex and the boundary between *a* and *b* containing *x* is the concave boundary. It is possible to have multiple concave vertices on the concave boundary. As shown in Figure 6b, both *x* and *y* are concave vertices. Our approach to obtain the shortest path in this kind of situation is to select one of the concave vertices on the concave boundary as an *intermediate point*, and partition the straight line into two segments. The partitioning process continues until all the generated segments do not intersect the boundary. For example, in Figure 5a, the straight segment connecting the access points *a* and *e* encounters the boundary. This segment is then partitioned into segments (a, v_{15}), (v_{15}, v_{14}) and (v_{14}, e) by the vertices v_{14} and v_{15} between them. Obviously, we can learn that a - v_{15} - v_{14} - e is the shortest path between a and e (shown in Figure 5b).

An open cell is different from a complex cell in the aspect of the open part of the boundary. In contrast to doors, it is difficult to determine the access points on the open boundary. Our approach to obtain the path segments in open cells works as follows:

Step 1: Combine all the open cells sharing open boundaries until the combined cell is a complex cell closed by walls.

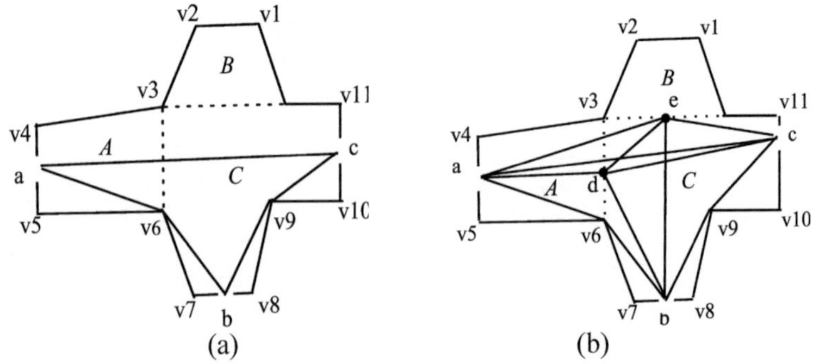

Fig. 7. The path segments in open cells

Step 2: Apply the same strategy of finding path segments in comple cells to this combined cell to obtain the path segments between the access points on the outer boundary.

Step 3: For all the open boundaries, select the center position of each open boundary as its access point. Then construct path segments between these new access points and all other existing access points.

There are two roles for a cell with multiple access points: a passage or a start/target object. The purpose of the combination in the first two steps is to construct all the shortest path segments when the cells functioned as passages. As shown in Figure 7a, *ac* is the shortest path when a user wants to go through these open cells. When an open cell is the target object in a query, the best position to lead users is the center of the open boundary. Thus, in Step 3, the center point of each open boundary is selected as an access point. As an example in Figure 7b, *d* and *e* are two access points of the region *A* and *B* respectively. When a user standing in the bottom of the combined cell wants to go to region *B*, the best way for her is the segment *be*.

There are only two directions in a connecter: up and down. Once a user knows the number of his current floor and the destination floor, he immediately knows the direction to the destination floor. Thus, we ignore the shape of connecters and assume that every two floors are straightly reachable. Then, path segments in a connecter are segments connecting each pair of access points (see Figure 4c).

3.3 Accessibility

An important issue of the way finding process is the aspect of accessibility of architectural cells in the indoor space. For example, while an employee in a building may have access to certain office rooms, these rooms are probably inaccessible for a customer. Another example is a construction site that prevents people from walking through a corridor and forces them to bypass it. In our model we control the accessibility of cells by assigning *accessibility* attributes to both access points and path segments.

The accessibility in an access point is controlled by a time stamp indicating when this access point is accessible. Taking an example from Figure 8a, assume the accessibility of the door "D3" is "8:00-17:00". Then only during this period of time, users can enter room103 or use the path (d_3, d_4). The reason why we also assign accessibility to each path segment is that the accessibility of the interior of a path segment may not be controlled by its two end points. For example, a path segment may not be accessible because of the construction site while its two end points are accessible. This means that an accessibility attribute only for access points or only for path segments would be insufficient. The relationship between the accessibility of access points and the accessibility of path segments is stated in the following observations.

Observation 1: If an access point is inaccessible, all its emanating path segments are inaccessible, and vice versa.

Observation 2: If an access point is accessible, there is at least one emanating path segment that is accessible, and vice versa.

Observation 3: If two end points of a path segment are accessible, the path segment can be inaccessible.

Observation 1 and Observation 2 are obvious. If an access point is inaccessible, then it does not make sense that any of its incident path segments is accessible since the access point can never be reached. An accessible access point must have at least one path segment that leads to it. Observation 3 indicates that a path segment can be blocked although its two end points are accessible due to other paths traversing them.

4 Constructing and Navigating the Direct Path Graph

Navigation is a process that successfully leads users from a source to a destination where they want to go. Usually, it should be able to find the optimal paths to destinations, which could be the shortest paths with respect to time, the shortest path with respect to distance, the path without paying fees, or any other paths according to users' requirements. The optimal route with respect to distance can be obtained by applying the shortest path algorithm on a path network that reflects the global path information of the entire indoor space. In this section, we will discuss how to build the path network according to the set of path segments we selected from each cell, and how to obtain the shortest routes from the path network.

4.1 Constructing the Direct Path Graph

The most efficient method to calculate paths is applying the shortest path algorithm to graphs. Therefore, we design a graph, named *direct path graph* (*DPG*), to support the shortest path algorithm.

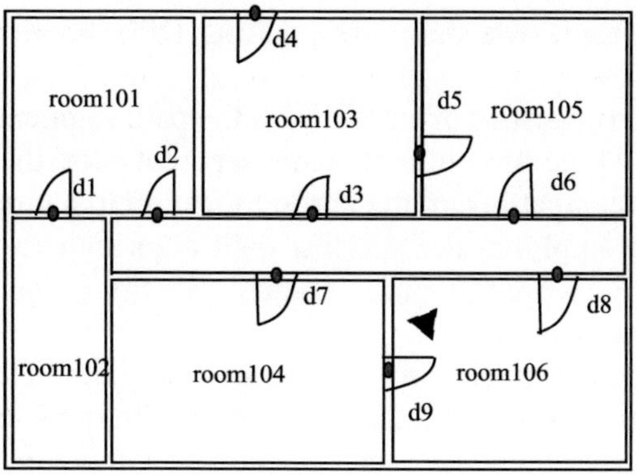

name	accessibility
d1	8:00-14:00
d2	8:00-17:00
d3	8:00-17:00
d4	8:00-17:00
d5	8:00-16:00
d6	8:00-17:00
d7	8:00-13:00
d8	8:00-17:00
d9	8:00-17:00

(a)

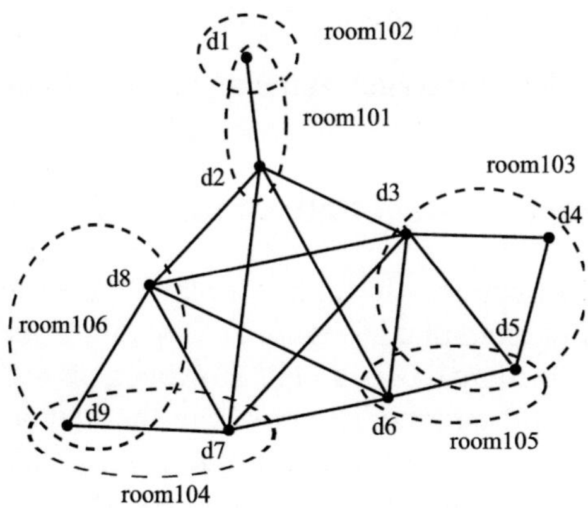

(b)

Fig.8. An example of navigation. Architectural map and the availabilities of doors(a) and its corresponding DPG(b)

Definition 1: A *direct path graph* $G := (V, E)$ is a graph which reflects all possible path constructions in a given indoor space scenario. V is a set of *access points* and *intermediate points*, and E is a set of *path segments* stored in the representations of the different cells in the indoor space.

In Section 3, we discussed how to determine the shortest path segments in different cells for routing. Obviously, the combination of these path segments from all cells constructs a path network which can support the shortest path search. A DPG is such a path network that is composed of access points, intermediate points, and path segments from different cells in this space. The reason why we call it *direct path graph* is because every edge in the graph represents a straight passage that is fully inside its corresponding cell. Figure 8b shows the corresponding DPG for the indoor space in Figure 8a.

There are several nice properties inherited from the path segments. First, a DPG represents the whole structure of path segments for the indoor space scenario. Therefore, a path from the current location to a target location can be obtained by applying the shortest path algorithm to a DPG. Second, edges in a DPG represent path segments in the indoor space, which are straight lines between access points. Therefore, every edge represents the shortest path between any two connected nodes. Third, any two locations between two connected nodes in the graph are visible from each other in the indoor space and can reach each other without encountering an obstacle like a wall. Fourth, the length of each edge in a DPG is the value of the attribute *length* stored with each path segment. It is calculated by using the Euclidean distance $\sqrt{(x_1 - x_2)^2 + (y_1 - y_2)^2}$ where (x_1, y_1) and (x_2, y_2) are the coordinates of the two access, or intermediate points.

4.2 Navigating the Direct Path Graph

Usually, users are interested in finding a path to a certain cell, such as to a store in a mall. For example, in Figure 8, they might ask "*How can I get to room103?*", where room103 is a cell. However, in a DPG, the nodes are the access points and intermediate points in the indoor space. There is no explicit cell information represented in this graph. Thus, a DPG cannot be directly used for answering navigation queries. In this subsection, we will show how we can nevertheless leverage a DPG to find the desired paths.

Originally, a DPG contains all access points and path segments for a certain indoor space. Since the current location and the target cell for a user might not be identical to any of the nodes in the DPG, our first step is to determine the starting node and the target node in a DPG. In Section 3, we have mentioned that the entire indoor space is composed of several non-overlapping cells, each of which contains its access points, intermediate points as well as path segments. Thus, from the current cell where the user is, we can determine all access points and intermediate points involved in this cell. In our model we choose the nearest access point or intermediate point to represent the user's current location and set it as the starting node for the shortest path algorithm. For example, Figure 8a is a typical indoor architectural map with the accessibility time for all doors. Its corresponding DPG is shown in Figure 8b, and the nodes and edges inside each dashed line cycle belong to one cell in the indoor space. Assuming your current location is the place marked by a triangle in Figure 8a, d_9 will be the starting point since you are in room106 and d_9 is your nearest access point in room106. The starting nodes are not necessarily the access points on the boundary of the source cell. They can also be intermediate points in the source cell. As shown in Figure 9, the intermediate points in this cell are v_3, v_6, v_9, v_{14}, and v_{15}. If your current location is the place marked by the triangle, then the starting point chosen will be v_6.

Choosing the starting node does not mean that users need to go to this starting node at the beginning. It is only used to represent the user's current location and determine the first access point that users need to go to. If the first path segment obtained from the shortest path algorithm is inside the cell where the user is, then users can directly go to the second point other than the starting point. For our example in Figure 8a, assuming that a user wants to go to room105 from his current location marked by the triangle in room106, and we learn that the shortest path from d_9 to room105 is "d_9 - d_8 - d_6". Then the start node is d_9, and the first access point the user need to take is d_8, since the path segment (d_9, d_8) is inside room106. If the user want to go to room101, and we learn that the shortest path to room101 is "$d_9 - d_7 - d_2$", then the first access point the user needs to take is d_9 because the first path segment (d_9, d_7) is not inside room106.

The way we determine the target node is different from the decision of the start node. Usually, if we want to know the way to a target place, we just need to find the way to any of its access points (e.g., doors and some openings) on its boundary. Thus, all the access points on the boundary of the target cell are our *potential target nodes*. For example, in Figure 8b, if our target cell is room103, then d_3, d_4 and d_5 are the potential target nodes. Because the shortest path algorithm will return the shortest paths from the starting node to any other node in a graph, we run this algorithm for all potential target nodes and determine that node (access point) as target node with the shortest distance from the source node.

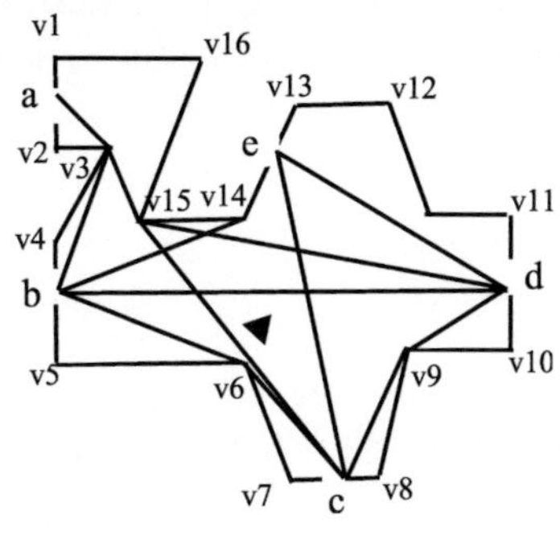

Fig. 9. An example that the starting node is an intermediate point

During the shortest path algorithm, we need to check the accessibility of path segments because some of the edges might not be accessible. For example, in Figure 8, assume the accessibility of all path segments is controlled by its two end points, and the current time is 15:00. Then the segment (d_7, d_9) will not be taken into account now since the value of accessibility of d_7 is 8:00-13:00.

In some models, each cell is represented by one node only, as shown in Figure 2a. This prevents them from providing shortest paths. As said before, in Figure 2a, the path from door2 to room106 these models will provide is shown by the solid line. This is a circuitous path, and the problem is serious when room105 is very large. Our model overcomes this problem by viewing each cell in general and by recording all possible path segments in it. Thus, our model can provide a more proper path from door2 to room106, as shown by the dashed line in Figure 2a.

5 Conclusions and Future Work

In this paper, we have proposed a model which supports length-dependent optimal routing for indoor navigation systems. We have explored how to select implicit path segments in cells with different shapes and access points. Then based on these path segments, a direct path graph is built to reflect the path network in indoor space. By using this graph, our model is able to provide the shortest path from the current location to the target location.

In the future, we plan to explore the hierarchical structure in the direct path graph. By considering hierarchical structures of indoor space, we can group some nodes and edges according to their relations with other nodes. Therefore the total nodes and edges can be reduced, and the efficiency of the path calculation can be improved. In addition, we will study how to generate nice descriptions for different routes, which should be clear and easy to follow.

References

1. B. Brumitt, S. Shafer (2001) Topological World Modeling Using Semantic Spaces. In UbiComp 2001 Workshop on Location Modeling for Ubiquitous Computing,
2. B. Lorenz, H. Ohlbach, E.-P. Stoffel (2006) A Hybrid Spatial Model for Representing Indoor Environments. Web and Wireless Geographical Information Systems 4295:102-112

3. B. Krieg-Brückner, H. Shi (2006) Orientation Calculi and Route Graphs: Towards Semantic Representations for Route Descriptions. International conf. on Geographic Information Science 4197: 234-250
4. E.-P. Stoffel, B. Lorenz, H. Ohlbach (2007) Towards a Semantic Spatial Model for Pedestrian Indoor Navigation. Advances in Conceptual Modeling-foundations and Applications 4802:328-337
5. F. Lyardet, D. W. Szeto, E. Aitenbichler (2008) Context-Aware Indoor Navigation. European Conf. on Ambient Intelligence, pp 290-307
6. F. Lyardet, J. Grimmer, M. Muhlhauser (2006) CoINS: Context Sensitive Indoor Navigation System. 8th IEEE Int. Symp. on Multimedia, pp 209-218
7. I. Park, G. U. Jang, S. Park, J. Lee (2009) Time-Dependent Optimal Routing in Micro-scale Emergency Situation. 10th Int. Conf. on Mobile Data Management: Systems, Services and Middleware, pages 714-719
8. J. Sakamoto, H. Miura, Noriyuki Matsuda, Hirokazu Taki, Noriyuki Abe, Satoshi Hori (2005) Indoor Location Determination Using a Topological Model. Knowledge-Based Intelligent Information and Engineering Systems, 3684:143–149
9. M. de Berg, O. Cheong, M. van Kreveld, and M. Overmars (2000) *Computational Geometry: Algorithms*
10. M. Raubal, M. Worboys (1999) A Formal Model of the Process of Wayfinding in Built Environments. Spatial Information Theory : Cognitive and Computational Foundations of Geographic Information Science, pages 381–399,
11. P. Hoppenot, G. Pradel, Catalin Caleanu, Nicolas Perrin, Vincent Sommeilly (2003) Towards a Symbolic Representation of an Indoor Environment. IMACS-IEEE Computational Engineering in Systems, Applications (CESA) Multiconference,
12. P.-V. Gilliéron, B. Merminod (2003) Personal Navigation System for Indoor Applications. Int. Association of Institutes of Navigation World Congress
13. R. Urs-Jako (2007) Wayfinding in Scene Space: Transfers in Public Transport. PhD thesis, University of Zürich
14. S. Werner, B. Krieg-Brückner, T. Herrmann (2000) Modelling Navigational Knowledge by Route Graphs. Integrating Abstract Theories, Empirical Studies, Formal Methods, and Practical Applications, pp 295–316

MIMEXT: a KML Extension for Georeferencing and Easy Share MIME Type Resources

Carlos Abargues, Arturo Beltran, Carlos Granell

Centre for Interactive Visualization, Universitat Jaume I, Castellón, Spain
abargues@uji.es, arturo.beltran@uji.es, carlos.granell@uji.es

Abstract. Georeferenced information is becoming more important in today's society leading to different techniques for georeferencing resources. Most of these techniques present some problems by the internal modification of the file and/or because of these are designed for very specific use cases, so supported formats are very limited. However, in this paper we present a new approach for a global solution to georeference and share any kind of Multipurpose Internet Mail Extensions (MIME) type resources. The proposed solution is based on the use of the Keyhole Markup Language (KML) and a new extension called MIMEXT for the annotation and georeference of these MIME type resources. The KML file containing the description for the resource and its geolocation is encapsulated within a KMZ file with the own resource and any other related resources such as licenses or thumbnails. This technique could facilitate to distribute, visualize and georeference a wide range of resource types not conveniently exploited yet in the GIS context.

1 Introduction

The geolocation of any kind of resource is acquiring a fundamental role in a wide range of applications. Examples of this trend are geotagged pictures: users are increasingly georeferencing their photos to position them geographically in virtual globes or map visualization services. One of the best-known examples in this context is the tandem formed by the virtual

M. Painho et al., (eds.), *Geospatial Thinking*, Lecture Notes in Geoinformation and Cartography,
DOI 10.1007/978-3-642-12326-9_17, © Springer-Verlag Berlin Heidelberg 2010

globe Google Earth[1] and the mapping service Google Maps[2] with the publishing and sharing pictures community Panoramio[3]. For instance, this enables sharing and searching of georeferenced images among users.

However geotagged images represent only one type of resources available in the collaborative environment empowered by social networks and Web 2.0 services. Essentially, users are demanding the possibility of georeferencing and sharing other types of resources, such as text files, videos, web pages or blog entries, through collaborative geospatial applications like virtual globes.

Given this context, the aims of this paper are two-fold. First we propose a solution to allow us to position geographically any kind of resource whatever its nature. This would break the trend of current tools that offer specific solutions addressed to the use case that are trying to tackle. Second, we provide a general way to easy share in one single file the data, its metadata and other related resources. For instance, this approach would enrich traditional data found in the Spatial Data Infrastructure (SDI) nodes (Nebert 2004) by incorporating new types of data from other domains not previously considered (i.e. spreadsheet documents, video or 3D models), however it can be applied in a wide range of scenarios.

The rest of this paper is structured as follows. In the following section we will see what options we currently have to incorporate metadata to resources. In section 3 the Keyhole Markup Language (KML) is presented as a basic element in our solution. Section 4 describes MIMEXT, the KML extension developed for annotating and georeferencing MIME type resources and in section 5 we mention how we achieve the encapsulation capability for exchanging georeferenced resources. Section 6 shows a proof-of-concept tool. Finally, conclusions and future work come in section 7.

2 Overview on Georeferencing Techniques

Georeferencing multimedia resources can be accomplished through the use of formats that natively support location information, such as JPEG2000 (Taubman and Marcellin 2002) (Schelkens et al. 2009) image format or MPEG[4] video format family. However, most applications cannot rely on the use of these formats, so they treat and store separately the resource they work with and its corresponding metadata or location information or

[1] http://earth.google.com/
[2] http://maps.google.com/
[3] http://www.panoramio.com/
[4] http://www.mpeg.org/

simply they do not consider this valuable information. The main reason for this behaviour is that, currently, many of the most used formats are long-standing and cannot include any georeferencing information, since they were not designed with this need in mind. Given the large amount of data already existent in these formats and their extensive use, we cannot ignore them in our solution.

On this basis, we will consider two families of solutions. On the one hand, the internal modification of the resource oriented solutions aimed to add the necessary information inside the own resource. On the other hand, the solutions aimed to encapsulate both the resource and its metadata as a unit, either at physical or logical level. In this section we will show some examples of both kinds of solutions and analyze their advantages and disadvantages.

2.1 Techniques for internal annotation

The internal modification of a file for adding metadata and georeferencing information is an extremely format-dependent approach. Basically, most of these approaches are based on the addition of certain metadata in some header fields. These fields inside file headers are usually considered as unused allowing other information to be placed there.

Adobe XMP (Extensible Metadata Platform) (Adobe 2007) is a labelling technique to incorporate metadata into the resource itself in terms of new labels attached as header fields. One limitation is that it can only be applied to certain formats (TIFF[5], JPEG[6], PNG[7], GIF[8] and PDF[9]) so it does not meet the needs of users who require other formats (i.e. audio or video). Furthermore, the added annotations are placed in different parts of the file depending on the target format, either including new labels or reusing existing ones designed for other purposes. This lack of consistency afterwards hampers the processing of metadata information in an automatic way. Although this technique does not damage the file in principle (Adobe 2007), we consider that modified (annotated) resources can become unsuitable for some applications.

A similar approach is the use of International Press Telecommunications Council (IPTC) (Löffler et al. 2007) tags. This format allows the insertion of metadata in some file types adding specific tags to the header of the file.

[5] http://partners.adobe.com/public/developer/tiff/index.html
[6] http://www.jpeg.org/
[7] http://www.libpng.org/pub/png/
[8] http://www.w3.org/Graphics/GIF/spec-gif89a.txt
[9] http://www.adobe.com/devnet/pdf/pdf_reference.html

The IPTC tags method represents an old meta-information format that is slowly being passed out in favour of XMP. An effort to collaborate between both produced the "IPTC Core Schema for XMP" allowing the merge of both approaches to embed metadata.

EXIF[10] (Exchangeable Image File Format) (JEITA 2002) is another interesting solution based on internal annotations that aims at incorporating labels in image files, especially those using JPEG compression. Again, the problem lies in the limited number of formats supported and the possible damage caused to the file.

2.2 Techniques for external annotation

Most techniques for external annotation are based on georeferencing resources from external files. Examples are SMIL[11] (Synchronized Multimedia Integration Language) (Bulterman and Rutledge 2008), and the generation of *".world"* files[12] that accompany the corresponding resource. With this kind of techniques, the resource remains intact, but loses the notion of unity that integrates data and metadata, making it hard to manage and share.

New formats are emerging to solve these problems. They encapsulate some of the external files with metadata, the data and other resources into one single file. This is the case of MEF format (Metadata Exchange Format) (Nottingham and Sayre 2008) or KMZ[13], a format broadly used in geobrowsers and web mapping services that we will explain in more detail later.

MEF was specifically created for the data and metadata exchange among different platforms and especially between GeoNetwork[14] nodes. MEF is focused on facilitating tasks such as the storage, transference and migration of spatial data, metadata, thumbnails, basic privileges and other related information. Its internal structure is composed of a *metadata.xml* file containing the metadata of the resources and an *info.xml file* with a specific format for GeoNetwork that gives extra information about resources and their metadata. They also contain the public and private folders to store resources depending on the privilege access intended for their content. The MEF files allow the transportation of files with geospatial information such as shapefiles (ESRI, 1998) or maps, encapsulated in one

[10] http://www.exif.org/
[11] http://www.w3.org/TR/REC-smil/
[12] http://en.wikipedia.org/wiki/World_file
[13] http://code.google.com/intl/en/apis/kml/documentation/kmlreference.html
[14] http://geonetwork-opensource.org

single file along with their own metadata. This encapsulated-oriented approach facilitates the exchange, distribution and reuse of resources.

Otherwise, the Web 2.0 promoted solutions oriented to georeference some resources via web. This is the case of some popular social networks like Panoramio or Flickr[15] used to share and possibly georeference pictures or Wikiloc[16] used with GPS tracks. These mashups usually store user's resources and annotate their geolocation inside their databases. The metadata included by this kind of solutions is very limited and furthermore they are oriented to their specific use case and format, so they are not applicable as a general solution.

2.3 Discussion

Table 1 compares the two georeferencing strategies described in this section: techniques for internal annotation and techniques for external annotation or encapsulation.

Table 1. Comparison of georeferencing strategies

	Internal Modification Techniques	Encapsulation Techniques
Advantages	Total integration of data and metadata in one single file	Integrates data, metadata and other resources into one unit. Metadata of unlimited size. Valid for any format. Original files remain intact. Compression capability.
Disadvantages	Can damage the file. Limited number of supported formats. Amount of information that we can add is limited. You cannot add other resources. Some formats are "closed".	Need to be expanded (divide and extract the different components). Metadata is not embedded in the resource itself.

As shown in the summary table, the main advantage of the solutions aimed at internal modification of the resource is the total integration of data and metadata in the same file. Thus, it greatly facilitates the transport, management and dissemination of the dataset along with its metadata.

[15] http://www.flickr.com

[16] http://www.wikiloc.com

However, this family of techniques has several drawbacks, arising mostly from the manipulation of the original format to add more information. The most obvious disadvantage is that the manipulation of the original may break them becoming unusable to other applications. In addition, we have to bear in mind that this family of techniques incorporates additional metadata in different places depending on the original format of the file, being really low the number of formats that support this kind of techniques. Furthermore, the amount of information we can add is limited in size because, in most cases, the information is incorporated into segments or labels with a bounded capacity. A notable drawback is the inability to add other resources such as licenses or thumbnails. Another disadvantage of internal change is that we depend on if the format is known and open to change, and if there are available drivers that provide us the read/write ability to change the metadata. In short, it does not exist the same freedom of manipulation for all formats.

The encapsulation-oriented techniques allow the integration of data and its metadata in one unit and, moreover, include other related resources without any limitation. Another advantage of this kind of techniques is that there are no restrictions concerning the size of the metadata or data; we can even make use of compression techniques. Encapsulation-oriented techniques are valid for any current and upcoming format furthermore do not alter the original files avoiding its associated kind of errors and problems. The main drawback of this family of techniques is that the resources are not directly available. It is required a pre-process to expand and extract the resources to operate with the files. Moreover, the data and metadata, yet encapsulated within a file are separated, making harder their management and synchronization.

In our opinion the use of one single unit to store data and metadata could facilitate their transportation and exchange. Furthermore, a generic solution is desirable to incorporate metadata information of current resources and be also flexible enough to accommodate new data formats. In the following sections we will explain in detail our approach based on the encapsulation-oriented solution using and extending the KML language.

3 Keyhole Markup Language

In this section, we describe briefly the relevant aspects of the KML language (OGC 2008), in which our solution relies on. Following we also analyze different ways for including metadata in KML to demonstrate that it is a suitable format for containing metadata descriptions.

3.1 Keyhole Markup Language for resources annotation

Virtual globes or geobrowsers are gaining momentum as platform for geospatial data visualization by professional and occasional users (Yamagishi et al. in press). In fact these specialized browsers force improvements in various aspects such as the 3D visualization or the approach of the geospatial world to the non-specialized public showing that the Geoweb (Scharl 2007) is becoming a reality. Most of them support KML, de-facto XML-based language for representing geographic information.

Recently adopted as OGC standard, KML has been broadly used to encode resources publicly available on the Web. Probably its popularity stems from its simplicity, inherent visualization and annotation capabilities (Wood et al. 2007), and the support offered by the most used geospatial tools and services (i.e. web mapping services, geotagging services).

KML can succeed as solution for georeferencing resources adding their metadata annotation and visualization details. Moreover this format can be processed by a large number of applications supporting it. However there exist other aspects to be considered that comprises conceptual and technical problems. Although KML offers a quite rich set of visualization options either in 2D and 3D environments, it seems important to consider the visualization of those kinds of resources not yet considered by the standard. KML offers a set of primitive geometries such as Point, Polygon or even COLLADA[17] models directly inherited from the OGC GML standard (OGC 2007). Beside these primitives it is possible to add different resources embedded in HTML code within the element *Description*. This is the case of images or Adobe Flash[18] video, usually georeferenced using a *Placemark* element with a simple geometry like *Point*. This geometry makes sense in most of the cases but others could be more suitable for concrete scenes (i.e. audio track describing a route represented by a lineal geometry, a polygon representing a terrain parcel with an associated PDF document with cadastral information, etc.). Unfortunately KML does not offer solutions for referencing resources of different types to those that can be embedded within the *Description* element.

3.2 Including Metadata in KML

In order to integrate new kinds of resources in SDI nodes using our proposed solution it is necessary to incorporate metadata inside the file de-

[17] http://collada.org
[18] http://www.adobe.com/devnet/swf

scribing the resources, in our case a KML file. To do so we propose two approximations using different KML elements. The first approach makes use of the KML standard element *ExtendedData* to include metadata in XML format. The second brings some concepts and techniques used in the Semantic Web (Berners-Lee et al. 2001) to embed metadata into the KML *Description* element.

3.2.1 The ExtendedData element for metadata containment

The *ExtendedData* KML element allows the insertion of XML content within a KML file. There exist three different methods to do so varying in the KML elements used within the *ExtendedData* element.

The first one consists on adding pairs of type name/value to any element derived from the KML element *Feature*. This method does not require the definition of any schema or data type. Derived *Feature* elements are for instance *Placemark*, *NetworkLink* or *Folder*.

The second method allows the definition of an arbitrary XML schema inside the KML code. This definition or pseudo-schema must be composed of simple elements without any type of nesting. Again any element derived from the *Feature* element can benefit of this method.

The last one allows the user to add metadata using a valid schema externally defined and referenced or imported using its URL. After referencing a given schema and assigning an arbitrary namespace to it, all the elements defined on it can be used inside a KML file. This last method offers a long list of advantages including the reuse of already existing schemas. These schemas can be as much complex as required including also the use of nested elements and complex types. One useful application of this method is the metadata addition to features in KML importing already existing and broadly used schemas such as ISO19115 (ISO 2003).

3.2.2 The Description element for Semantic annotation

The purpose of the *Description* element is to add some textual description to any of the elements derived from the element *Feature* within a KML file. This element commonly contains the information in plain text without any given structure however KML also allows the insertion of XHTML code via CDATA tags.

The use of XHTML offers several possibilities concerning the addition of metadata mostly based on the inclusion of microformats (Suda, 2006) and RDFa[19]. Both have the goal of encoding semantic information into

---

[19] http://www.w3.org/TR/xhtml-rdfa-primer/

XHTML documents so that the same content can be processed by humans and automatic agents. Microformats exploit specific XHTML attributes to put in place metadata to indicate the meaning of data. Currently it is possible to find a variety of defined structures for representing events, social relationships, or addresses and locations among others. On the other hand, RDFa format assigns XHTML elements (i.e. *div*, *p* or *span*) with RDF classes using the *meta* and *link* elements to express structured metadata within XHTML pages. Then it is possible the creation of RDF classes instances inside XHTML.

Embedding the metadata inside the *Description* element within a KML file presents several advantages. First the metadata would present not just a structured format facilitating its automatic processing but also a human readable format thanks to the use of XHTML. Another asset is that the content present in the *Description* element in KML files is specially used for content indexing by some search engines such as Google. This means that the content is analyzed when performing searches on some of the mapping services offered by these search engines. Unfortunately the search engines do not yet recognize the structure given by the use of RDFa or microformats in KML as it starts to happen in HTML.

3.3 Missing functionality in KML

KML offers a suitable format for annotation and visualization of information with geographic component. Some methods to add metadata have been presented as well demonstrating that this format could be successfully used inside an SDI.

Despite all these advantages, KML still miss some important aspects for its use as format to georeference resources. Probably the most important is the lack of methods to annotate or refer to these resources. Currently it is just possible to use some resources embedded inside the *Description* element for a given *Feature*. However the type and use of these resources is usually limited to elements such as videos or photos. Furthermore this method of embedding the resource inside KML elements that are not originally designed for such a task is clearly not the best solution.

To go through these and other types of limitations the KML standard defines a set of rules to extend the language adding more functionality. In section 4 we explain how, following these rules, we have created a new KML extension by which references to any type of MIME type resource can be performed. This extension allows georeferencing any type of resource using KML.

4 The MIMEXT KML extension

We are trying to georeference a wide range of kinds of resources that cannot be effectively annotated within a KML file for their processing. A similar situation took place with the HTML and the web browsers. In that case it was required to install different plug-ins or applications to visualize some media types directly in the web browser.

All this complications in HTML try to be avoided with the release of the HTML 5 specification[20] that includes new tags (video, audio, canvas, etc.). These new tags simplify the content creator's task shifting most of the work to web browsers since they should be able to reproduce the content according to these tags. Maybe it is too early to obtain measures about the success in the use of these tags however it is true that this approach intends to facilitate, among other things, the content creation in the web.

Considering the similarities between the Web and the Geoweb and between HTML and KML as their de-facto languages, a similar approach could extend KML to facilitate the integration of new kinds of resources as it already happened with the HTML in the last years.

4.1 Extending the KML standard

KML functionality can be increased by the use of extensions, adding new elements to those defined by the OGC standard. In our case, extending KML would allow its use for georeferencing and representing resources of any MIME type (Borenstein and Freed 1993) (Freed and Borenstein 1996a) (Freed and Borenstein 1996b) on those applications capable of handling such a format and the following proposed extension.

The standard OGC KML schema proposes several mechanisms to extend or restrict the format for specific purposes in what is known as *Application Profiles*. All these profiles must follow a list of requisites, which make reference to the use of appropriate namespaces, the correctness of the new schema and the reuse of the existing KML schema. Other requirements deal with dependencies related with the inheritance and extension of new or already existing elements.

Basically KML presents two methods for its extension: extension by inheritance and extension by composition. The former defines a method for adding new schemas derived by core abstract base types. The latter method is based on the substitution of already existing elements and presents two options based on simple and complex elements.

[20] http://www.w3.org/TR/html5

As we proposed in section 3.1 the georeferenced resources should be able to become associated to any geometrical element that can be represented in KML. Following this premise, a hypothetical language extension could effectively associate these geometries with resources of MIME types that are not currently supported by the KML standard. All these geometry features derive from the KML element *Geometry*. As a first approach, our solution consists of associating any resource with any *Geometry* derived element.

Besides the reference to the resource content, the proposed extension could also offer information about its file type. This information could become useful, not just for the end user but also for applications capable of offering a direct visualization of these resources. This visualization could be performed over a virtual globe or opening the resource with the corresponding application associated to the file type or extension in a similar way as it is done in most of the current operating systems. This resource type description should follow a specification or standard such as the one used to define the MIME types. An extensive list of MIME types is defined, each one specifying their file type, subtype and extension. This has been taken as base for our approach in order to give information about the resource in a way as standardized as possible.

4.2 MIMEXT structure

Taking into account the above requirements a first version of a model extending the KML specification to annotate MIME type resources in virtual globes has been developed. We name this new extension as MIMEXT whose purpose is the annotation in KML of all those resources with a registered MIME type. It will allow the georeferencing of the resource, the addition of metadata and at the same time its representation and use whenever KML and this new extension are recognized.

Figure 1 shows all the elements used to build up this new extension, their type and the hierarchy of elements. Three new simple types have been created: *MimeTypeEnumType*, *MimeSubTypeEnumType* and *MimeFileExtensionEnumType*. These new element types serve to specify the MIME type, subtype and file extension for each resource defining string enumerations with the corresponding values for each category. *MimeTypeEnumType* contains a collection of 7 elements with the values *application*, *audio*, *image*, *message*, *text*, *video* and *x-world*. *MimeSubTypeEnumType* contains a list with more than a hundred elements with the subtypes defined for each one of the *MimeTypeEnumType* categories. Some values found in this enumeration are *pdf*, *msword*, *x-latex* or *mpeg*.

Finally the *MimeFileExtensionEnumType* lists all the file extensions corresponding to the previously defined MIME types and subtypes.

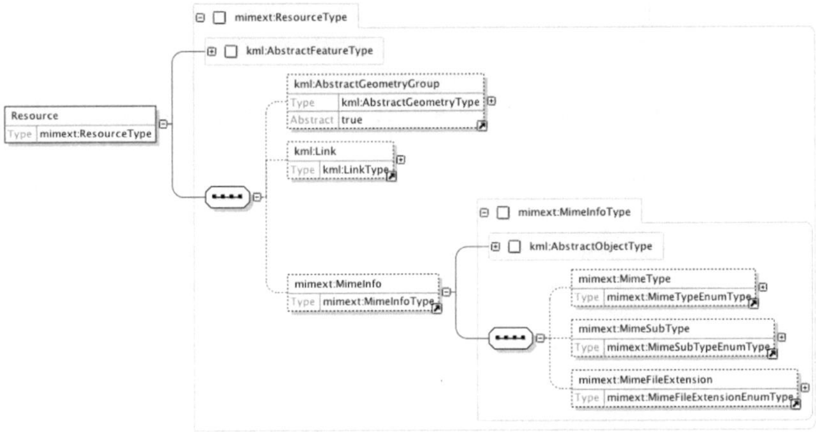

Fig. 1. MIMEXT extension structure

MIME simple types are used to define the simple elements *MimeType*, *MimeSubType* and *MimeFileExtension* that will hold the information about the type, subtype and file extension for a given resource using the corresponding value. These three simple elements are used to define the complex type *MimeInfoType* that serves for grouping, as a sequence of the above simple types, all the information concerning the resource. This complex type serves to define the complex element *MimeInfo* that at the same time is used to define the complex type *ResourceType*. This complex type defines the basic element for the extension, the complex element *Resource*.

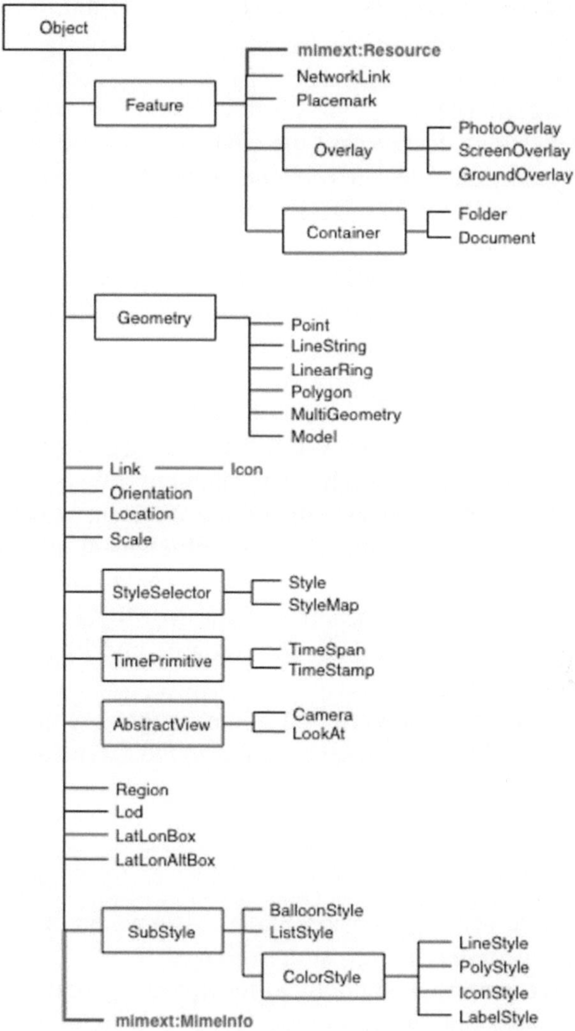

Fig. 2. KML hierarchy including MIMEXT extension elements

Figure 2 shows how the element *Resource* derives from the abstract element *Feature* and by inheritance the new element has the same elements that *Feature* has. The addition of other metadata for the referenced resource could be accomplished by using any of the techniques explained in section 3.2. This aspect allows the insertion of metadata within the resource in order to be used in SDI. Besides the *Feature* inherited elements, others defined in the complex type *ResourceType* compose the *Resource* element:

- *AbstractGeometryGroup*: The use of this abstract element allows the association between elements, in our case the element *Resource*, and any of the geometry types defined in KML (*Point*, *LineString*, *LinearRing*, *Polygon*, *Multigeometry* or *Model*). Besides georeferencing a resource by its coordinates as a point, complex geometries could add additional geometric information associated with it (i.e. land extension referenced in a PDF document and represented by a *Polygon*, a race's length which video is georeferenced by using a linear geometry)
- *Link*: This KML element permits specify the location and some handling information for a given resource. The location of the resource is usually given by its Unified Resource Location (URL)[21]. Parameters of *Link* are usually used when loading content such as the output of a Web Mapping Service (WMS) (OGC 2006) request, specifying different parameters such as the refreshment period. Although the Link element allows the referencing of both remote and local resources, the latter are the ones considered in our approach. The benefits of the MIMEXT extension could be easily appreciated where both the resource and its georeferenced information coexist together.
- *MimeInfo*: It facilitates the type description of the resource by the use of information about its MIME type, subtype and also the extension of the file representing it. This information intends to be processed and used by applications such as virtual globes to visualize or work conveniently with the resources.

To clarify the structure of our extension Figure 3 shows the necessary code for annotating and georeferencing a simple image.

At this point we have proposed solutions to incorporate georeferencing information and other metadata into resources. However a solution for integrating both the resource and its associated information is still missing. In the following we describe how to encapsulate MIMEXT files.

[21] http://www.w3.org/Addressing/URL/url-spec.txt

```
<?xml version="1.0" encoding="UTF-8"?>
<kml xmlns="http://www.opengis.net/kml/2.2"
    xmlns:mimext="http://www.geoinfo.uji.es/kml/ext/mimext">
    <Document id="d1">
        <mimext:Resource>
            <name>Image Resource #1</name>
            <description>
                This is a test for image resources
            </description>
            <Point>
                <coordinates>
                    26.0822035425683,37.42228990140251,0
                </coordinates>
            </Point>
            <Link>
                <href>http://www.example.com/fire.jpg</href>
            </Link>
            <mimext:MimeInfo>
                <mimext:MimeType>image</MimeType>
                    <mimext:MimeSubType>jpeg</MimeSubType>
                        <mimext:MimeFileExtension>jpg</MimeFileExtension>
            </mimext:MimeInfo>
        </mimext:Resource>
    </Document>
</kml>
```

Fig. 3. MIMEXT extension example

5 KMZ Encapsulation

One of the requirements of our approach has been the encapsulation of the resource, its metadata including georeferencing information and other related resources. The use of KML facilitates this task introducing the use of KMZ files. Basically a KMZ is a compressed (zipped) file containing a KML file and a folder with resources referenced in this KML. These resources use to be graphics such as icons, images or photos.

Thanks to its simple structure most of the applications that work with KML also work with KMZ making of this a perfect solution for transporting and sharing geographic information. In this sense KMZ represents a simple but powerful solution for the encapsulation into one single file of any type of resource along with its geospatial location and other descriptive information expressed in KML as shown in Figure 4. This approach offers the same benefits listed for KML plus a simple way of combining both the resource and its associated information.

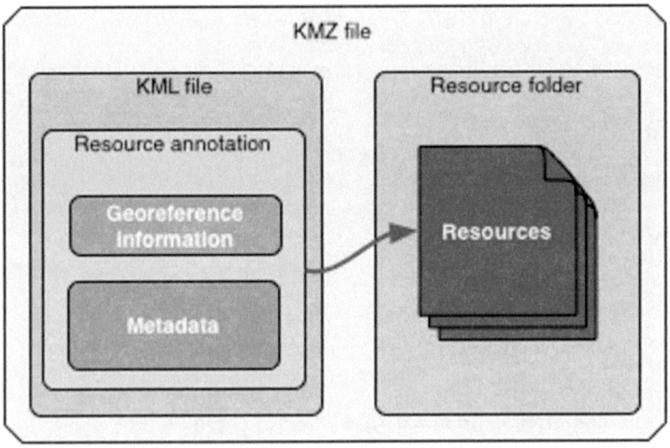

Fig. 4. KMZ encapsulation file content.

6 A Practical Approach

In order to test our approach and prove the concept a small application has been build. Its main objective is to visualize within a virtual globe a MIME type resource encapsulated in a KMZ file and described using the MIMEXT extension.

This application can demonstrate that our approach could become an easy way to add geographic information and also metadata to any resource. The tool also tries to demonstrate how a geobrowser could become a multi-purpose and multi-format browser with effective markup and data processing. Ideally, a user with a HTML5 enabled web browsers can browse the Web without worrying too much about some content type (i.e. audio, video) that could be found. The same could happen with the users browsing the Geoweb with a geobrowser. In this sense MIMEXT could facilitate the integration of existing data types into the emerging Geoweb.

The application is divided into three main modules: the object data model, the parser and the visualization module. The first one comprises a set of Java classes and interfaces representing the data model specified in the KML standard and in the MIMEXT extension. Not all the classes, interfaces and data types are implemented but all those necessary for visualization and metadata annotation.

The parser module allows the reading of a KML file and its transformation into data model objects in memory. The reading of a KML or KMZ file allows the recognition of all those elements implemented in the object data model including the MIMEXT extension elements.

Finally the purpose of the last module is the visualization of all the elements defined in the KML file over a virtual globe using the WorldWind Java SDK[22]. This NASA's free SDK allows the rapid creation of Java applications or applets that require the visualization of data over a virtual globe. It also allows the use of annotations to visualize information on windows or balloons over the globe.

The first version of our application is able to represent, visualize and when required, reproduce images and audio clips annotated in KML with MIMEXT over the virtual globe. Figure 5 shows the result of processing the code presented in section 4.2.

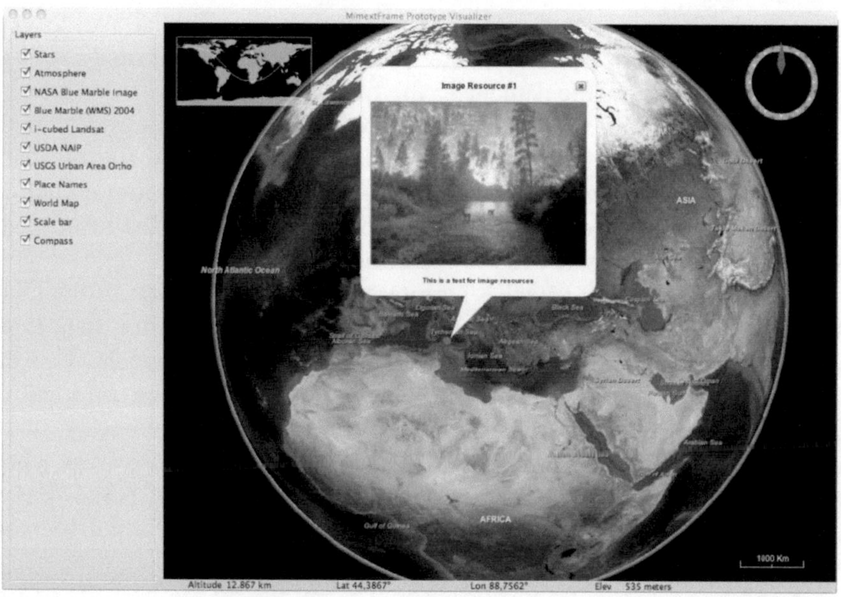

Fig. 5. MIMEXT demo application processing an image resource

Since the MIMEXT extension allows the annotation of any MIME type resource the interpretation and processing of more types only depends on the implementation and improvement of the processing application (i.e. geobrowser). In future releases of this tool, capabilities for reproducing video and visualize different text-based documents as well as typical GIS formats are planned.

[22] http://worldwindcentral.com/wiki/WWJava_FAQ

7 Conclusions

The provision of data and metadata in one single file has a number of advantages. One of these is to facilitate and to improve the management and sharing of resources. Current techniques aimed at internal modification of the resource represent many complications especially that can cause damage to the file, leaving it unusable for some applications. Furthermore, the creation of a georeferencing model based on the metadata addition at internal level would require the individual examination of each and every one of the data types and formats to georeference. These techniques involve a considerable workload that does not guarantee the development of an effective model for both existing and future formats. However, this is not the case with techniques aimed at encapsulate the original file with their metadata and related resources into a new file. This allows to georeference any resource since this method is independent of format avoiding at the same time possible damage to the file.

The solution proposed in this paper presents a method for georeferencing any kind of MIME type file, integrating in one single file the data, metadata and any other related resources. This implies that future applications can use at their discretion the geospatial metadata of the resource or simply the resource itself without its metadata.

The KML extension MIMEXT represents a valid extension of the OGC standard KML version 2.2 and a first step towards integrating data types and resources that are not totally exploited in GIS environments. This extension creates a new element that allows easily annotating information about any MIME type resource, including its associated geometry, metadata and geolocation. Thus MIMEXT can be still used in those areas where the KML format is used. In addition, we exploit the use of KMZ files to encapsulate as a unit geographical information in KML format together with the resources that they reference.

In order to experience and to observe more clearly the benefits of this approach, as shown in Section 6, we developed a simple application. It is a virtual globe application capable to interpret the KML format and the MIMEXT extension and operate on any kind of resources.

Our future plans include the extension of the application to support more types of resources, in order to validate its benefits. If accomplished, the next step would be to promote its use for georeferencing and share MIME type resources.

Acknowledges

This work was partially funded by the project “CENIT España Virtual”, funded by the CDTI in the program “Ingenio 2010” through *Centro Nacional de Información Geográfica* (CNIG).

References

Adobe (2007). Extensible metadata platform (XMP). http://www.adobe.com/products/xmp/overview.html.

Berners-Lee T, Hendler J, Lassila, O (2001). The Semantic Web – A new form of Web content that is meaningful to computers will unleash a revolution of new possibilities. Scientific American, 284(5): 34-43.

Beard K, Smith T, Hill L (1997). Meta-information models for georeferenced digital library collections. IEEE Computer Society metadata conference No2, Silver Spring MD, ETATS-UNIS (16/09/1997), pp. 1-9.

Borenstein N, Freed N (1993) MIME (Multipurpose Internet Mail Extensions) Part One: Mechanisms for Specifying and Describing the Format of Internet Message Bodies. RFC 1521, Bellcore, Innosoft.

Bulterman DCA, Rutledge L (2008). SMIL 3.0: Interactive Multimedia for the Web, Mobile Devices and Daisy Talking Books. Springer.

ESRI (1998). ESRI Shapefile Technical Description. An ESRI White paper, July 1998. http://www.esri.com/library/whitepapers/pdfs/shapefile.pdf.

Freed N, Borenstein N (1996a). Multipurpose internet mail extensions (mime) part one: Format of internet message bodies, RFC 2045. http://tools.ietf.org/html/rfc2045.

Freed N, Borenstein N (1996b). Multipurpose internet mail extensions (MIME) part two: Media types, RFC 2046. http://tools.ietf.org/html/rfc2046.

ISO (2003). Geographic information - Metadata. ISO/TS 19115:2003, International Organization for Standardization (ISO).

JEITA (2002). Exchangeable image file format for digital still cameras: Exif Version 2.2. Technical Standardization Committee on AV & IT Storage Systems and Equipment. Specification by JEITA.

Löffler H, Baranger W, Steidl M (2007). Photo Metadata White Paper 2007. IPTC, the International Press Telecommunications Council.

Nebert D (2004). Developing Spatial Data Infrastructures: The SDI Cookbook v.2.0. http://www.gsdi.org: Global Spatial Data Infrastructure (GSDI).

NISO (2007). The Dublin Core Metadata Element Set: ANSI/NISO Z39.85-2007. American National Standards Organization.

Nottingham M, Sayre R (2008). Geonetwork opensource: The complete manual. http://www.fao.org/geonetwork/docs/Manual.pdf.

OGC (2006). OpenGIS Web Map Server Implementation Specification, Version 1.3.0. Open Geospatial Consortium Inc (Open GIS Consortium Inc). http://www.opengeospatial.org/standards/wms.

OGC (2007). OpenGIS Geography Markup Language (GML) Encoding Standard, Version 3.2.1. Open Geospatial Consortium Inc (Open GIS Consortium Inc). http://www.opengeospatial.org/standards/gml.

OGC (2008). OpenGIS Keyhole Markup Language (KML) Implementation Specification, Version 2.2.0. Open Geospatial Consortium Inc (Open GIS Consortium Inc). http://www.opengeospatial.org/standards/kml.

Scharl, A. (2007). Towards the GeospatialWeb:Media Platforms for Managing Geotagged Knowledge Repositories. In A. Scharl, K. Tochtermann (Eds.): The GeospatialWeb - How Geo-Browsers, Social Software and the Web 2.0 are Shaping the Network Society. London: Springer, pp. 3-14.

Schelkens P, Skodras A, Ebrahimi T (2009). The JPEG 2000 Suite. Wiley, Wiley-IS&T Series in Imaging Science and Technology.

Suda, B. (2006). Using microformats. O'Reilly Media, Inc.

Taubman D, Marcellin M (2002). JPEG 2000: Image Compression Fundamentals, Standards and Practice. Springer: The Springer International Series in Engineering and Computer Science, vol. 642.

Yamagishi Y, Yanaka H, Suzuki K, Tsuboi S, Isse T, Obayashi M, Tamura H, Nagao H (in press) Visualization of geosciences data on Google Earth: Development of a data converter system for seismic tomographic. Computers & Geosciences. DOI: 10.1016/j.cageo.2009.08.007.

Wood J, Dykes J, Slingsby A, Clarke K (2007) Interactive visual exploration of a large spatio-temporal dataset: Reflections on a geovisualization mashup. IEEE Transactions on visualization and computer graphics. 13(6): 1176-1183.

Schema Mapping in INSPIRE - Extensible Components for Translating Geospatial Data

Chunyuan Cai[1], Sven Schade[2], Thimmaiah Gudiyangada[3]

[1] Institute for Geoinformatics, University of Muenster, Germany
cagehouhou@gmail.com
[2] Institute for Environment and Sustainability,
European Commission - Joint Research Centre, Ispra, Italy
sven.schade@jrc.ec.europa.eu
[3] Centre for Interactive Visualization, Universitat Jaume I, Castellón, Spain
thimmaiah21@hotmail.com

Abstract. The general situation on geospatial information is one of fragmented data sets and sources, gaps in availability, lack of harmonization between data sets at different geographical scales, and duplication of information collections. Addressing these issues, providers of geospatial information in Europe have to deliver their contents relating to specifications emerging in for the context of INSPIRE (Infrastructure for Spatial Information in Europe). Schema mapping plays a critical role in this process. We suggest an approach that supports the data translation procedure using Web Service technology. Efficient and sufficiently loosely coupled components for rule management and translation execution were designed and implemented. We present them in this paper. Our solution guarantees extensibility in mapping rule structure and of translation operators. We allow users to customize geospatial information and thus enable flexible information exchange. We suggest and compare two options of using the developed software.

M. Painho et al., (eds.), *Geospatial Thinking*, Lecture Notes in Geoinformation and Cartography,
DOI 10.1007/978-3-642-12326-9_18,

1 Introduction

The general situation on geospatial information is one of fragmented data sets and sources, gaps in availability, lack of harmonization between data sets at different geographical scales and duplication of information collection (Masser 2005; Vandenbroucke et al. 2007). The Infrastructure for Spatial Information in Europe (INSPIRE) is established as a means for addressing these issues in respect to environmental data within the European Union (INSPIRE 2004). Existing spatial data infrastructures (Nebert 2004) should support INSPIRE by following its recommendations to share geospatial data between various users. To achieve, national data models (also called schemas) have to be mapped to INSPIRE Data Specifications (INSPIRE 2009) and data sets have to be translated accordingly. Supporting tools are topic to current practices (Curtis and Müller 2006; SafeSoftware 2008; Madsen 2009) and research (Donaubauer et al. 2006; Lehto 2007; Beckman et al. 2009).

Management of mapping rules and execution of these rules has been identified as central requirements for data translation (Schade 2009). Our research concentrates on these two critical issues. We aim at an extensible framework for managing rules, as well as an extensible and loosely coupled translation execution service. We propose a framework running on the level of download services and consider two possible implementations. While INSPIRE motivates our work and serves multiple use cases, detailed elaborations on INSPIRE-required types of translation operators are out of scope. We also do not address elaborated translation tests, as for example presented in (Östman et al. 2009).

In the next section we present required background. The third section deals with the rule language and component design followed by two alternate options for implementing translation with web service technology (fourth section). The fifth section describes the implementation of the component architecture. The final section includes the summary and an outline for future work. For illustrations, we use data models for transport networks. We focus on road networks and particularly on forest roads. We work on the German NavLog data model (NavLog 2005) and its mapping to INSPIRE road networks (INSPIRE 2009b).

2 Background and Preparatory Work

Data translation requires the definition of mapping rules of source data model(s) to a desired target schema and execution of these rules on source data set(s). Input data may have to be collected from heterogeneous sources. In this section, we introduce a running example, give a brief introduction to the general topic of (geospatial) data translation, and define the relation to INSPIRE. We also visit existing solution for data translation, and analyze their capabilities.

2.1 Schema Mapping Example

We selected NavLog (NavLog 2005), a data model from the forestry domain, as relatively small test case. NavLog was developed as an operator model for the capturing, maintaining, and distributing of forest road data in Germany. It consists of two layers; one contains lines of forest road network, the other points of other road features such as bridges and curves. Within this example we focus on linear data. As the NavLog data model is initially specified in German language, we provide English translations of required attribute definitions below (Figure 1).

NavLog Attribute	Description
GEOMETRY	Center line of the road.
NAME	Name of the road.
LANEWIDTH	Restricted width of the roadway in meters (physical road width on the ground for trucks to pass).
CLEARWIDTH	width restricted by a clearance gauge in meters (visible width for driving).
CLEARHEIGHT	height restricted by a clearance gauge in meters (visible height for driving).
WAYCLASS	Classification of the road according to capabilities for wood removal (support for trucks with or without steering function).
CONCR	Pavement (road paved or not).
GRADIENT	Maximum slope of the road in percent.
BLOCK	Blockage (road blocked/closed).
COMMENT	Free text message.

Fig. 1. Description of NavLog line attributes

INSPIRE in aims at using maintained local data sets for serving geospatial data with European coverage (EC 2007). In the medium term INSPIRE targets data provision for 34 different themes, ranging from address data through transport networks to species distribution (INSPIRE 2004). The themes are separated into three annexes, where currently only the themes of annex one have been fully specified. All INSPIRE Data Specifications for the first annex make use of the Geography Markup Language (GML) in its current version, 3.2.1 (OGC 2007). The third version of data specifications has just been released (INSPIRE 2009b). We identified elements related to NavLog as parts of the INSPIRE Data Specification for Transport Networks. The relevant attributes are listed in Figure 2.

INSPIRE Attribute	Description
centerlineGeomerty	Centre line of the road.
RoadName	Name of the road.
RoadWidth	Road width value.
FormOfWay	Classification of the road based on its physical attributes.
FunctionalRoadClass	Classification of the road based on its importance to the connectivity of the transport network.
RoadSurfaceCategory	State of the surface of the road (paved or unpaved).

Fig. 2. Description of INSPIRE road link attributes

Schema mapping can provide instructions on translating NavLog data sets according to INSPIRE Data Specifications, so that the attributes of Nav-Log refer to appropriate attributes of INSPIRE. Elaborated matches are given in form of a mapping table (Figure 3). Mappings, which require further elaborations, are indicated by dashed lines.

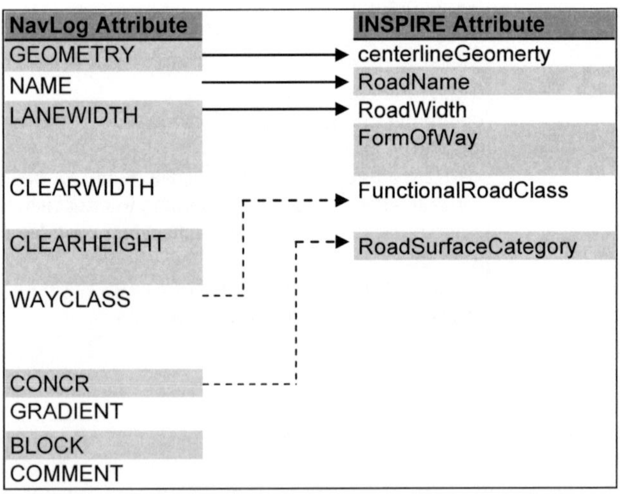

Fig. 3. Identified schema mapping from NavLog to INSPIRE

2.2 Translation Operators and Procedures

The example above contains simple one-to-one mappings only. In general, more complex (one-to-many and many-to-many) mappings may occur.

Some data models, for instance, separate X and Y coordinates of point geometries. In the INSPIRE context, such have to be mapped to a point geometry in a two-to-one mapping.

Additionally, translation operators may be required in order to translate a source data set to a desired target. For example, if the source data model only provides a geometry attribute, but the target requests length information, a spatial calculation has to be applied. Categorizations of the large amount of translation operators have been provided previously (Lehto 2007; Schade 2010). Most common categorizations distinguish between complexities. Following Schade (2010), renaming is the most simple category while filtering, reclassification, value conversion, augmentation, merging/splitting, and morphing have increasing complexity.

Once the source and target schema have been analyzed and mapping rules are available, source data sets have to be retrieved, the rules have to be executed on them, and the newly created data set has to be delivered. Many conceptual approaches have been proposed. Especially, one has been defined on high-level (Lehto and Sarjakoski 2004). The authors suggest a five-level view, separating the provision of source data, data integration capabilities, processing, portal functionality, and the application layer.

2.3 Translation as an INSPIRE Service

A Draft Implementing Rule for INSPIRE Transformation Services has been released recently (INSPIRE 2009c). The document suggests three architectures for transformation services and identifies possible transformation types, which may be supported in the context of INSPIRE. Those can be seen as concretizations of the first three layers of Lehto's and Sarjakoski's suggestion. At this moment, coordinate and data model (or schema) transformations are foreseen. Detailed guidance is provided only for the former (INSPIRE 2009d), while work on the latter is ongoing. In INSPIRE terms, schema transformation is the main topic of our work.

INSPIRE basically distinguishes between three types of architectures for transformation services (INSPIRE 2009c). The transformation workflow may be driven by the:

- *Application/Client*: In this case the client has to contact a service for data and subsequently one for transforming the retrieved data set.
- *Transformation Service*: The client requests a transformation from a service, which itself requests the source data set and returns the data set complying to the target schema.

- *Download Service*: The client sends a data request to an INSPIRE Download Service, which internally contacts another service for transformation.

In our work, we provide basic components, which could potentially be used in many of these settings. Opposed to other work, which considers the use of decoupled transformation services (Lehto 2007), we focus on download services as the primary contact point.

The type of download service (INSPIRE 2009e) can be used for further structuring. Using a Direct Access Download Service requires translation of the data request (including possible filters), while this is not the case for Pre-Defined Data Set Download Services. We illustrate the application of our components in two scenarios. The first can be seen as a realization of a Download Service-driven architecture for direct access, while the second goes beyond the current suggestions of INSPIRE. In addition, we concentrate on the access to and exchange of mapping rules, which is not explicitly mentioned in the INSPIRE documents.

2.4 Analysis of Mapping Tools and Languages

Several tools already implement geospatial data translation. We revisit an earlier analysis of ours (Beckmann et al. 2009) and summarize their characteristics in respect to the following criteria (Table 1):

- *Support of GML as output format*: Does the tool allow for producing GML data sets?
- *Service front-end*: Does the tool offer its functionality as a web service interface, preferably via a standard interface for download services, such as an OGC Web Feature Service (WFS) (OGC 2005)?
- *Graphical User Interface (GUI) for rule generation*: Does the tool offer a GUI for creating even sets of complex mapping rules?
- *Support for mapping rules*: Which types of operators can be defined and executed? What are the possibilities for extension with external translation operators? Does the tool support re-use of rules?
- *Type of software*: Is the product is commercial or open-source?
- *Comment*: Including any other specific information about the tool.

Table 1. Comparison of existing schema mapping tools

	FME	GoPublisher	Spatial Data Integrator	GeoXSLT
GML support	in-built	in-built	in-built	Possible, but must be user defined
Service	WFS	WFS	not included	not included
GUI	sophisticated, graph-based	table-based	sophisticated, graph-based	not included
Rule support	more then 600 operators in-built, user may add self-defined operators	simple filters in-built, user-defined rules can be added using XSLT or Java code	many operators in-built, user may add self-defined operators (using Java)	standard XSLT operations with extensions to include spatial processing from the GeoTools library
Software	commercial	commercial	open source	open source
Comment	use of the GUI has steep learning curve	table-based GUI has quite steep learning curve, hard to include own operators	already not performing well with small data sets	generation of XSLT code inconvenient in general

Feature Manipulation Engine (FME) (Safe Software 2008), GoPublisher (Snowflake Software 2008), Spatial Data Integrator (Camp to Camp 2008), and GeoXSLT (Klausen 2006), listed above, make use of proprietary formats for storing mapping rules. In order to achieve re-use and exchange, we envision an independent rule language. The Web Ontology Language (OWL) (Bechhofer et al. 2004), Ontology Mapping Language (OML) (Scharffe and de Bruijn 2005), and extensible Ontology Mapping Language (XeOML) (Pazienza et al. 2004) are considered as candidates. We came to the following findings:

- *OWL*: a popular language to describe ontologies, but it lacks the completed capabilities for representing mappings.
- *OML*: a language for describing mapping rules between ontologies. It is not based on XML format. Related Application Programming Interfaces

(APIs) have to implement an individual compiler for parsing the language.

- *XeOML*: an extension of OML and based on XML format. Unfortunately, the mapping execution tool can not provide a fine grained access capability. Moreover, we find the four basic elements specified by it are not all necessary for the mapping issues we encountered so far.

According to this brief analysis, we conclude that there are a variety of translation tools and a bunch of potential language available. However, we envision a more open and independent solution that can support any users to integrate it into their own applications. The fact that the existing solution can not be seamlessly assembled into service architectures and the need in extensibility motivated us to propose an alternative. In the next section, we will introduce our solution from a comprehensive designing view. Our goal is constructing a completely loosely-coupled structure that does not depend on specific products.

3 Rule Language and Supporting Components

In this section, we concentrate on design of the rule language and of management components. Rule management provides a feasible and convenient tool to freely define the structure of mapping rules, which is referred to in subsequent translation processing. Inside this component, we iteratively implement support for the proposed rule language.

3.1 Rule Language

We reviewed existing solutions and conclude that most of them lack extensibility either in programming level, or in rule structure (section 2.3). For supplying a basic capability of mapping description, we analyzed XeOML and reserve two critical elements of rules (*Class* and *Attribute*). Both are respectively corresponding to the XML element node and attribute node. We develop the rule language along these lines. A segment of mapping rule for translating NavLog into INSPIRE is shown in Listing 1. It describes three basic mapping possibilities: changing attribute name, changing element name and adding a new element. We make use of XPath, which is used to navigate through elements and attributes in an XML document, for navigating between nodes.

```
01<?xml version="1.0" encoding="UTF-8"?>
02 <MappingRule id="1" version="1.0">
03  <SimpleMapping type="AttributeMapping" id="1">
04   <Source>
05    <AttributeElement name="NAME" parentElement="//topp:HE_F_WAY"/>
06   </Source>
07   <Target>
08    <AttributeElement name="RoadName" parentElement="//topp:HE_F_WAY"/>
09   </Target>
10   <Operation expression="AttributeNameChange"/>
11  </SimpleMapping>
12  <SimpleMapping type="ClassMapping" id="2">
13   <Source>
14    <ClassElement name="LANEWIDTH" id="//featureMember/topp:HE_F_WAY/ "/>
15   </Source>
16   <Target>
17     <ClassElement name="RoadWidth id="//featureMember/topp:HE_F_WAY/ "/>
18   </Target>
19   <Operation expression="ClassNameChange"/>
20  </SimpleMapping>
21  <SimpleMapping type="ClassMapping" id="3">
22   <Source/>
23   <Target>
24     <ClassElement name="FormOfWay" id="//topp:HE_F_WAY"/>
25   </Target>
26   <Operation expression="ClassCreateForFormOfWay"/>
27  </SimpleMapping>
28 </MappingRule>
```

Listing 1. Segment of a mapping rule

The element *MappingRule* (line 02) is the root element and two required attributes should be specified: *version* and *id*. Now version is marked as '1.0' mostly, and every rule should a unique identifier that is used to identify the rule file later. Each *SimpleMapping* (lines 03-11) records all necessary information for one translation processing, which, except for a required attribute type, holds three essential children elements: *Source* (lines

04-06), *Target* (line 07-09), as well as *Operation* (lines 10-11). The first two are used for specifying the resource nodes and target nodes. They describe the elements needed to translate and how the result looks like. The Operation element assigns an appropriate operator for translation execution. According to the type of the translation operator clarified in SimpleMapping, sets of elements are defined in Source and Target.

The proposed language also supports one-to-one, one-to-many, and many-to-many mappings by using customized operation classes. Taking the SimpleElement identified as ’1‘ as an example (lines 03-11), it is a kind of translation for attribute node. An operation named ’AttributeNameChange‘ implies that we will convert the attribute’s name defined as ’NAME‘ in Source into ’RoadName‘ in Target. ’parentElement‘ is used to navigate these attributes through XPath syntax. A many-to-many example can be found in SimpleElement with ’id=3‘. In the example, we add an element to the target data set, which does not exist in source. Accordingly, the Source element is kept empty (line 22). Depending on the algorithm implementing the translation operator, a new attribute named ’FormOfWay‘ will appear in the translated data set. It can be considered as a special case of many-to-many mapping. In our case, we handle processing of the complex mapping over to concreted operation classes, rather than completely depend on the rule definition. It guarantees that our solution can easily handle with complex mapping situation.

3.2 Rule Managing Components

In order to support the previously introduced language for mapping rules, we suggest an extensible rule management, which comprises two building blocks; *Rule Analyzer* and *Rule Generator*. These components guarantee extendable rule structure, and that rules in form of XML documents can be converted into programming objects and vice versa. Users can benefit from this bi-direction rule generation in two aspects:

1. The rules stored into files can be re-used. Generating programming objects in real time from rule files guarantees that this component can sufficiently loosely coupled with translation executing component. Any possible component or system, which applies the rule to execute the concreted translation, only need to retrieve the rule files and are able to obtain related rule programming objects for executing. We become able to reduce the amount of consuming internet resources, because the only element which needs to be transported amongst distinct translation platforms is a rule file. Furthermore, the components become platform-independent.

2. The rule structure itself can be extensible and modified freely without making any changes in components. Due to applying reflection (Bates and Sierra 2005), the processing of rule generation is completely automatic. For instance, if the user needs to extend the rule structure, they only add a simple JavaBeans (SUN 2009) as a standard means for presenting objects, which will represent rule's element and register the new element in the rule structure configuration file. This registration includes a set of descriptions including how the rule file looks like.

The interplay of the rule management components is shown in Figure 4. In the beginning of starting up this component, Rule Analyzer prepares the necessary *Rule Objects* for the Rule Generator. After that, Rule Generator can communicate with users (programmers or terminal client) in two possible phases; generating rules either in file, or in programming instances.

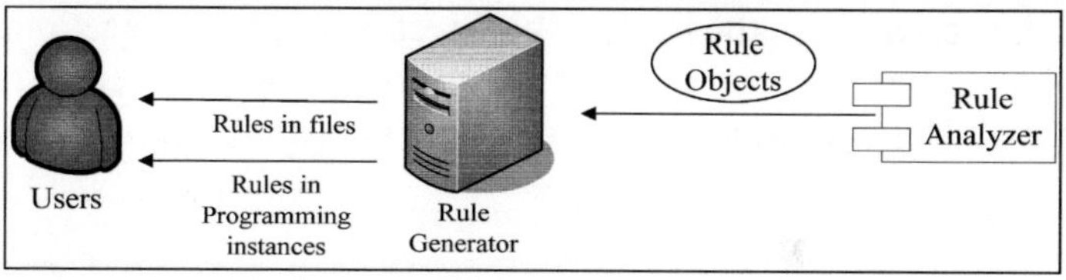

Fig. 4. Interplay of rule management components

Rule Analyzer guarantees the extensibility in rule structure. Every time, while starting this service, the analyzer will first automatically analyze the supported rule structure, which is described in form of a configuration file.

According to the description in the rule structure property file, the Rule Analyzer will choose a corresponding implementing class file from the repository of rule elements and loads them into system as run-time instances (Figure 5). These classes are implemented as JavaBeans, which includ-ing a set of configurations and related 'set' and 'get' operations. Extending the supported structure only requires provision of the according Java-Beans.

Rule Generator provides means to communicate with another components or users in the system. Communication happens in two modes. A translation service needs run-time instances of mapping rules. Hence, it should be able to obtain mapping rules through operations implemented in Rule Generator. The Rule Generator loads a file tat contains a set of mapping rules, and then chooses rule elements instances offered by Rule Analyzer. On this basis it can finally generate that rule instance.

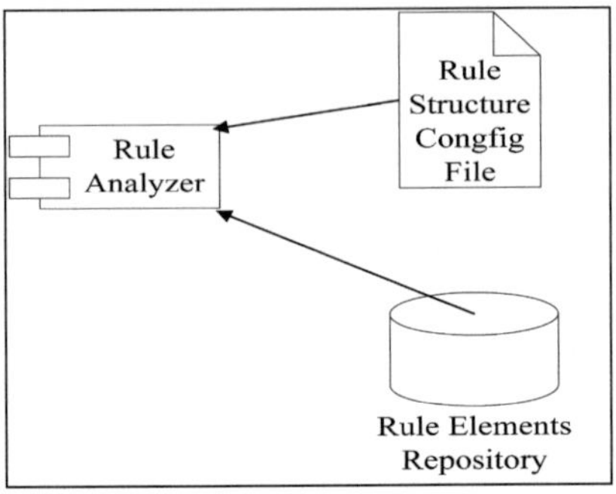

Fig. 5. Structure of Rule Analyzer

In order to guarantee the mapping rule can be re-used in future translation, the Rule Generator should also support to store rules on persistent media. We call the related storage *Rule Repository* in the following. We equip the Rule Generator with functionality to store rule instance as XML files. This capability allows for 'bi-directions' communication, because Rule Genera-tor can convert rule file into rule programming instance, and vice versa. The core functions of the Rule Generator are shown in Figure 6.

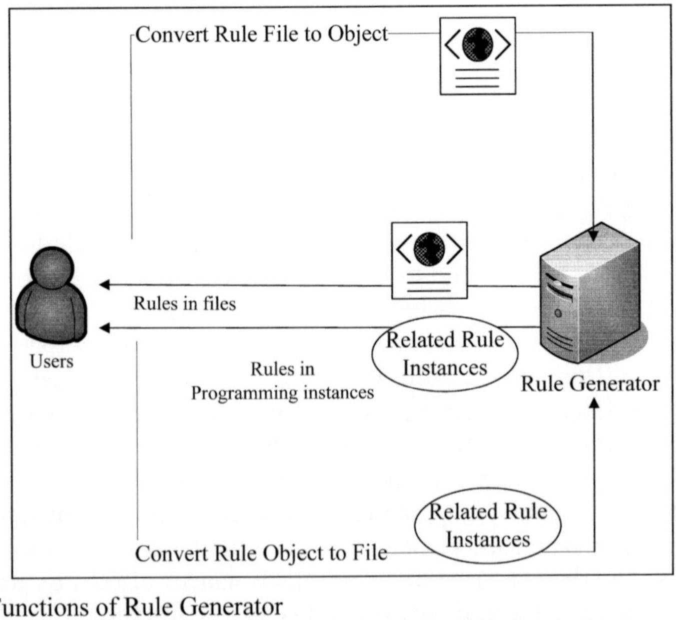

Fig. 6. Functions of Rule Generator

4 Design of Translation Architecture

Having rule support available, we now provide a method for executing translation of geospatial data. For this purpose, we modify and extend a Web Service-based solution that we developed previously (Beckman et al. 2009). We analyze two architectural options; both could be applied for implementing the 5-layered approach as proposed by Lehto and Sarjakoski (2004) (see also section 2.2). In respect to the first option, we emphasize a classification of participants. We distinguish between common and advanced users, and give the latter more freedom in rule customization. The second option, which we name *translation encoding*, is based on the idea that schema mapping is considered as an accessory function of download services. It implies that users can trigger the translation function while invoking a WFS.

4.1 Translation – Option One

The architecture representing the first option is given in Figure 7. In respect to the INSPIRE Implementing Rules (section 2.2), this corresponds to a realization of a Direct Access Download Service. Consequently, this central component redirects the mapping rule and the data set to a *Translation Component* in order to obtaining target data sets.

In the following we define the workflow of translation. We, realize the download service as a separated component (*Central Controller Component*). A possible application for converting NavLog data to INSPIRE data could be made up with six steps (indicated by the numbers in Figure 7):

(1) A user directly sends the request for NavLog data to Central Controller Component. He or she should specify a desired output, for example INSPIRE transport network, as part of the request.
(2) Central Controller Component accesses the source data (NavLog in this case) from a WFS.
(3) Central Controller Component retrieves a suited set of rules from Rule Repository.
(4) Central Controller Component requests the execution of the rules on the data set from the translation component, which provides execution environment for the rule language as defined above (section 3.1).
(5) After execution, the desired data set is passed to the control component.

(6) Now the Central Controller Component can reply to the user by sending the final data (INSPIRE transport network data in this case).

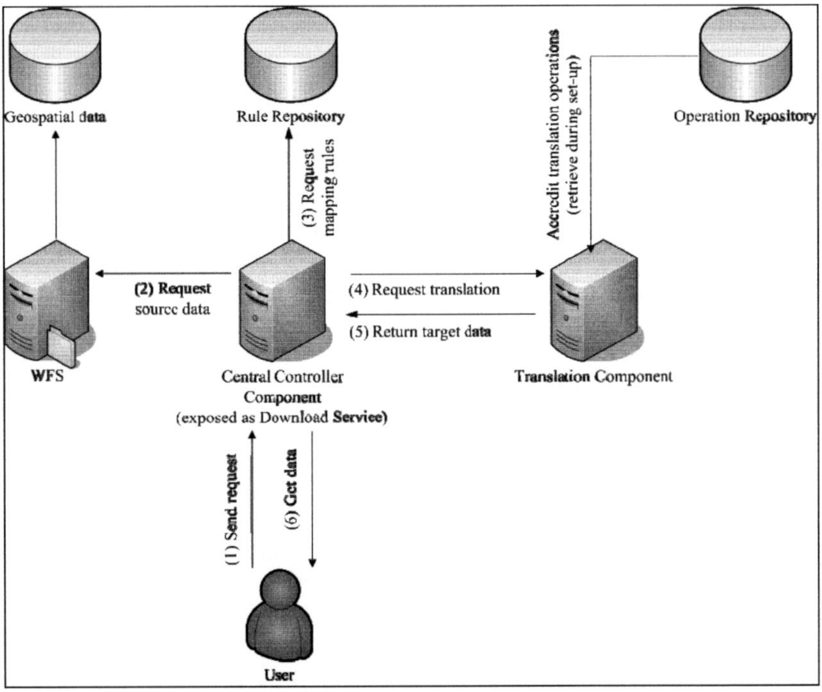

Fig. 7. Proposed architecture of Translation Execution Service

With our architecture, we support two user roles; *Common Users* and *Ad-vanced Users*. Common users can search for available pre-defined map-ping rules in Rule Repository and for geospatial data that they want to translate. In addition, Advanced Users are qualified to customize transla-tion operations. For instance, users may want an attribute creation opera-tion but it has not been implemented in standard operations library yet. In addition, new rules can be defined based on operations available in the Operation Repository. These rules can be added to the Rule Repository for later re-use.

Notably, translation is provided as an independent component. A de-coupled component executes translation depending on the translation op-erations defined in Operation Repository. These operations are constructs for rule representation. Independent of the rule structure, the Translation Component only determines whether or not operations have already de-fined in repository. Due to exposing an access to specific users for updating operations, it guarantees capabilities of rules extension and customization.

4.2 Translation – Option Two

Opposed to the option outlined above, we may implement a *translation encoding* directly on top of the download service, in our case on top of an OGC WFS. Following this option, mapping rules are directly included in a data request. A possible architecture is shown in Figure 8. Notably, this solution differs from what has been introduced for INSPIRE (section 2.2). We suggest a way of sending flexible rules together with a request for data instead of simply executing previously defined set of mapping rules.

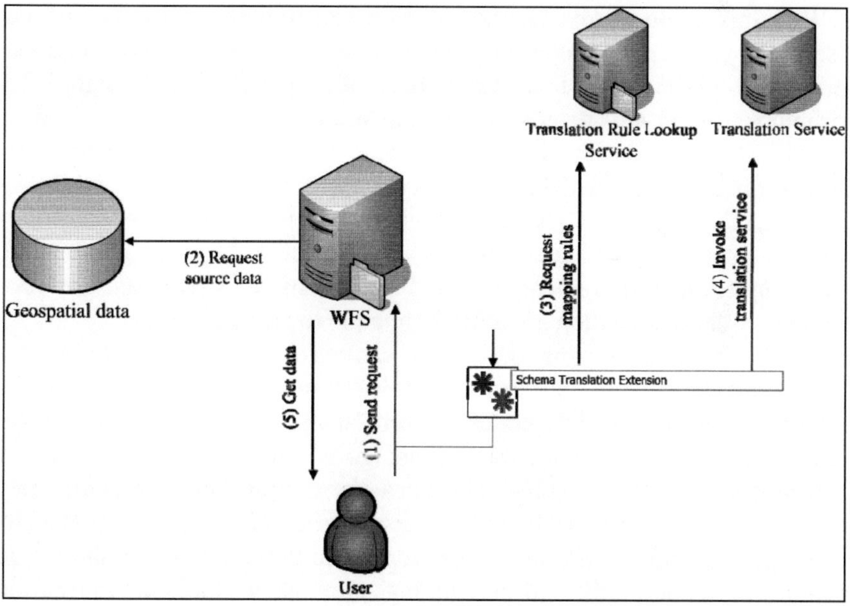

Fig. 8. Possible architecture for Translation Encoding

Requests are processed as follows (numbering corresponds to labels in Figure 8):

(1) Users request the extended WFS to deliver a specific data set after a distinct set of mapping rules has been executed.
(2) The WFS connects to a data base offering source data.
(3) Depending on users' requirement, the extension named 'Schema Translation Extension' will search appropriate mapping rule support.

(4) The extension will choose an appropriate translation service, according to the rule format, and executes the translation.
(5) The WFS redirects the resulting data set to the requester.

Some obvious changes occur in this architecture. First, it simplifies the WFS part in option one, now which WFS service will be retrieved data is not an issue any more, instead, we make the translation function as an optional processing for every WFS requests. Second, we reduced the complexity of the rule management component by decoupling functions for system extensibility. We change our strategy to make the system use different translation tools automatically, according to uses' requirements. Accordingly, searching mapping rules and the manner to invoke them become crucial facets.

With this solution, the manner of data consumption does not change. Users still only need to communicate with WFS to retrieve encoded data in for-mat of GML. Data Provides publish the spatial data through WFS without knowing any details about translation. Schema mapping is integrated with WFS as an optional function. Whether to execute practical translation de-pends on whether the user triggers this function while requesting data. Once it is activated, it will search for feasible mapping rules and corre-spondingly pick up appropriate translation service for execution according to the requirement specified by user in WFS requests. The procedures of data translation are entirely opaque to terminal users.

Another advantage is that the system can make a set of different translation products cooperate in translation processing. A third is the flexible integra-tion with established concepts from spatial data infrastructure (Nebert 2004). A fourth remarkable advantage is that the Central Controller Com-ponent is not necessary. The occasion of translation is completely under control within every time the users request WFS. This solution guarantees applicability of the structures mentioned in section 4.1, while it also pro-vides the possibility of integrating previously available translation tools, such as FME, GoPublisher, Spatial Data Integrator, and GeoXSLT.

5 Component Implementation with Example

According to the research above, we implement an example including the rule management components (Rule Analyzer and Rule Generator) and the translation component using Java (Bates and Sierra 2005). These components can cooperate with the two optional settings discussed above. We also applied these components for implementing the first option of data translation (section 4.1). The implementation of the second solution (section 4.2) is topic to ongoing work. The concreted implementation will not be presented in this paper. We expect a first prototype by next month (February 2010).

5.1 Components for Rule Management

A set of classes corresponding to rule elements is created as simple JavaBeans. All element class should extend *SimpleMappingElementWrapper* class. It handles the general processing of bi-directional mapping by implementing the adapter-pattern (Gamma et al. 1998).

A configuration file (named 'rules-config') holds the information about how to organize the rule elements. RuleXMLFileMapper is responsible for mapping processing and offering services. The main interfaces of the service are list in Table 2 (the overloaded methods are not listed).

Table 2. Function list of Rule Component

Function	Description
getCapabilities	Return a set of available operations for rule definition.
createXMLFile	Generate the rule in file.
loadSimpleRule-FromXML	Mapping rule into object from file.

5.2 Translation Component Framework

In this part, we implement a framework for execution of mapping rules on source data sets. The distinct translation algorithms are capsulated into a set of Java classes with a suffix 'Action' and registered in 'action-config' file. The *SchemaTranslator* will determine which actions will be used to execute translation and then invoke them in run-time based on Java reflect technique.

The actions can be classified into two categories: common action and spatial action, which respectively refer to ordinary textual translation and spatial calculation. *SpatialAction* and *CommonActionAdapter* are the parents for spatial operation and common translation. SpatialAction firstly extract the elements from input GML file and creates related Java Topology Suite (JTS) Objects (VividSolutions 2009). JTS is a Java API that implements a core set of geospatial data operations using an explicit precision model and robust geometric algorithms. Spatial actions only need to hand over processing tasks to JTS spatial functions. Own (spatial) algorithms can be added at will.

CommonActionAdapter is a centralized element navigator based on XPath syntax defined in rule file. So its subclass can simply simulate the business logic to complete translation processing avoiding XML technical handling.

Besides that, a user interface based on web-page was developed. It plays the role of Central Controller Component (section 4.1). It exposes an interface to users for manually controlling the main workflow.

5.3 Example Walkthrough

We created a web-based user interface to present the system workflow. A screenshot is shown in Figure 9, where we apply the developed components to the NavLog example.

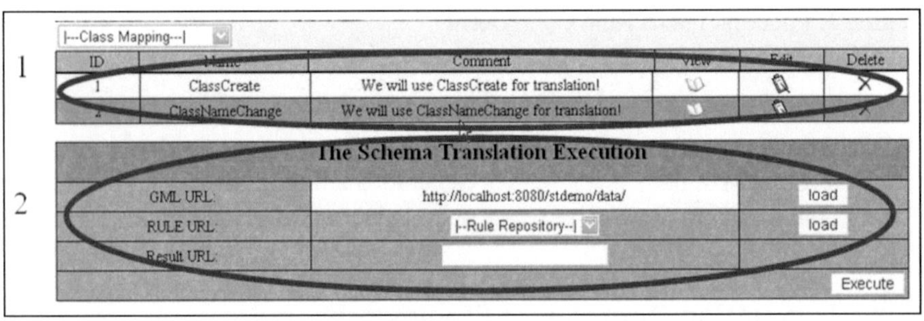

Fig. 9. User interface for Rule Management

In the first part of the interaction (Figure 9, part 1), user can obtain a metadata about available operations. Rule elements can be selected. The second part of interaction considers a WFS services manager, which decides on the source data provider, how to retrieve the rule file and where to dispatch the translated data within system's real time lifecycle. Here we load a NavLog data named 'HE_F_WAY' from the service that is located at http://localhost:8080/stdemo/ (Figure 10, part 1). Users should choose an existing rule from repository for translation execution (shown as Figure 10, part 2). If they fail in finding an appropriate rule, they can use Rule Generator to create a new customized one. As example, we choose the rule defined above, whose name is 'Navlog2INSPIRE' (Figure 10, part 1).

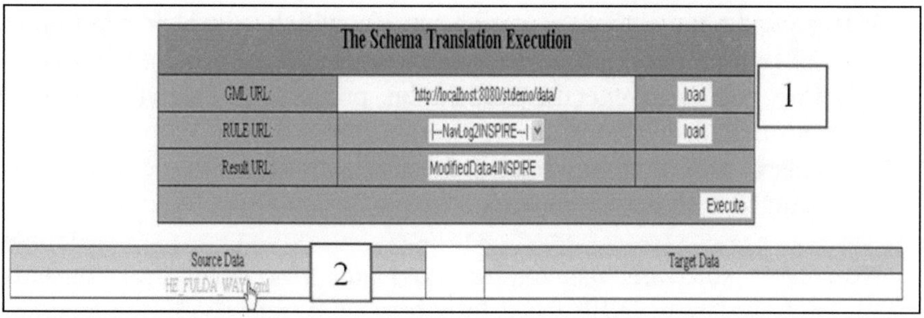

Fig. 10. Translation Prepare

Finally, users specify the output URL to either redirect translated data to another service or send it back to source URL (Figure 11). We keep the translation results in local file system in order to directly check them in target data repository.

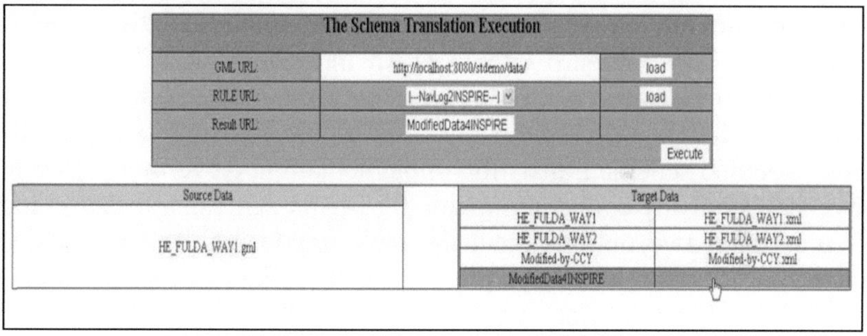

Fig. 11. Specifying Output URL and execute translation

6 Conclusion and Future Work

Schema mapping plays a critical role in geographic information harmonization. In this paper, we implement translation procedure in Web Service level. In particular, we:

1. Specified the rule structure, where the structure was simplified into only two categories such as class (elements of XML) and attributes (attributes of XML).
2. Designed a rule management tool using the Java programming language. This tool can efficiently convert mapping elements in Java objects into XML elements and vice versa. Also the tool has a feasible extensibility that allow user to modify rule structure, add new mapping elements without changing programming codes.

3. Designed a translation execution service, which is implemented into a flexible framework. Users are able to use a set of standard actions to define rule and execute translation processing through invoking encapsulated functions, or they can customize their actions based on standard programming interfaces and later they can be seamlessly interacted with our framework.

Executable software was implemented and a scenario that translates from NavLog data to INSPIRE data is presented. Specifying the mapping rules based on XeOML. We are in the second iteration of the rule language and supporting tools. Currently, we implement a solution that addresses the complex mappings. It has the capability for spatial functions processing by integrating JTS. It is adapted in practical applications where we translate further data models into INSPIRE Data Specifications.

Comparing to other existing products, our solution guarantees the extensibility in definition of mapping rule structure and translation operation. Users can flexibly customize rule structure and translation operation through extending standard interface. An individual workflow definition component provides the possibility of integration between our service and others. The developed components offer a good design spectrum for a loosely coupled schema translation implementation. Nevertheless, we have not yet achieved any cooperation with third-party schema translation tools.

We consider two options for using the components for geospatial data translation. The first is fully implemented. It realizes one of the architectures for transformation as suggested by INSPIRE. Work on the second setting is close to a first release. Defining a specification for schema mapping encoding, which can standardize the manner of invoking schema translation function in order to integrate with OGC Services goes beyond the architectures proposed by INSPIRE. Our future research will focus on definition this specification, and then implement it as an extension of existing WFS products. The issues of interoperation amongst different schema translation tools will also be addressed with the defining standard implementation specification. As we only implemented mapping language support as far as required in respect to the illustrating example, the iterative development is still ongoing. We are far from completion, but strongly believe that we are on the right track.

Acknowledgements

The research presented in this paper is based on a schema mapping project, which was carried out at the Institute for Geoinformatics of the WWU Münster during 2008 and 2009. We deeply acknowledge the contributions of all other participants. We would also like to thank four anonymous reviewers for their valuable feedback on an earlier version of this paper. The work has been supported by the European Commission, Erasmus Mundus Programme, M.Sc. in Geospatial Technologies, project no. 2007-0064.

References

Bates, B. and Sierra, K. (2005). Head First Java. O'Reilly, Sebastopol, CA.

Bechhofer, S., van Harmelen, F., Hendler, J., Horrocks, I., McGuinness, D.L., Patel-Schneider, P.F., Stein, L.A. (2004). OWL Web Ontology Language Reference, http://www.w3.org/TR/owl-ref/, last date accessed: 11.2009.

Beckmann. O., Blut, C., Deelmann, T., Michels, H., Osmanov, A., Roth, M., Weerasinghe, H., Wilden, M., Schade, S. (2008). Getting Inspired? Geoinformatik 2009. Osnarbrück, Germany.

Camp to Camp (2008). http://www.spatialdataintegrator.com/, last date accessed: 11.2009.

Curtis, E. and Müller, H. (2006). Schema Translation in Practice. White Paper. http://83.138.131.106/eurosdr/workshops/models_2006/discussion/Schema%20Translation%20in%20Practice.pdf, last date accessed: 11.2009.

Donaubauer, A., Fichtenberger, A., Schilcher, M., Straub, A. (2006). Model Driven Approach for Accessing Distributed Spatial Data Using Web Services - Demonstrated for Cross-border GIS Applications. XXIII International FIG Congress, Munich, Germany.

EC (2007). Directive 2007/2/EC of the European Parliament and of the Council of 14 March 2007 establishing an Infrastructure for Spatial Information in the European Community (INSPIRE). European Commission (EC), Official Journal of the European Union.

Frankel, D.S. and Parodi, J. (2004). The MDA Journal: Model Driven Architecture Straight From the Masters. Tampa, FL, Meghan Kiffer Press.

Gamma, E., Helm, R., Johnson. R. and Vlissides, J.M. (1998). Design Patterns: Elements of Reusable Object-Oriented Software. Addison-Wesley, Boston, MA.

INSPIRE (2004). INSPIRE scoping paper, http://www.ec-gis.org/inspire/reports/inspire _scoping24mar04.pdf, last date accessed: 11.2009.

INSPIRE (2009). D2.5: Generic Conceptual Model, Version 3.2. INSPIRE Data Specifications Drafting Team.

INSPIRE (2009b). INSPIRE Data Specification Transport Networks – Guidelines (Version 3.0.1). INSPIRE Thematic Working Group on Transport Networks.

INSPIRE (2009c). INSPIRE Draft Implementing Rules for INSPIRE Transformation Services (Version 3.0). INSPIRE Networking Services Drafting Team.

INSPIRE (2009d). INSPIRE Draft Technical Guidance for INSPIRE Coordinate Transformation Services (Version 2.0). INSPIRE Networking Services Drafting Team.

INSPIRE (2009e). INSPIRE Draft Implementing Rules for Download Services (Version 3.0). INSPIRE Networking Services Drafting Team.

Klausen F.M. (2006). GeoXSLT: GML processing with XSLT and spatial Extensions. Department of informatics. University of Oslo.

Lehto, L. (2007). Schema translations in a Web service based SDI. 10th AGILE International Conference on Geographic Information Science, The European Information Society: Leading the way with geo-information, Aalborg, Denmark.

Lehto, L. and T. Sarjakoski, 2004. Schema Translations by XSLT for GML-Encoded Geospatial Data in Heterogeneous Web-Service Environment. Proceedings of the XXth ISPRS Congress, July 12-23, 2004, Istanbul, Turkey, International Archives of Photogrammetry, Remote Sensing and Spatial Information Sciences, XXXV(B4:IV): 177-182.

Madsen, M. (2009). The Role of Open Source Data Integration - An Analyst White Paper, http://www.talend.com/library/reflibrary.php, last date accessed: 11.2009.

NavLog (2005). Formatbeschreibung zur Erstellung einer forstspezifischen Navigationsdatenbasis - Pragmatisches Shape Forst, http://www.kwf-online.de/deutsch/arbeit/geodat/Spezifikation_pragmatischesshapeForst_3_0.pdf, last date accessed: 11.2009, (German).

Nebert, D. (2004). Developing Spatial Data Infrastructures: The SDI Cookbook, Version 2.0. Global Spatial Data Infrastructure.

OGC (2005). OpenGIS® Web Feature Service Implementation Specification 1.1, Open Geospatial Consortium Inc.

OGC (2007). OpenGIS® Geography Markup Language (GML) Encoding Standard - Version 3.2.1. The Open Geospatial Consortium.

Östman A, Abugessaisa I, Tanzilli S, He X, El-Mekawy M, 2009. GeoTest: A

Testing Environment for Swedish Geodata. Paper presented at the GSDI 11 World Conference, Rotterdam, June 15-19, 2009.

Pazienza M.T., Stellato, A., Vindigni, M., Zanzotto, F.M. (2004). XeOML: An XML-based extensible Ontology Mapping Language, in Proceedings of Workshop on Meaning Coordination and Negotiation, held in conjunction with 3rd International Semantic Web Conference (ISWC-2004) Hiroshima, Japan, November 2004.

SafeSoftware (2008). Spatial ETL: Making Spatial Data Accessible. White Paper, http://www.safe.com/technology/whitepapers/spatial-etl.php, last date accessed: 11.2009.

Schade, S. (2009). Translation of Geospatial Data: Challenges, Solution and Vision. AGILE Workshop Challenges in Geospatial Data Harmonisation, Hannover, Germany.

Schade, S. (2010). Ontology-Driven Translation of Geospatial Data. AKA, Heidelberg, Germany.
Scharffe, F. and de Bruijn, J. (2005). A language to specify mappings between ontologies. SITIS6: IEEE Conference on Internet-Based Systems.
SUN (2009) JavaBeans SUN, http://java.sun.com/docs/books/tutorial/javabeans/whatis/ index.html, last date accessed: 11.2009.
VividSolutions (2009) JTS Topology Suite - Version 1.4 Developer's Guide.
W3C (2009) XPath - W3School, http://www.w3schools.com/XPath/default.asp, last date accessed: 11.2009.

Automated Image-Based Abstraction of Aerial Images

Amir Semmo, Jan Eric Kyprianidis, Jürgen Döllner

Hasso-Plattner-Institut, University of Potsdam,
doellner@hpi.uni-potsdam.de

Abstract. Aerial images represent a fundamental type of geodata with a broad range of applications in GIS and geovisualization. The perception and cognitive processing of aerial images by the human, however, still is faced with the specific limitations of photorealistic depictions such as low contrast areas, unsharp object borders as well as visual noise.

In this paper we present a novel technique to automatically abstract aerial images that enhances visual clarity and generalizes the contents of aerial images to improve their perception and recognition. The technique applies non-photorealistic image processing by smoothing local image regions with low contrast and emphasizing edges in image regions with high contrast. To handle the abstraction of large images, we introduce an image tiling procedure that is optimized for post-processing images on GPUs and avoids visible artifacts across junctions. This is technically achieved by filtering additional connection tiles that overlap the main tiles of the input image. The technique also allows the generation of different levels of abstraction for aerial images by computing a mipmap pyramid, where each of the mipmap levels is filtered with adapted abstraction parameters. These mipmaps can then be used to perform level-of-detail rendering of abstracted aerial images.

Finally, the paper contributes a study to aerial image abstraction by analyzing the results of the abstraction process on distinctive visible elements in common aerial image types. In particular, we have identified a high abstraction potential in landscape images and a higher benefit from edge enhancement in urban environments.

M. Painho et al., (eds.), *Geospatial Thinking*, Lecture Notes in Geoinformation and Cartography,
DOI 10.1007/978-3-642-12326-9_19,

1 Introduction

Since its first advent in 1858, when the French photographer and balloonist Gaspard-Félix Tournachon captured aerial images over Paris, aerial photography has evolved to a fundamental tool for geodata capturing, processing, and visualization.

With today's usage of aerial photography in photorealistic visualization, aerial images have become a viable medium in applications such as in cartography, photogrammetry, construction, planning, and marketing. They are also used as popular source of geoinformation in today's geographic information systems (GIS) and by the general public, in particular since the wide spread application of aerial photography in Internet mapping services such as GoogleEarth or Microsoft Bing Maps.

Furthermore, the technical advances in the field of oblique imagery (Pictometry, 2009) have opened new avenues for efficient 3D geoinformation acquisition.

Simultaneously, with today's technical advances in remote sensing for aerial photographs (e.g., aerial images with a resolution of 10 cm per pixel Blom Pictometry, 2009), aerial images have been established as a fundamental data source in 3D geovisualization for constructing realistic, detailed models of reality.

1.1 Non-Photorealistic Spatial Communication

Aerial images show the richness of visual details and the natural appearance of our environment. Insofar, they serve as important geodata source for photorealistic visualization. Non-photorealistic visualization, however, can be a better choice for spatial perception, analysis, understanding and knowledge discovery (Meng, 2002; Nienhaus, 2006) in many areas of application. Those computer-generated presentations basically build on the intention for communicating highly complex contents by emphasizing subtle attributes and features and omitting extraneous information. A typical approach is the imitation of classical depiction techniques in terms of illustrative and aesthetically pleasant visualizations. For example there has been extensive research on visualizing images using computer-generated watercolor effects (Curtis et al., 1997), pencil or pen and ink drawings (Gooch and Gooch, 2001).

Fig. 1. Examples of abstracted aerial images created using our framework

Another classical area of research is the stylization and abstraction of photographs using edge-preserving smoothing and enhancement filters (e.g., Kyprianidis and Döllner, 2008; Kyprianidis et al., 2009; Winnemöller et al., 2006) for depicting bold edges and large regions of constant color. Research in visual perception argues that non-photorealistic rendering should take into consideration the specific nature and type of objects displayed, since "meaningful abstraction clearly affected viewers in a way that supports an interpretation of enhanced understanding" (Santella, 2005). Additional evaluations proof that non-photorealistic visualization has the ability to direct a viewers gaze and focus the interest on particular areas. These are not just applicable to art, but also to wider problems of graphical illustration and visualization (Santella and DeCarlo, 2004), e.g. 3D geovirtual environments. Earlier research on non-photorealistic rendering of 3D geovirtual environments involves the non-photorealistic rendering of expressive representations of 3D city models (e.g., Baumann et al., 2005; Döllner et al., 2005; Döllner, 2007; Glander and Döllner, 2009) and how users can leverage from these expressive depictions in graphics design, primary valuable in application areas like city and landscape planning or tourist information systems

1.2 Abstracted Aerial Images for Non-Photorealistic 3D Geovirtual Environments

Research in the fields of non-photorealistic rendering primarily focuses on rendering of 3D models such as terrain models and 3D building models

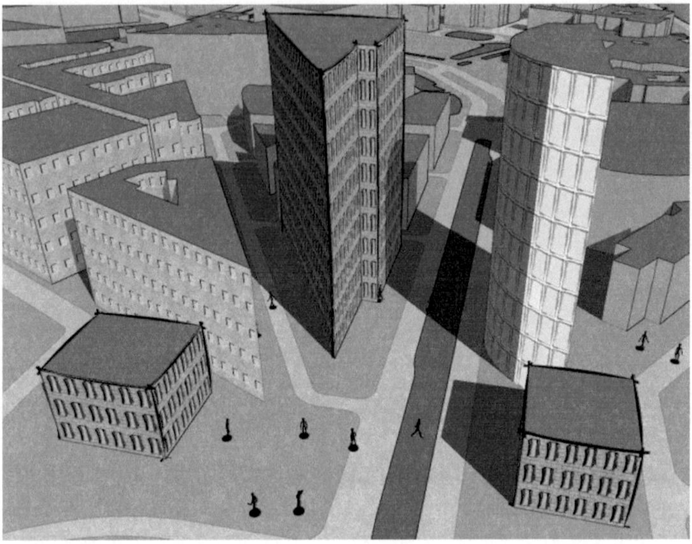

Fig. 2. Illustrative visualization of a 3D city model with stylized edges, facade textures and regular shadows as described by Döllner et al. (2005)

(see Figure 2), but has not developed specific solutions for a proper stylistic representation of the terrain with projected aerial images.
In this work we present a technique for automatically abstracting aerial images in a non-photorealistic way (see Figure 1); the results supply a concise presentation of the terrain in non-photorealistic renderings of 3D geovirtual environments.

Additionally, common image abstraction techniques focus on image filtering that can be processed as a whole by the graphics processing unit (GPU). Our framework, in contrast, contributes an approach to filter massive image data on the GPU in order to allow for post-processing of very large aerial images in a seamless way. Technically we make use of image tiling, which we arrange as main and connection tiles for preventing visible artifacts at junctions. With this procedure we facilitate pre-processing of stylistic image representations such that rendering is feasible on low-cost computing devices or devices with limited computing power, e.g., smart phones or PDAs. To show the potential of abstraction, we analyze the results on distinct common elements that regularly occur in aerial images, e.g., vegetation, buildings, and transportation objects. Our analysis focuses on the abstraction potential in detailed structures and the benefits of edge enhancement that occurs in image discontinuities.

For filtering, we use as initial position primarily those aerial photographs that correspond to two classes: Oblique photos and orthophotos.

Oblique photos are images that are captured at an angle of about 40 degrees from four principle directions. Orthophotos are most valuable to cartography and city planning and are the primary source for GIS to create maps because of their nature, i.e., having a uniform scale and providing undistorted geometry.

2 Related Work

The presented image filtering technique is based on work by Kyprianidis and Döllner (2008), who describe automatic non-photorealistic image processing for creating stylistic illustrations from color images. The technique extends the approach of Winnemöller et al. (2006) to use iterated bilateral filtering for abstraction and difference-of-Gaussians (DoG) for edge extraction by adapting these filters to the local orientation of the input image. A detailed survey on bilateral filtering can be found in (Paris et al., 2007). The Difference-of-Gaussians (DoG) filter is an approximation of the Laplacian-of-Gaussian (Marr and Hildreth, 1980).

A standard approach for dealing with the rendering of large textures is texture tiling. Tanner et al. (1998) introduce the technique of texture clipmaps, a mechanism that virtualizes mipmap textures and is able to manage high-resolution terrain textures by clipping the mipmaps into regions that are updated according to the viewer's movements. A more general approach is introduced by Döllner et al. (2000), who present a texture hierarchy based on a multiresolution model that incorporates for each texture layer an image pyramid and texture tree. They utilize multi-pass rendering and multi-texturing to get real-time performance and texture layer combinations.

3 System Overview

From a technical perspective, there are two crucial aspects for implementing post-processing techniques for aerial images: First the massiveness of the data and second limited hardware resources for processing these data. For example, if the city of London (2.90 km^2) would be captured at a resolution of 10cm per pixel (24-bit), we would have to deal with an uncompressed dataset of about 20GB. And as the image post-processing is usually carried-out by the GPU in order to take advantage from parallel

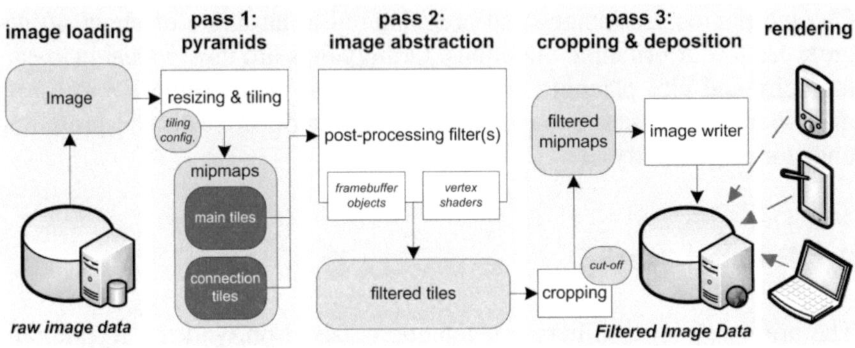

Fig. 3. Overview of our abstraction pipeline. Raw image data (e.g. GeoTIFFs) is loaded and pre-processed by image resizing and tiling. A filtering is conducted on each of the tiles, which are cropped and written back into memory. Finally, the filtered mipmaps are stored onto disk, from where they can be requested for rendering onto different devices

processing, the GPU itself (respectively the texture memory) turns out to be the main bottleneck. Typically GPUs of today's generation can address and process images with 8,192 x 8,192 pixels at the same time.

In our approach, the technique is designed in such a way that these GPU limitations can be by-passed by means of image tiling. Figure 3 illustrates the abstraction pipeline, which we also use for the analysis given in Section 4. The system starts to incrementally load a given aerial image into main memory. Next, the image is tiled in evenly sized main parts such that the GPU can process them as a whole (see Section 3.1). Each of these tiles (or the whole image) are resized down to 1 x 1 pixels, capturing the halve side length images of the upper levels. Those layers correspond to the mipmaps of a usual mipmapping procedure (Williams, 1983), with the difference that we are able to achieve a level of abstraction as each layer will be filtered apart from the other layers (see Section 3.2). After this, the images tiles are loaded into texture memory and filtered by the abstraction stages as described by Kyprianidis and Döllner (2008). For the abstraction process we use off-screen buffers using GLSL shaders with OpenGL. Additional (connecting) tiles that overlap the main tiles are taken into account. This is done to avoid visible artifacts at the junctions of the main tiles during the rendering process (see Section 3.1). This includes the cropping of each of the tiles in the third pass of the pipeline (Figure 3).

The storage of the filtered mipmaps can be handled differently. We compress the filtered mipmaps by the S3 Texture Compression algorithm (S3 Corporation) and store them as DDS (Direct Draw Surface) file such

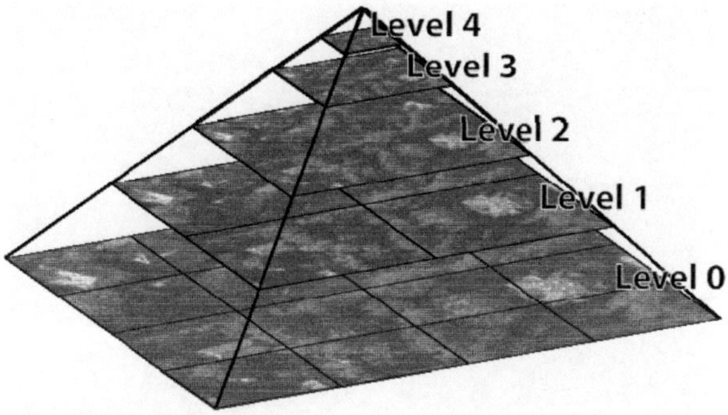

Fig. 4. Overview of an image pyramid consisting of a base image (level 0) and a series of successively smaller sub-images, each at half the resolution of the previous image

that the images/tiles can be rapidly loaded into texture memory and rendered (see Section 3.3). An alternative method could base the storage on a streaming procedure that makes use of the TIFF file standard, which supports storage of mipmaps.

The abstraction pipeline can be handled by a system that provides the abstracted images pre-tiled and filtered on each of the mipmap levels. Computing devices that do not have the power to compute these images in an adequate time span, or do not provide the required hardware, can request these images and render them directly on-screen.

3.1 Image Resizing & Tiling

After loading, the system scales down the aerial image on each mipmap level and tiles those layers in smaller parts. Most web services providing aerial images already have them split in evenly sized tiles; mapping services for example prefer a tile dimension of 256 or 512 pixels. In our approach, we do not assume to have a pre-tiled image and partition the scaled mipmaps in tiles with a size of 1,024 x 1,024 or (if possible) 2,048 x 2,048 pixels. Image tiling and resizing are processes that are not performed in a distinctive order, but usually images are tiled first and resized afterwards. In any case, this results in an image pyramid that looks like the one in Figure 4.

The plain abstraction of image tiles by filtering in local image domains, however, will cause visible artifacts at the junctions as soon as the scene is being reconstructed and rendered (Figure 5). By taking additional tiles into

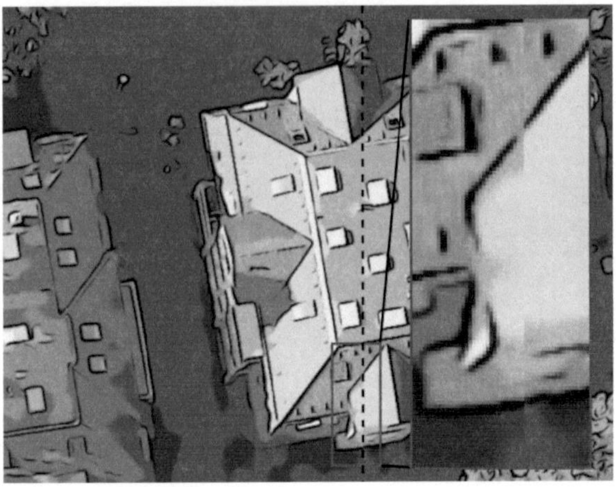

Fig. 5. A visible junction between two tiles in case a plain filtering is conducted. Edges are apparently interrupted and colors abruptly change

account that overlap the junctions, we are able to deal with this problem. Conceptually, each one of these connection tiles overlap two adjacent main tiles in each direction by a cut-off value that is adapted to the domain range of the bilateral filter, which is defined by its standard deviation σ_d. For calculating the lowest cut-off, we will also need to consider the number of iterations of the bilateral filter n_a:

$$\lceil 2 \cdot \sigma_d \rceil \cdot n_a$$

The connection tiles can occur in each main direction of the area as well as in the center of 4 adjacent main tiles, where we take an additional smaller squared tile into account. The dimensions of these tiles can be seen in Figure 6.

3.2 Image Abstraction

For image abstraction, we load each of the tiles into texture memory and run the following post-processing stages on input and temporary generated data. For further details on each of the stages and optimizations we refer to Kyprianidis and Döllner (2008):

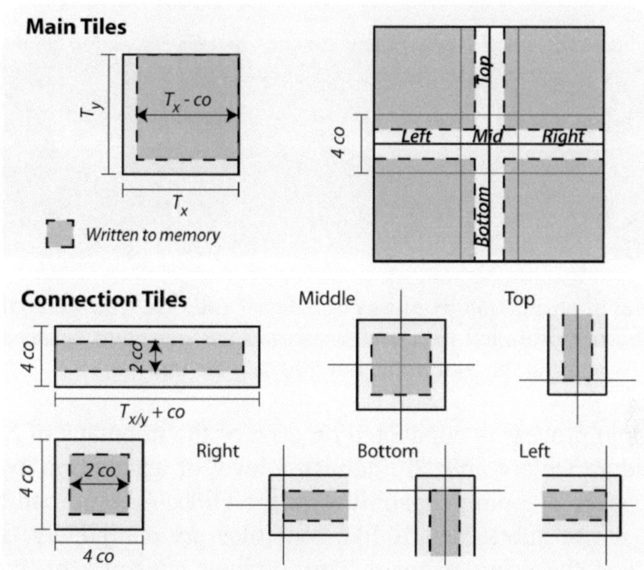

Fig. 6. Overview of the cropping of main tiles and connection tiles for encountering the problem of visible artifacts at junctions. $(T_x \times T_y)$ denotes the size of image tiles

1. Local Orientation Estimation: The estimation of the local orientation of an image allows us to obtain information about the dominant orientation of a region. This information is used for optimizing the bilateral and difference-of-Gaussians filter. It is mathematically retrieved from the structure tensor and its eigenvectors, which encode information about a region's structure.
2. Bilateral Filter: We use the bilateral filter as the main construct for our image abstraction. It replaces the pixel's value by a weighted average of its neighbors in both, space and intensity, and has the characteristic to smooth an image while preserving edges.
3. Edge Detection: A flow-guided anisotropic difference-of-Gaussians kernel, whose shape is defined by the local orientation, is used for edge detection. As first pass, we apply a one-dimensional DoG filter in gradient direction followed by a second pass that applies smoothing along the flow curves of the vector field induced by the smoothed structure tensor.
4. Color Quantization: As last step, we quantize the color range of the abstracted image by using the smoothed step function from (Winnemöller et al., 2006). The final output of this stage is then combined with the product of the edge detection.

Fig. 7. Level of abstraction by means of trilinear filtering. The viewpoint distance decreases linearly from left to right. The greater the viewpoint distance, the fewer details are depicted at the roof of the building and the tree

The whole process is conducted on each of the mipmaps of a tile. With this procedure we are able to establish a level of abstraction by using the trilinear texture mapping capabilities of the GPU. A result can be seen in Figure 7, where subtle details like roof tiles are completely filtered out from 25% of the original image size, outlines shadows are increasingly coarsened and complex objects shapes like bushes or treetops are combined to simpler structures. This is due to the nature of the bilateral filter, having much less pixels of a distinct structure in a resized image left for weighting the color values into the local neighborhood.

3.3 Image Compression & Storage

For image compression and storage we make use of the nVidia Texture Tools library (see NVIDIA Texture Tools 2) for storing the filtered mipmaps as DDS (Direct Draw Surface) files and stitching them to a single DDS file. The DDS scheme is an optimized data format for storing mipmaps and DXT compressed textures. This allows us to render the image tiles quickly with moderate memory consumption.

4 Analysis of Abstracted Aerial Images

To show the feasibility of automated aerial image abstraction, we analyze the results of our approach by distinguishing between the visible objects in aerial images. This is done because the process of abstraction and edge enhancement scales differently among the distinct visible objects, whose characteristic illustration mainly depict similar structures across different images.

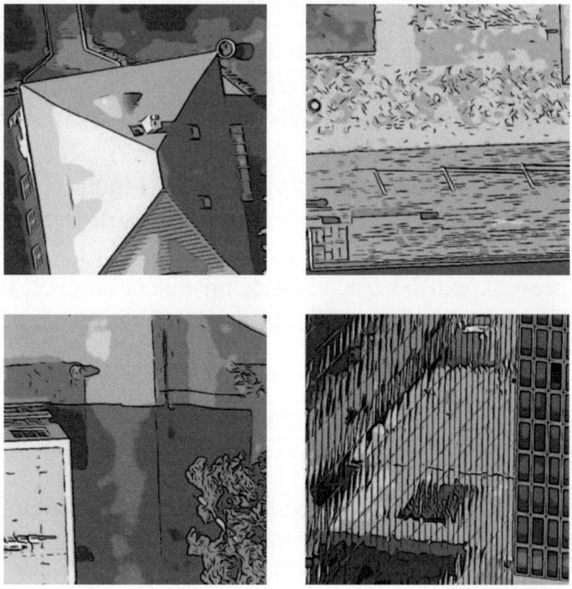

Fig. 8. The four categories of discontinuities that can occur in aerial images and affect the edge detection

On the one hand, this analysis focuses on the abstraction potential in fine granular structures for filtering extraneous information. On the other hand, our analysis identifies benefits from edge enhancements that are derived from four different discontinuities that regularly occur in aerial images: Discontinuities in depth, surface orientation, illumination and reflectance. Those discontinuities primarily highlight well-defined boundaries of objects and allow, due to contrast differences, a detection of edges and enhancement for a non-photorealistic representation (see Figure 8).

Our analysis is based on the thematic model of the CityGML (OpenGIS City Geography Markup Language) specification and the categorization of Döllner et al. (2005). For an overview, we refer to Figure 9 and Figure 10.

4.1 Relief & Terrain Objects

Relief and terrain represent the essential element of aerial images, especially in landscapes photographs. We count everything as plain terrain that reveals the rock of the earth and does not feature unnatural structures, e.g., sand, split, or gravel. Dependent on the granularity of the terrain, the abstraction may unfold a high potential for filtering extraneous information. For example, sandy soils can mainly be depicted as single-colored planes, especially when no edge detection is conducted and the color quantization is limited to few colors. On the other hand, structures with a high granularity like mountainous areas benefit from their three-dimensional structure as light spots get caught and accordingly strengthen the color differences of the terrain's surface.

Additionally, lighting plays the main role for highlighting variations in the slope of the surface and correspondingly supports the process of edge enhancement for depicting the characteristic shapes like trenches and crests. This process can be traced back on discontinuities in the surface orientation, for example when different rock faces meet or at the boundary of snow lines in mountainous areas.

4.2 Vegetation Objects

Vegetation objects always have been a challenge for modeling in 3D geovirtual environments due to their complex structure and fuzziness (see Foody, 1996; Muhar, 1999). Basically we distinguish between the range and 3-dimensional structure (physiognomy) of plants, e.g., trees and bushes as rangy plants and grass as open and plane vegetation. With this categorization we achieve considerable variations in the depiction of abstraction. Lawn for example can induce the highest potential of abstraction just as in the case of fine granular terrain structures. On the other hand, treetops are structures that can highly leverage from the level of abstraction concept. In this case, the edge enhancement turns out to be the main depiction technique, since sub-branches of the tree and their foliage are mainly depicted as partial clouds that primarily differ from their surrounding by their difference in height - an effect that is reinforced by the actual direction of lighting. As the viewing distance increases, the differences in the structures decrease and sub-leafages more and more merge.

4.3 Water Surface Objects

Water surfaces can be described as boundaries that occur between water and air. We count water surfaces as the third element with high abstraction potential, because generally they can be depicted as single-color planes without any edge enhancement. Nevertheless, due to high specular characteristics of water surfaces, they tend to feature areas with high reflectivity, whose abrupt color differences to the normal surface color might influence an edge enhancement process.

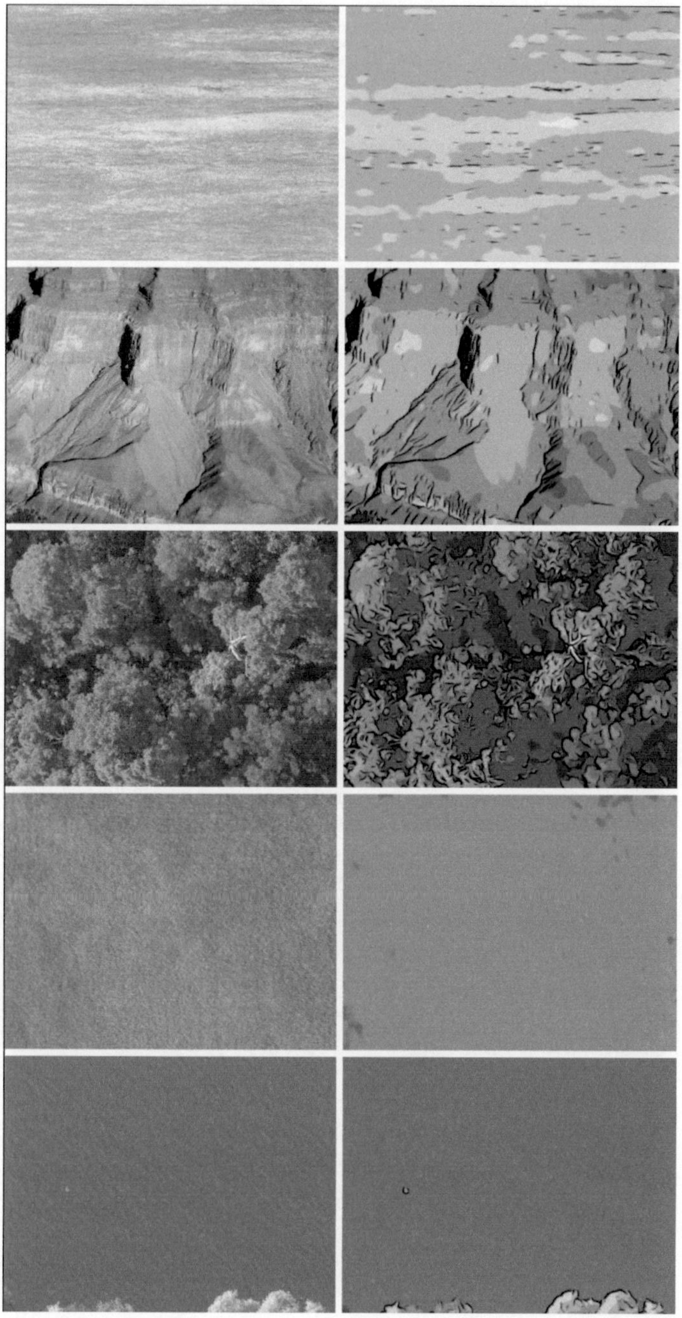

Fig. 9. Common object types that occur in aerial images. Top: Relief and terrain. Middle: Vegetation objects. Bottom: Water surfaces

Fig. 10. Common object types that occur in aerial images. Top: City furniture. Middle: Building and sites. Bottom: Transportation and population objects

4.4 Building & Site Objects

The major illustrated elements of a building are the roofs, facades and installations, whose richness of detail highly depends on the viewing distance and lighting condition. Regarding the roofs, usually each one of the sides is depicted by its own major color tone. The edge enhancement highlights the roof-edges and illustrates a suggestion of the roof tiles due to discontinuities in the surface orientation. In any case the exterior shell and distinctive shape of the roof is reflected due to its difference in elevation from the surrounding (depth discontinuities, see Figure 8). Regarding the facades we usually have the same effect as for the roof: Plastering or fine granular textures are depicted as single colored areas and objects like windows are enhanced as squares.

4.5 Transportation Objects

Transportation objects refer to all objects that are dedicated for the movement, traffic and transport, e.g., roads, rails or pavements. Usually roads and pavements can be abstracted with wide filtering kernels and a high color quantization process. The difficulty occurs when those objects are roughly structured, for example just like in the case of cobbled streets. Here the abstraction needs to operate on smaller local areas and the edge enhancement ideally takes a higher abstracted buffered image as input in order to operate more abrasive but with the characteristic to not enhance noise that occur on the surface, especially in the case of rapid changes that occur in the surface orientation (Figure 8).

4.6 City Furniture Objects

City furniture denotes an inhomogeneous group of objects that are statically placed with special function related to traffic, transportation, decoration, or advertisement. Examples are street lights, traffic signs and lights, benches, garbage cans and billboards for advertisement. Due to their architectural structure, they similarly behave like the abstraction of buildings and sites.

4.7 Population Objects

We count for example cars or people as population objects. These elements are usually the least important details in aerial images that should be kept in an abstraction process. Unfortunately due to its size, objects like cars will be obtrusively linked with the surrounding. Recent work studied the recognition of those objects (Leberl et al., 2007; Ram et al., 2001). Results in this area might facilitate a pre-processing that is able to remove those objects in certain conditions.

5 Performance

We used the following default values for the filtering process, whose configuration is feasible for pleasant results of any type of aerial image. The flow field of the local orientation estimation is smoothed by a standard deviation of 2.0 in order to remove discontinuities. For the bilateral filter the first iteration is used for edge extraction and the fourth iteration as input for the color quantization. The size of the kernel has a standard deviation of 3.0, whereas the amount of smoothing the bilateral filter applies at edges is 4.25%. For the edge detection we conducted one iteration. To balance bandwidth and sensitivity of the difference-of-Gaussians filter, standard deviations 1.0 and 1.6 are used, as recommended by Marr and Hildreth (1980). We used a standard deviation of 3.0 for the smoothing filter applied along the streamlines. Our color quantization outputs at most 8 distinct colors for a local region. We additionally conducted a usual Gaussian-smoothing on the final output with a kernel size of 3x3 pixels. Figure 11 depicts an output that is based on these values.

The percentage of connection tiles in relation to the input size of an image can be calculated as follows. Let $I_x \times I_y$ be the dimension of the input image and let $T_x \times T_y$ be the dimension of main tiles. Then the number of main tiles in x- and y-direction are given by

$$n_{Tx} = \frac{I_x}{T_x} - 1 \quad \text{and} \quad n_{Ty} = \frac{I_y}{T_y} - 1 \, .$$

The total size in pixels for the connection tiles is:

$$p_t(I_x, I_y) = n_{Tx} \cdot n_{Ty} \cdot 4 \cdot co^2 + \left(n_{Tx} + n_{Ty} + 2 \cdot n_{Tx} \cdot n_{Ty}\right)\left(4 \cdot co \cdot (I_x + I_y + co)\right)$$

The number of these connection tiles increase nonlinear with the dimension of the input image $I_x \times I_y$. Nevertheless, in case of a tiling size of

2,048 x 2,048 pixels and cut-off of $co = 24$ pixels, the percentage levels off at a limit of approximately 18.9%. GPUs that would be able to post-process image tiles of 8,192 x 8,192 pixels would end up with an off-cut of 4.7% (2.3% with $co = 12$).

An empirical analysis of the run-time emphasizes the decreased overhead if larger main tiles are conducted. As filtering settings we used the default configurations described previously on a test system with a Core2 Duo E8400 3.0GHz and a nVidia GeForce 9600 GT. Images with a size of 8,192 x 8,192 pixels took in mean 21.1 seconds to process for a tiling of 1,024 x 1,024 pixels (16.5s with $co = 12$), whereas the same image took 17.8 seconds to process for image tiles of 2,048 x 2,048 pixels (15.4s with $co = 12$).

6 Conclusions & Future Work

In this paper, we have presented an approach for automated, image-based abstraction that can handle massive, tiled aerial photography and that takes advantage of a GPU for efficiently performing image filtering.
We have shown that the abstraction of aerial images result in visually pleasant depictions. By analyzing seven different object types and identifying the impact of a filtering configuration on each of the elements' structure, we have shown how different content types of aerial photography can be graphically abstracted. In addition, we have shown how a concept of levels of abstraction can be established. Overall, our work contributes to the feasibility of non-photorealistic, automated aerial image abstraction as a pre-processing stage, required by applications such as for mobile devices with a low computing power and limited display size.

Primary application areas include non-photorealistic representations of 3D city models, which can take advantage from proper terrain stylization. As future work, we would like to explore the acceptance of abstracted aerial images as orientation guidance in mapping services like Google Maps or Microsoft Bing Maps, especially in areas that are little-known to a user. A second topic could deal with the quality of abstraction by conducting a dynamic filtering, which could be based on the actual mapped content. We could do so by distinguishing between primarily landscape or aerial images, or tie it to the wide topic of automatic object recognition (e.g. Mayer, 1999; Zhao et al., 2008; Idbraim et al., 2008) for combining it with our object categorization from Section 4. In this context it would also be worthwhile to investigate other abstraction filters that preserve edges, e.g., the anisotropic Kuwahara filter by Kyprianidis et al. (2009).

Fig. 11. An example created by our abstraction framework. Image tiles: 1,024 x 1,024 pixels

Figure 8 (bottom-right) original photo by Clearly Ambiguous - http://www.flickr.com/photos/clearlyambiguous
Figure 9 (top-left) original photo by mistress_f - http://www.flickr.com/photos/mistressf
Figure 9 (second from top-left) original photo by <<graham>> - http://www.flickr.com/photos/schnappi
Figure 10 (top-left) original photo by Daquella manera - http://www.flickr.com/photos/daquellamanera
Figure 11 (top-left) original photo by IRRI Images - http://www.flickr.com/photos/ricephotos

References

Bigün, J. and Granlund, G. H. (1987) Optimal orientation detection of linear symmetry, Proceedings of the IEEE First International Conference on Computer Vision, London, Great Britain, pp. 433-438.

Blom Pictometry (2009) http://www.blompictometry.com, Last date accessed 11.2009.

Buchholz, H., Döllner, J., Nienhaus, M. and Kirsch, F. (2005) Real-Time Non-Photorealistic Rendering of 3D City Models, 1st International Workshop on Next Generation 3D City Models, EuroSDR.

Curtis, C. J., Anderson, S. E., Seims J. E., Fleischer, K. W. and Salesin, D. H. (1997) Computer-generated watercolor, SIGGRAPH '97: Proceedings of the 24th annual conference on Computer graphics and interactive techniques, New York, pp. 421-430.

Döllner, J. (2007) In: Non-Photorealistic 3D Geovisualization, Multimedia Cartography, Springer, pp. 229-240.

Döllner, J., Baumann, K., Hinrichs, K. (2000) Texturing techniques for terrain visualization. In: Proceedings of IEEE Visualization, pp. 227-234.

Döllner, J., Buchholz, H., Nienhaus, M. and Kirsch, F. (2005) Illustrative Visualization of 3D City Models, In: Visualization and Data Analysis, Proceedings of the SPIE, International Society for Optical Engine (SPIE), pp. 42-51.

Foody, G. M. (1996) Fuzzy modelling of vegetation from remotely sensed imagery, In: Ecological Modelling, vol. 5669, pp. 3-12.

Glander, T. and Döllner, J. (2009) Abstract representations for interactive visualization of virtual 3D city models, In: Computers, Environment and Urban Systems

Gooch, B. and Gooch, A. A. (2001) Non-Photorealistic Rendering, AK Peters Ltd.

Idbraim, S., Mammass, D., Aboutajdine and Ducrot, D. (2008), An automatic system for urban road extraction from satellite and aerial images, In: WSEAS Trans., Sig. Proceedings, vol. 4, no. 10, pp. 563-572.

Open Geospatial Consortium (O.G.C.) (2008) OpenGIS City Geography Markup Language (CityGML) Implementation Specification, http://www.opengeospatial.org/standards/citygml, http://www.citygml.org, Last date accessed 11.2009.

Kuwahara, M., Hachimura, K, Eiho, S. and Kinoshita, M. (1976) Digital processing of biomedical images, Plenum Press.

Kyprianidis, J. E. and Döllner, J. (2008) Image Abstraction by Structure Adaptive Filtering, In: EG UK Theory and Practice of Computer Graphics, Eurographics Association, pp. 51-58.

Kyprianidis, J. E., Kang, H. and Döllner, J. (2009) Image and Video Abstraction by Anisotropic Kuwahara Filtering, Computer Graphics Forum, vol. 28, no. 7.

Leberl, F., Bischof, H., Grabner, H. and Kluckner, S. (2007) Recognizing cars in aerial imagery to improve orthophotos, In: GIS '07: Proceedings of the 15th annual ACM international symposium on Advances in geographic information systems, New York, pp. 1-9.

Marr, D. and Hildreth, E. (1980) Theory of edge detection, RoyalP, vol. B-207, pp. 187-217.

Mayer, H. (1999) Automatic object extraction from aerial imagery - A survey focusing on buildings, Computer vision and image understanding, vol. 74, no. 2, pp. 138-149.

Meng, L. (2002) How can 3D Geovisualization Please Users Eyes Better?, Geoinformatics Magazine for Geo-IT Professionals, Emmeloord, The Netherlands, vol. 5, pp. 34-35

Muhar, A. (1999) Three-dimensional modeling and visualization of vegetation for landscape simulation, Institute for Landscape Architecture and Landscape Management, Ascona, Switzerland, Technical report.

Nienhaus, M. (2006) Real-Time Non-Photorealistic Rendering Techniques for Illustrating 3D Scenes and their Dynamics, Ph.D. dissertation, HPI, Universität Potsdam, Germany, 2006.

NVIDIA Texture Tools 2 - GPU-accelerated Texture Tools with support for DirectX 10 texture formats (2009) http://code.google.com/p/nvidia-texture-tools, Last date accessed 11.2009.

Paris, S., Kornprobst, P., Tumblin, J. and Durand, F. (2007) A gentle introduction to bilateral filtering and its applications, In: SIGGRAPH '07: ACM SIGGPRAPH 2007 courses, New York, p. 1.

Pictometry - The Aerial Oblique Photography Company (2009) http://www.pictometry.com, Last date accessed 11.2009.

Ram, T. Z., Zhao, T. and Nevatia, R. (2001) Car detection in low resolution aerial images, In: Image and Vision Computing, pp. 710-717.

Santella, A. (2005) The Art of Seeing: Visual Perception in Design and Evaluation of Non-Photorealistic Rendering, Ph.D. dissertation, Rutgers University, New Brunswick, New Jersey, USA.

Santella, A. and DeCarlo, D. (2004) Visual interest and NPR: An evaluation and manifesto, In: NPAR '04: Proceedings of the 3rd international symposium on Non-Photorealistic animation and rendering, New York, USA, pp. 71-150.

Tanner, C. C., Migdal C. J. and Jones, M. T. (1998) The clipmap: A virtual mipmap, In: SIGGPRAPH '98: Proceedings of the 25th annual conference on Computer graphics and interactive techniques, New York, USA, pp. 151-158.

Tomasi, C. and Manduchi, R. (1998) Bilateral filtering for gray and color images, IEEE International Conference on Computer Vision (ICCV), p. 839.

Williams, L. (1983) Pyramidal parametrics, In: SIGGRAPH '83: Proceedings of the 10th annual conference on Computer graphics and interactive techniques, New York, pp. 1-11.

Winnemöller, H., Olsen, S. C. and Gooch, B. (2006) Real-time video abstraction, In: ACM Trans. Graph., vol. 25, no. 3, pp. 1221-1226.

Zhao, F., Zhang, H., Li, Z. and Pang, Y. (2008) The extraction of individual treecrown in aerial digital camera imagery, In: FSKD (3), pp. 183-188.

Geospatial Annotations for 3D Environments and their WFS-based Implementation

Jan Klimke, Jürgen Döllner

Hasso-Plattner-Institute, University of Potsdam,
Prof.-Dr.-Helmert-Strasse 2-3,
14482 Potsdam, Germany
{jan.klimke, juergen.doellner}@hpi.uni-potsdam.de

Abstract. Collaborative geovisualization provides effective means to communicate spatial information among a group of users. Annotations as one key element of collaborative geovisualization systems enable comprehension of collaboration processes and support time-shifted communication. By *annotations* we refer to user-generated information such as remarks, comments, findings and any other information related to the 3D environment. They have to be efficiently modeled, stored and visualized while precisely retaining their spatial reference and creation context. Existing models for annotations generally do not fully support spatial references and, therefore, do not fully take advantage of the spatial relationships associated with annotations. This paper presents a GML-based data model for geospatial annotations that explicitly incorporates spatial references and allows different types of annotations to be stored together with their context of creation. With this approach annotations can be represented as first-class spatial features. Consequently, annotations can be seamlessly integrated into their 3D environment and the author's original intention and message can be better expressed and understood. An OGC Web Feature Service is used as standardized interface for storage and retrieval of annotations, which assures data interoperability with existing geodata infrastructures. We have identified three types of *annotation subjects*, namely geographic features, geometry, and scene views, represented by their corresponding 2D/3D geometry. The model also defines a point-based approximation for complex geometry, such that annotations can also be used by client application with limited abilities regarding display size, band-

M. Painho et al., (eds.), *Geospatial Thinking*, Lecture Notes in Geoinformation and Cartography,
DOI 10.1007/978-3-642-12326-9_20, © Springer-Verlag Berlin Heidelberg 2010

width or geometry handling. Furthermore we extended our model by annotations that can contain 3D geometry besides textual information. In this way the expressiveness of annotations can be further enhanced for communicating spatial relationships such as distances or arrangements of geographic features.

1 Introduction

Collaborative geovisualization provides effective means to communicate spatial information among a group of users for sharing knowledge and information. This kind of communication occurs in a variety of applications such as public participation in planning projects, city management (i.e., complaint management), security monitoring or disaster management. To enable a comprehensible, potentially time-shifted communication of spatial information a user should be able to create, store, display and analyze *geospatial annotations* as pieces of information that are connected to geospatial objects, structures or regions. These annotations can represent, e.g., opinions, remarks, hints, explanations, or questions regarding a spatial subject. Contents and spatial references of annotations should be as flexible as possible to allow users to precisely, directly, and efficiently express their thoughts. Beside textual and multimedia contents, we propose freehand sketches as expressive type of annotation for visually communicating fuzzy, sketchy or vague information. Using sketches, for example, feature arrangements or change proposals in planning scenarios can be effectively communicated.

To provide a common understanding of geospatial annotations, a model is required that is general enough to serve as basis for data integration into heterogeneous service-based software systems and applications. Especially the definition of a model for an annotation's *spatial reference* is important to prevent loss of information concerning the annotation's spatial subject. Such spatial references are typically specified explicitly using tools provided by an annotation authoring system to avoid non-georeferenced, purely textual descriptions of spatial subjects that may lead to ambiguities. The comprehension of such descriptions depends on a user's context like skills or current tasks (Cai et al., 2003). Explicit specification using georeferenced geometry obviates the use of specialized language to draw a reader's attention to an annotation's spatial subject (Hopfer and MacEachren, 2007).

Our annotation model is designed for 3D geovirtual environments (3D GeoVE) such as 3D virtual city and landscape models. In this paper we assume in the following an urban area as the scope of a collaboration. Simple 2D geometries are not fully sufficient for describing an annotation's subject geometry due to the nature of features in such areas. For example, underground structures or indoor references for certain parts of a building cannot be expressed unambiguously using 2D geometry as spatial reference. Our annotation definition and implementation uses 3D georeferenced geometries for spatial reference specification. The unambiguously specified spatial reference geometry is particularly important to enable automated analysis of larger amounts of annotations using spatial parameters. Using our annotation model, for example, to gather and afterwards manage and visualize annotations in a public participation scenario, such analysis can help to improve the process of evaluation and processing of issues expressed by annotations.

Besides supporting a clear and flexible specification of an annotation's spatial reference, our model supports capturing the *creation context* of an annotation. The collaboration context includes metadata such as creation time and author information but also the author's 3D view on model data visualization. This view bears information that helps a later reader to comprehend the meaning of an annotation.

The purpose and applicability of geospatial annotations is widespread. They may be used, for example, to collect information concerning urban planning scenarios for public participation purposes or for persisting agreements on problem solving during remote or local meetings using a virtual 3D city model. Such annotations can afterwards help to review findings and therefore help to recall key aspects of a collaborative work process (Shrinivasan and van Wijk, 2009). Annotation data created during collaboration processes must be widely usable in heterogeneous software environments. When using the same open and standardized data encoding and service interface that is used for geodata itself, annotation functionality can be embedded into a variety of applications that are already capable of dealing with such data.

In this paper we introduce an object oriented model of geospatial annotations in connection with its implementation using the *Geography Markup Language* (GML) (Portele, 2007) as data exchange format between a transactional *Web Feature Service* (WFS-T (Vretanos, 2005)) and clients creating and visualizing annotation data in 3D geovirtual environments. For this purpose an annotation's spatial references are modeled as distinct objects describing 3D geometries. By doing so, those reference objects can be shared throughout annotation objects to explicitly share spatial references.

The rest of this paper is organized as follows: Section 2 provides a short overview of related work. Section 3 introduces our model of geospatial annotations. The design and implementation of the collaborative annotation system is presented in Section 4. A short discussion including the limitations of our approach is given in Section 5. Section 6 summarizes the paper and proposes some additional research directions to take.

2 Related Work

Schill et al. (2008) introduce in the context of the Virtual Environment Planning System project (VEPs) a model of geospatial comments for public participation in urban planning projects, using GML for data encoding and an OGC Web Feature Service for storage and retrieval. Text is used as annotation contents, and object URLs for each annotation can be stored to reference multimedia objects. An annotation's spatial reference is modeled as point, which is interpreted differently depending on the type of the annotation. An identifier of a parent annotation can be set to create annotation chains as discussions. The approach is limited regarding the definition of multiple objects, for example feature groups, or more complex geometries as spatial reference for annotations.

An interactive geocollaboration framework supporting geographic annotations is introduced by Mittlböck et al. (2006), which combines data from heterogeneous sources for presentation and analysis. A user is able to vote and to comment on geospatial subjects visualized by maps. Annotations are georeferenced using 2D coordinates. As real-time visualization component Google Earth[1] is used. Unlike our implementation, a separate service combines data from different sources (for example WFS and WMS) for generating output of annotation data in KML (Wilson, 2008) format.

Several researchers worked on supporting geo collaboration using maps. Yu and Cai (2009) propose *GeoAnnotator* as a service-oriented system for map based public participation. They outline requirements of such a system to provide necessary features for annotation of geospatial objects as well as for encouraging people to provide their opinions. A many-to-many relation between annotations and spatial references is considered to be important to support, e.g., comparison arguments as annotated information. Further they outline the need for multi-modal multimedia annotations to support sharing geographical information more easily. Rinner (2001, 2005) introduced Argumentation Maps to support discussions on planning activities by connecting discussion contributions to geographic features or geometries. This object-based model is used to store discussion information in

[1] http://earth.google.com

databases. In contrast to Argumentation Maps, our model aims at a more general approach for annotation of geographic areas and features, which can be used in many application domains.

Hopfer and MacEachren (2007) investigate the use of geospatial annotation for collaboration using map-based displays, analyzing how annotations facilitate decision making in groups. They recommend to avoid introducing knowledge that is already known to each participant (shared knowledge) into decision making processes and outline the importance of flexible annotation systems (query, analysis and access possibilities).

Text and sketch annotations in 3D virtual environments for architectural design are presented by Jung et al. (2002). They outline the demand for non text annotations in an earlier user study Jung et al. (2002a).

Tohidi et al. (2006) report on the usage of user created sketches during user interface design processes. They state the advantage of providing a user with communication means to propose own ideas or proposals beside, e.g., textual comments or questionnaires. Sketches are also used frequently for describing intentions in the field of human-computer-interaction, e.g., for navigation (Igarashi, et al., 1998, Hagedorn and Döllner, 2009) or 3D modeling (Karpenko and Hughes, 2006). We are using a sketch-based approach for communicating visual information to provide equally expressive communication tools, which allow more useful annotations, i.e., to express alternate approaches or change requests.

Heer et al. (2009) deal with asynchronous (time shifted) collaboration on data visualization using annotations. They provide tools for diagram annotation. Additionally they conducted a user study to analyze the usage of these tools. Drawing sketches on top of the visualization is seen as expressive means especially for pointing: It turned out that 88.6 % of all sketch annotations involved pointing. In contrast to our drawing approach more tools are provided for drawing complex shapes like arrows or boxes, while our client implementation does exclusively support free-hand sketching.

Isenberg et al. (2009) conducted a user study on usability of a collaboratively retrofitted information visualization system. They introduced collaborative interaction and did changes concerning the data representation to enable collocated collaborative work. One improvement requested by several participating groups was to integrate explicit ways to ensure that decisions would not get lost in the collaboration process, which is also motivation for annotation in collaborative processes in 3D GeoVEs.

3 Modeling Geospatial Annotations

This section presents a model for geospatial annotations that concentrates on precise, creation context aware storage of information concerning a spatial subject. Annotations are used as means to make knowledge or information persistent and are intended for later access and analysis. We distinguish three types of annotations by their contents: textual information, multimedia contents (e.g., images, videos, or audio records), and additional geometry visually communicating a concept or proposal.

3.1 Spatial References

Spatial references define the location and extent of an annotation's subject in 3D space. To ease sharing of those between annotation objects, we model spatial references as separate first-class features, which does also allow us to define groups of spatial references to be the subject of an annotation. By marking an annotation's spatial subject area using our model of spatial reference, specialized language to communicate the spatial reference in annotation contents can be obviated (Hopfer and MacEachren, 2007).

Our model for spatial references is partitioned into two parts (Fig. 1): `SpatialReference` and specialized reference types. The `SpatialReference` class defines basic parameters, which every reference type must have. A point as location indicator facilitates using an annotation's spatial references for clients that have very limited capabilities concerning computational power, display size, bandwidth or geometry handling. This especially eases the implementation of web-based clients for annotation exploration and creation, without having to implement the full support for GML geometry needed for precise and complete handling of complex reference geometry. The `modelId` attribute identifies the model data set in the database. This data set describes parameters of the city model that is used to create the `SpatialReference` object. The information about the used model can be retrieved from the WFS if information about the overall spatial extend or additional information like access parameters for model data are required.

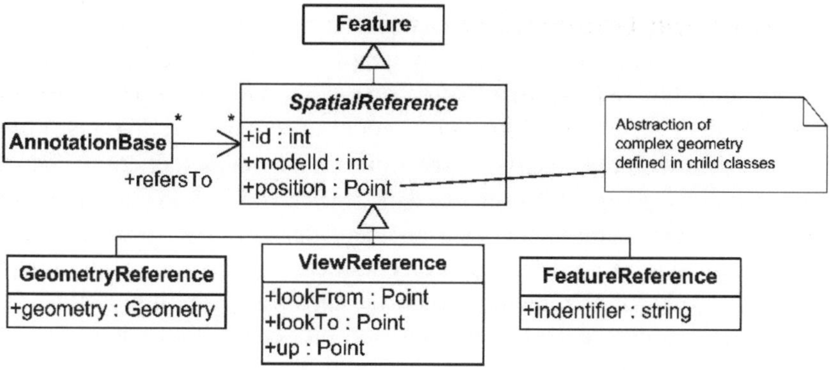

Fig. 1. Reference types as UML class diagram. Every geographic reference is a unique feature which can be referred

We define the following three types of spatial reference objects (Fig. 1):

- **Geometry**: This is the most explicit type of spatial reference. It contains geometry (e.g., point, polygon or box) defined in real-world coordinates. It is encoded using the GML 3 geometry model, which supports 3D geometries. The point geometry defining the approximate location is set depending on the type of geometry the references holds. If the geometry is a point, it is set as position property equally. For lines or line strings the center of the line, defined by the client creating the reference, is used as position marker. For areas or volumes the center of the bounding box is used to provide the value of the position property.
- **Scene Views:** A scene view is the second type of an annotation's spatial subject. A large amount of information is included in a user's current view of the scene through many perceptional impressions like the current line-of-sight or visible parts of certain structures are view dependent. A `ViewReference` instance is specified by three point properties: look-from position, look-to position and up-position. The up-position determines in conjunction with the look-from position camera's up direction. The look-from position also defines the position property of this type of spatial preference.
- **Geographic Features:** In contrast to references containing explicitly defined complex geometries, a reference to a model object is connected to a geographic feature (e.g., building, square, or street). This provides possibilities to use topological, hierarchical and other relations defined by the city model for computation like positioning calculation for annotation visualization elements or further analysis of larger numbers of annotations. Large amounts of annotations can occur, e.g., in public participation scenarios or planning activities. The indirection of those

references allows us to follow changes in the feature geometry providing the possibility to, for example, annotate features that do not have a fixed location. A `FeatureReference`, therefore, defines a link to a feature data set included in a city model. The `identifier` string in connection with the `modelId` identifier must enable a client to retrieve the complex geometry from the data source defined in the model description. An example for such an identifier is a URI used as gml:id attribute value in a GML-based city model. At least the client creating this type of spatial reference must be capable of getting feature data from the model to calculate the position property. Other clients can use the precomputed position property instead. By default the position property is set to be the center of the referenced feature's bounding box.

•

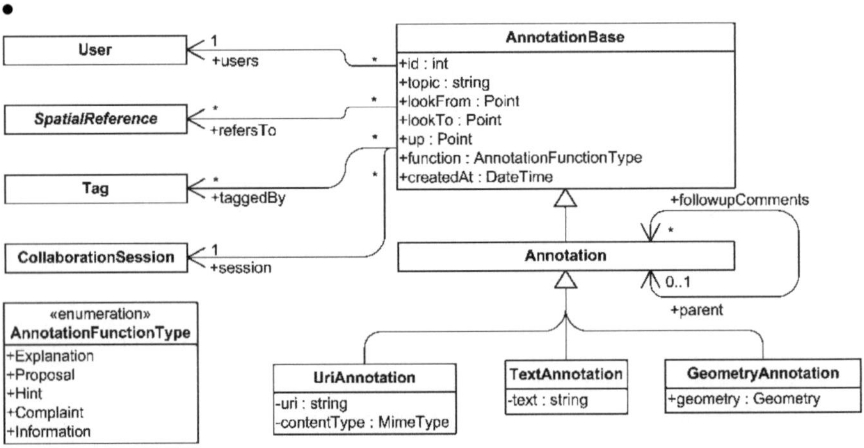

Fig. 2. UML class diagram for geospatial annotations. Metadata like geospatial annotation subjects or the author of an annotation is defined at the base class. The Annotation class adds the possibility to define annotation chains for discussions

3.2 Annotation Contents

An annotation's content defines the information associated to the spatial reference. To provide the users with a wide range of possibilities to express their opinions, remarks, or proposals we define three types of annotations according to their type of contents (Fig. 2):

- **Text:** A user can state opinions or other information by giving textual descriptions. Because of the well defined spatial references, users may refer to those objects easily. Although being quite expressive, text is not the optimal means for communicating information that refers to spatial relations.

- **References to Multimedia Contents:** This annotation type enables a user to connect multimedia contents to a spatial reference. The contents themselves are not stored together with the annotation data but are referenced using an URL. To support clients to handle the playback or display of the linked contents, information about the type of the referenced media is stored (see `contentType` property). Through using this quite flexible form of annotation contents, a wide range of media (e.g., audio recordings, videos, or images) can be associated with spatial references.
- **Geometry:** The third type of annotation contents is either 2D or 3D geometry that is used to annotate the city model by using direction indicators (i.e., arrows), measurement indicators, sketches, or extensions to existing object geometries like, e.g., lines as proposal for routes (Strobl, 2007) (Fig. 3). Those geometries are means to communicate spatial information like object arrangement, object size or design ideas. The communication of such visual forms of information through non visual (verbal, textual) means involves a loss of information due to the necessary mental translation effort (Yao, et al., 2005). To help to avoid such a translation loss, we allow the creation of free-hand sketches as special case of geometry annotation. A sketch is an intuitive and efficient way for communicating information or concepts (Stefik, et al., 1987). The user is free to express its own concepts or proposals. Due to the creative freedom a resulting sketch serves as a basis for later analysis and interpretation (Tohidi, et al., 2006), which may help to improve planning.

Fig.3. An example of view-plane sketches for communicating proposals, ideas, or spatial relations. The sketches are connected to one viewpoint, but the camera orientation can be changed while the sketch's position is maintained

3.3 Expressing Uncertainty for Spatial References

If there is no precisely definable subject geometry for an issue, further means for expressing spatial vagueness are needed. Imprecisely known subject geometry can be necessary, e.g., when assumptions or guesses concerning spatial issues shall be made. Our concept of an annotation's uncertainty extent provides means for specifying further spatial attributes than the spatial reference as annotation subject only. Annotation's contents may refer to this geometry to express an alternative concerning a spatial extent. An extent geometry can be defined in two ways:

- Indirectly by using an offset given in meters which enlarges the spatial reference geometry
- Directly through defining a separate explicit extent geometry

By taking the geometry specified by the annotation extent into account for search and analysis, the scope of a search request can be broadened to include annotations that are possibly related to the geometry defined as search parameter.

3.4 Annotation Metadata

Basic annotation attributes describe metadata concerning the annotation's contents. They can help to comprehend the author's original intention and message when annotations are explored. The following 6 items are stored alongside with every annotation for that purpose:

- **Scene View Specification:** The parameters describing an author's current scene view are stored together with annotation data providing the reader with information about the creation context. When the annotation was created using an interactive 3D client, we assume the author chose a viewpoint in such a way that objects that are important for understanding the spatial situation are visible and properly aligned concerning the message that is intended to be communicated.
- **Annotation Function:** As shown in Fig. 2 each annotation can have a function assigned that describes what the author has intended to express. The categorization, which is possible through this function attribute, can be used for annotation visualization and analysis. E.g., where and how many complaints or proposals have been given as annotations to identify problematic areas. By now, the annotations functions have been defined exemplarily for the use case of public participation in urban planning or city management scenarios. They may have to be adapted or extended to

serve for other application areas. The function is also a good criterion for grouping of annotations especially for visualization purposes.

- **Session:** Annotations can be grouped by sessions that describe the occasion for annotation creation, i.e., a team meeting or a planning project. An annotation's session does also describe the geographical extent of the overall area of interest using a bounding box. A model description associated with a session holds information about the model that is used for annotation authoring. A session provides a short description of the overall topic (e.g., project name or activity description). By assigning a session id to an annotation, they are assigned to a session dataset.
- **Tags:** Keywords (*Tags*) can be assigned by a user to briefly describe what an annotation is about. Through using tags for annotation description groups are created, each containing annotations that hold the same tag. A user is free to assign arbitrary unstructured keywords to his annotations. The keywords may define a broad range of annotation attributes like, e.g., contents or intended function. The meta-information provided by such a keyword set per annotation can be used for searching or filtering of annotation objects (Xu et al., 2006).
- **Author:** An annotation holds information (e.g., id, name, color) about its author to enable to tracing of annotations created by a certain user or a group of users which matches given parameters.
- **Discussion:** Parent-child (`followupAnnotations`) relations introduced by the Annotation class definition in Fig. 2 allow to create tree structures of annotations. Those structures can be used to model discussion threads. The spatial reference of a parent annotation is implicitly inherited by child annotations, which do not have to have a `SpatialReference` object assigned. This also provides the possibility to define further `SpatialReference` objects to broaden the spatial scope of a follow-up annotation. Since we can model discussion threads using our geospatial annotation model it can be applied for issue-based information systems (Kunz and Rittel, 1970) for applications like map-based argumentations (Rinner, 2005).

4 Implementation

This section provides an overview of our prototypic implementation of a system that supports authoring, storage and visualization geospatial annotations using 3D GeoVEs.

4.1 System Architecture

Our system architecture consists of three major parts (Fig. 4): A geo-enabled database management system, a WFS implementation, and client applications. The data back end is provided by a PostGIS[2] spatially enabled PostgreSQL database which is encapsulated using a transactional WFS-implementation. This supports usage of the annotation data by a variety of client applications through offering an open, standardized interface for data retrieval, insertion and update. OGC Filter (Vretanos 2005), as query language for WFS requests, allows to define queries using spatial and non spatial predicates for restricting search requests in a standardized way.

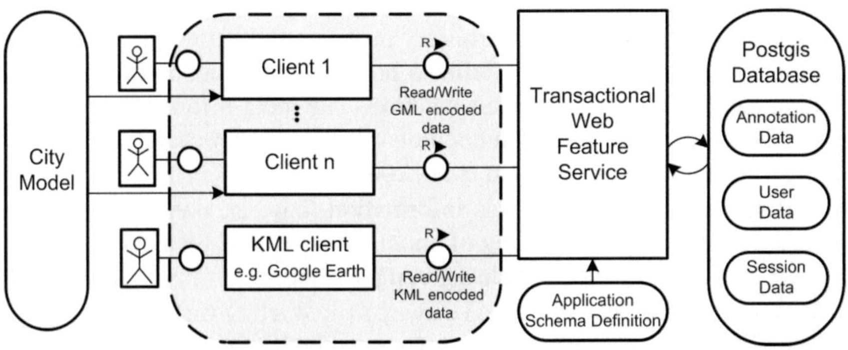

Fig. 4. Architectural overview of the annotation system. The client is responsible for user interaction, visualization and conversion to the respective GML-based data encoding while the WFS provides the interface for data storage using the PostGIS database. Through the standard WFS interface and its variable output format different types of clients are possible

Several types of clients may be used together with our data back end. We have implemented a C++ application that can create and visualize annotations for virtual 3D city models. It uses GML for encoding of annotation data. For annotation exploration, a KML enabled client like Google Earth can be used to visualize annotations by using the KML encoded output generated by the WFS implementation through transforming the GML encoded output to KML using XSL transformations.

The city model must be shared throughout all clients creating annotations to be able to create and resolve `FeatureReferences`, e.g., through distributing the model data file throughout all client instances.

[2] http://postgis.refractions.net

Another possibility for data distribution is a service-based access to a CityGML (Gröger, et al., 2008) encoded model also using a WFS.

4.2 Data Storage

Data retrieval, creation and update is encapsulated by a transactional WFS. Due to our requirements regarding support for 3D geometries and transactional service functionality (insert, update and delete features), we use a service implementation that supports the WFS specification version 1.1., which defines GML version 3.1.1 as mandatory data exchange format and allows encoding of 3D geometries. Since annotation data creation and update of data properties are inevitable functionalities for our purposes, a WFS implementation is used which supports the *Transaction* operation, defined by the WFS specification to be optional.

We selected the WFS implementation of the deegree project[3] version 2.2 to be used with a PostGIS spatially enabled database. The WFS is configured using a GML application schema that defines XML elements and XML Schema types according to our model presented in section 3. In the schema definition we made extensive use of complex typed child elements to implement our annotation model. Each complex typed element is mapped onto a database table in our relational database model. Inheritance relations (e.g., the one between `SpatialReference` and `GeometryReference` depicted in Fig. 1) are defined in both, the XML schema definition using the type extension mechanism and the relational database model using the table inheritance feature of PostgreSQL. All XML schema types that extend other types are mapped to an own database table that inherits from the table of the extension's base type. This eases a consistent handling of child types regarding id generation, feature update, or deletion behavior down to database level.

Data Export as KML

A system for annotations should enable the largest possible public to use it (Strobl, 2007). Virtual globe tools like Google Earth, Microsoft's Bing Maps 3D[4] or NASA World Wind[5] are very popular and many web users are familiar with using such virtual globes. All those applications mentioned support KML as input or output format. We defined a XSL trans-

---

[3] http://www.deegree.org
[4] http://www.bing.com/maps
[5] http://worldwind.nasa.gov

formation that is used as output Filter for our WFS instance to enable annotation exploration via afore mentioned KML enabled clients (Fig. 4). It translates annotation data encoded in GML according to our application schema into KML placemarks. Therefore the position property held by `SpatialReference` instances provides a placemark's location. Equally the definition of an input transformation from KML to our GML dialect would be possible to allow creation of annotation features using KML-encoded input data.

4.3 Interactive 3D Client

For annotation authoring and visualization we have implemented a C++ client application that uses the GML-encoded data provided by the WFS. It provides a user interface for exploring of a 3D city model. Users are enabled to define spatial references visually by selecting features or defining reference geometries. Annotation metadata is captured implicitly on annotation creation.

Annotations are visualized using 3D display elements, which are embedded into the scene and additionally through icons on a mini map. Annotations are grouped by their spatial references to limit the amount of 3D annotation items being displayed. Different interaction and visualization strategies have to be applied depending on annotation types. Sketches concerning scene views can be created as `GeometryAnnotation` using the mouse or a tangible display. To view such sketches the user takes the author's camera position and sketch geometry is displayed front of the user's viewpoint. This creates the impression of the sketch being projected into the scene (see Fig. 3).

The process for annotation creation has to assure data integrity, which also includes avoidance of double entries, especially for `SpatialReference` instances, where possible. Partly this is ensured by the deegree WFS implementation by checking input data for validity according to the application schema. It does also check for doublets using predefined equality criteria for feature types. Unfortunately, the check for duplicates does not take complex typed child properties and geometry properties into account. So this has to be implemented in the client application. We avoided this missing feature by creating the `SpatialReference` features independently before creating the annotations objects.

5 Discussion

Currently the client annotation-system is implemented to prove the applicability of our annotation model. No larger user tests have been done right now. In the following we will discuss the current client/server-system and model.

Screen overlay sketches are currently defined using a set of curves given in 3D real world coordinates which have been calculated depending on near clipping plane of the client's current camera projection settings. For it, screen coordinates of each point of the sketch are unprojected to 3D coordinates situated on the camera's view plane. By saving those coordinates in that way, we enable a camera look around while maintaining the alignment of the sketch in connection to the city model scene (see Fig. 3). While this is sufficient for a fixed viewpoint in connection with sketches, it is not possible to display sketches that are associated with objects or certain areas independently from viewpoints. Sin et al. (2006, 2006a) provide possibilities to use object surfaces as sketch canvas. They adjust the sketch display depending on the orientation and viewpoint dependence of the information contained in such sketches.

Our representation of overlay sketches and camera viewpoints defined for annotation features are based on GML geometries. From a semantic point of view the geographical bounding box of those features should be the bounding box of their spatial reference. Unfortunately this is not possible at the moment because the WFS takes every geometry property of a feature into account for bounding box calculations. This falsifies the bounding box of such features. The bounding box definition is also a problem for view references and other types of annotations because of the definition of a camera position using three georeferenced points. Those points are also contributing to the bounding box calculation of the deegree WFS.

The client introduced in this paper provides basic functionality to explore a city model, to select (also multiple) spatial references and to visualize annotations in 3D scenes. The methods for alignment and display of such metadata elements are not subject of this paper. But there are some restrictions concerning the usage of our annotation model's expressiveness. For example the ability to define a `GeometryReferences` is limited for simplicity to points, boxes, and 2D polygons. Also the client is not yet able to create or visualize annotation chains (discussions).At the moment KML output is restricted to a placemark for each annotation which are localized using the position property of the `SpatialReference` type. Also `GeometryAnnotation` objects are not covered by the XSL transformation by now. Complex reference geometries could be translated into KML geometries using more sophisticated transformations. Those would have to include additional functionalities implemented in Java classes providing for example coordinate transformations or other computations. This would also enable geometry processing for `GeometryAnnotations` and enable sketch display in KML enabled 3D clients through KML-encoded complex geometries.

The types of annotation content are defined exemplarily. All metainformation that is needed for annotation handling are defined independently from the type of the information contained in an annotation. This way the model can be extended towards, e.g., composite types of annotation contents. For each new type of annotation content client applications have to be adapted to enable authoring and visualization of those new content types.

6 Conclusions and Future Work

In this paper we have shown a model for geospatial annotations that pays special attention on modeling the data structures for referencing geospatial objects or geometries using GML-features. A transactional WFS provides a standardized interface for data access and storage, which allows embedding annotation functionality into a variety of GML compatible applications. Metadata supports comprehension of annotation meanings especially with regard to 3D GeoVEs. We further propose an annotation type containing geometry as intuitive communication tool. We have outlined the possible value of sketches for communication of proposals or ideas.

Concerning geospatial annotations some further work may be applying geospatial ontologies for defining a more precise and semantically valuable georeferencing. The additional information provided by such an ontology may be used for a more subject specific visualization (arrangement, appearance) when interactive 3D client applications are used. Further, additional annotation types with regard to their contents could be defined. For example a type for representing questionnaires for guided data acquisition by users could be evaluated in connection with virtual tours through the area of interest.

We do not use a definition of what equality of spatial references mean, except for equal-valued object attributes. Defining variable criteria concerning the degree of containment of a reference's geometry in other reference geometries could help to visualize and analyze larger amounts of annotations.

A web based approach for interactive creation and visualization of annotations would lower barriers that are posed through installation requirements and dependencies of the current client system. Therefore other OGC web services may be used to provide map-based (Web Map Service - WMS (de la Beaujardiere, 2006)) or 3D visualization and interaction. For the three dimensional case, a *Web Perspective View Service* (WPVS) could be used to generate images of a virtual city model. Using such images for presentation, very thin client applications are possible without losing the possibility to give an impression of the author's context when creating the annotation (Hagedorn, et al., 2009). When additional interactivity is introduced using image data delivered by a WPVS, users would be able to define all necessary attributes for annotation creation using a web browser. Such a technique could enable the usage of annotation with precise geospatial references for a far broader audience and therefore enable using our annotation model in connection with a virtual 3D city or landscape model for large scale public participation applications.

An important next step is the acquisition of a larger amount of user-generated data to find requirements for annotation analysis; user tests should show how users can handle annotation tools and which kind of information they want to store as annotation.

References

de la Beaujardiere, J. (2006) OpenGIS® Web Map Service (WMS) Implementation Specification 1.3. *Opengeospatial Consortium.*. http://www.opengeospatial.org/standards/wms (accessed 11 05, 2009).

Cai, G., Wang, H., and Maceachren, A. M. (2003) Communicating Vague Spatial Concepts in Human-GIS Interactions: A Collaborative Dialogue Approach. Proceedings of the Conference on Spatial Information Theory, COSIT 2003, Springer. 287-300.

Gröger, G., Kolbe, T. H., Czerwinski A. and Nagel. C. (2008) OpenGIS® City Geography Markup Language (CityGML) Encoding Standard. *Opengeospatial Consortium.* http://www.opengeospatial.org/standards/citygml (accessed 11 05, 2009).

Hagedorn, B. and Döllner, J. (2009) Sketch-Based Navigation in 3D Virtual Environments. *it - Information Technology 03/2009*, 163-170.

Hagedorn, B., Hildebrandt, D. and Döllner, J. (2009) Towards Advanced and Interactive Web Perspective View Services. Developments in 3D Geo-Information Sciences. Springer: 33-51

Heer, J., Viégas, F. B. and Wattenberg, M. (2009) Voyagers and voyeurs: Supporting asynchronous collaborative visualization. *Communications of the ACM* (ACM) 52 (2009): 87-97.

Hopfer, S., and MacEachren, A. M. (2007) Leveraging the potential of geospatial annotations for collaboration: a communication theory perspective. *Int. J. Geogr. Inf. Sci.* (Taylor & Francis Inc.) 21 : 921-934.

Igarashi, T., Kadobayashi R., Mase, K., and Tanaka, H. (1998) Path drawing for 3D walkthrough. *UIST '98: Proceedings of the 11th annual ACM symposium on User interface software and technology.* ACM. 173-174.

Isenberg, P., Carpendale S., Bezerianos, A., Fekete, N., and Henry, J. (2009) CoCoNutTrix: CollaboartiveRetrofitting for Information Visualization. *IEEE Computer Graphics and Applications*: 44-57.

Jung, T., Gross, M. D., and Do, E. Y. (2002) Annotating and Sketching on 3D Web Models. *IUI '02: Proceedings of the 7th international conference on Intelligent user interfaces.* New York: ACM. 95-102.

Jung, T., Gross, M. D., and Do, E. Y. (2002a) Sketching Annotations in a 3D Web Environment. *CHI '02: CHI '02 extended abstracts on Human factors in computing systems.* New York: ACM. 618-619.

Karpenko, O. A. and Hughes, J. F. (2006) SmoothSketch: 3D free-form shapes from complex sketches. *SIGGRAPH '06: ACM SIGGRAPH 2006 Papers.* ACM, 589-598.

Kunz, W. and Rittel, H. W. J. (1970) Issues as Elements of Information Systems, Working Paper No. 131, Studiengruppe fur Systemforschung, Heidelberg (July 1970)

Mittlböck, M., Resch, B., and Eibl, C. (2006) geOpinion: Interaktives geo-Collaboration Framework - 3D-Visualisierung in Google Earth mit OGC WMS- und WFS- Diensten. *Angewandte Geoinformatik 2006. Beiträge zum 18. AGIT-Symposium.* Salzburg: Herbert Wichman Verlag, Heidelberg 464-469.

Portele, C. (2007) OpenGIS® Geography Markup Language (GML) Encoding Standard. *Opengeospatial Consortium.* http://www.opengeospatial.org/standards/gml (accessed 11 05, 2009).

Rinner, C. (2001) Argumentation maps: GIS-based discussion support for on-line planning, *Environment and Planning B: Planning and Design*, vol. 28, 2001, 847-863.

Rinner, C. (2005) Computer Support for Discussions in Spatial Planning. *GIS for Sustainable Development*. Taylor & Francis, 167-180

Schill, C., Koch, B., Bogdahn, J., and Coors, V. (2008) Public Participation Comment Markup Language and WFS 1.1. *Urban and Regional Data Management.* Taylor & Francis. 85-92.

Shrinivasan, Y. B., and van Wijk, J. J. (2009) Supporting Exploration Awareness in Information Visualization. *IEEE Computer Graphics and Applications*: 34-43.

Sin, E., Choy, Y., and Lim, S. (2006) A Study on Sketch Input Technique by Surface of 3D Object for Collaboration. *Proceedings of the 2006 International Conference on Hybrid Information Technology.* IEEE Computer Society: 623-628.

Sin E., Choy Y., and Lim, S. (2006a) Content based sketch annotations for collaboration. *SIGGRAPH '06: ACM SIGGRAPH 2006 Research posters.* ACM. 21.

Stefik, M., Bobrow, D. G., Foster, G., Lanning, S., and Tatar, D. (1987) WYSIWIS Revised: Early Experiences with Multiuser Interfaces. *ACM Trans. Inf. Syst.* (ACM) 5 : 147-167.

Strobl, J. (2007) Visual Interaction: Enhancing Public Participation ? Edited by Strobl Buhmann, Ervin. *Trends in Knowledge-Based Landscape Modeling Proceedings at Anhalt University of Applied Sciences 2006..*

Tohidi, M., Buxton, W., Baecker, R., and Sellen, A. (2006) User sketches: a quick, inexpensive, and effective way to elicit more reflective user feedback. *NordiCHI '06: Proceedings of the 4th Nordic conference on Human-computer interaction.* ACM. 105-114.

VEPs. VEPS project website. http://veps3d.org (accessed 11 05, 2009).

Vretanos, P. A. (2005) OpenGIS® Web Feature Service (WFS) Implementation Specification. Vers. 1.1. *Opengeospatial Consortium.* http://www.opengeospatial.org/standards/wfs (accessed 11 05, 2009).

Wilson, T. (2008) OGC® KML. *Opengeospatial Consortium.* http://www.opengeospatial.org/standards/kml (accessed 11 05, 2009).

Xu, Z., Fu, Y., Mao, J., and Su, D. (2006) Towards the Semantic Web: Collaborative Tag Suggestions. *Proceedings of Collaborative Web Tagging Workshop at 15th International World Wide Web Conference.*

Yao, J., Fernando, T., Tawfik, H., Armtiage, R., and Billing, I. (2005) A VR-centred Workspace for Supporting Collaborative Urban Planning. *Proceedings of the Ninth International Conference on Computer Supported Cooperative Work in Design.* 2005. 564-569.

Yu, B., and Cai, G. (2009) Facilitating participatory decision-making in local communities through map-based online discussion. *Proceedings of the 4th international Conference on Communities and Technologies.* ACM, 2009. 215 221.

Towards Spatial Data Infrastructures in the Clouds

Bastian Schäffer[1], Bastian Baranski[1], Theodor Foerster[2]

[1] Institute for Geoinformatics, University of Münster, Germany
{schaeffer,baranski}@uni-muenster.de

[2] International Institute for Geo-Information Science and Earth Observation, Enschede, The Netherlands
foerster@itc.nl

Abstract. Cloud Computing is one of the latest hypes in the mainstream IT world. In this context, Spatial Data Infrastructures (SDIs) have not been considered yet. This paper reviews this novel technology and identifies the paradigm behind it with regard to SDIs. Concepts of SDIs are analyzed in respect to common gaps which can be solved by Cloud Computing technologies. A real world use case will be presented, which benefits largely from Cloud Computing as a proof-of-concept demonstration. This use case shows that SDI components can be integrated into the cloud as value-added services. Thereby SDI components are shifted from a Software as a Service cloud layer to the Platform as a Service cloud layer, which can be regarded as a future direction for SDIs to enable geospatial cloud interoperability.

1 Introduction

Cloud Computing is one of the latest trends in the mainstream IT world (Gartner 2009a). A cloud metaphor is used to represent large networking and computational infrastructures. From a provider perspective, the key aspect of the cloud is the ability to dynamically scale and provide computational and storage capacities over the internet. From a client perspective, the key aspect of a cloud is the ability to access the cloud facilities on-demand in a cost efficient way without managing the underlying infrastructure and dealing with the related investments and maintenance costs.

M. Painho et al., (eds.), *Geospatial Thinking*, Lecture Notes in Geoinformation and Cartography,
DOI 10.1007/978-3-642-12326-9_21,

In this regard, Spatial Data Infrastructures (SDIs) undergo a transition from providing geodata towards providing web-based geoinformation (GI) (Kiehle 2007, Schaeffer et al. 2009). To provide this web-based geoinformation, massive processing tasks are required in a cost efficient way to maintain sustainability. In the past, the processing of geodata has been performed mostly on desktop machines and mainframes. Due to this requirement for massive processing capabilities, Cloud Computing is a promising approach. Additionally, this novel technology is beneficial to sufficiently scale these processing tasks on the organization's infrastructure or within an SDI. The problem of scaling can be demonstrated for the example in disaster management scenarios, which requires a large-scale computational infrastructure for extensive computations only for a short period of time. Another aspect related to scaling is the coupling of SDIs with the mass market domain such as the integration of volunteered geoinformation (i.e. collected via mobile phones) in SDIs. In this case many users create and share their geodata on-demand concurrently, which is seen as a beneficial application for SDIs to enrich existing databases in real-time. The risk management scenario as well as the volunteered geoinformation do not follow a fixed schedule (such as for instance the periodically update of data in an agency) and therefore require new approaches to technically meet the requirements and to limit the infrastructure costs. Therefore, Cloud Computing is a technical and economic opportunity for SDIs to support future geospatial applications. Moreover, it is also an approach for novel business to create, operate and utilize SDIs. All these aspects motivate to investigate the potentials of Cloud Computing for SDIs.

Thus, this paper presents a cloud-enabled SDI addressing some of the current obstacles of SDI development. Section 2 reviews the related concepts of Cloud Computing and SDIs. The cloud-enabled SDI is described in Section 3. The application of the risk management use case is presented in Section 4. In addition, Section 5 validates the scalability promise of the cloud computing paradigm with regard to the presented use case. Finally, Section 6 gives an outlook and concludes the findings.

2 Review of Relevant Concepts

This section provides a review of relevant concepts in the context of Cloud Computing and SDIs.

2.1 Cloud Computing

Cloud Computing is one of the latest trends in the mainstream IT world (Gartner 2008) (Gartner 2009a). Several IT companies such as Amazon,

Google, Microsoft and Salesforce have already built up significant effort in this direction (see Section 2.3.1). The term Cloud Computing describes an approach in which the storage and computational facilities are no longer located on single computers, but distributed over remote resources facilities operated by third party providers (Foster 2008).

Cloud Computing overlaps with some concepts of Distributed Computing and Grid Computing (Hartig, 2008). Both, grid and cloud environments provide a network infrastructure for scaling applications by sufficient storage and computational capabilities. However, Grid Computing is applied by the scientific community for large-scale computations (e.g. a global climate change model or the aerodynamic design of engine components). Whereas Cloud Computing enables small and medium-sized companies to deploy their web-based applications in an instant scaleable fashion without the need to invest in large computational infrastructures for storing large amounts of data and/or performing complex processes (Myerson 2008). As a consequence, national and international grid infrastructures (for example the Worldwide LHC Computing Grid[1]) are typically funded by the government and operated by international joint research projects, whereas cloud infrastructures are operated by large-sized enterprises under economic aspects, such as Amazon or Google, enabling smaller companies to use their infrastructure (e.g. WeoGeo).

In essence, Cloud Computing is not a completely new concept, it moreover col-lects a family of well known and established methods and technologies under the umbrella of the term Cloud Computing. These well known methods and technologies are for example Software as a Service (SaaS) as a model for software deployment and virtualization as an efficient hosting platform (Sun Microsystems Inc. 2009). Besides, it describes a paradigm of outsourcing applications and specific tasks to a scalable infrastructure and therefore consequently enabling new business models with less up-front investments.

The following sub-sections describe the paradigm of Cloud Computing grouped by its characteristics and anatomy.

2.1.1 Characteristics

The key characteristics of Cloud Computing are the ability to scale and provide computational power and storage dynamically in a cost efficient and secure way over the web (ANSI 2009). Besides, a client application is able to use these resources without having to manage the underlying complexity of the technology. These characteristics lead to the following benefits:

- **Efficiency**
  From a provider perspective, Cloud Computing enables IT companies to

[1] http://lcg.web.cern.ch/LCG/

increase utilization rates of their existing hardware significantly. Existing infrastructures such as large data centers are now able to utilize their hardware infrastructures more efficiently by dynamically distribute their applications and processes to free available resources in an on-demand fashion. From a client perspective, the client's infrastructure can be utilized to the maximum and whenever more resources are needed, additional resources could be provided by the cloud.

- **Outtasking**
  By outtasking software and data to computational facilities operated by third parties, clients do not need to operate their own large-scale computational infrastructure anymore. Therefore, enterprises of any size - from Web 2.0 start-up companies to global enterprises - can decrease their costs for initial infrastructure and maintenance significantly. Thereby, fixed costs can be transformed into variable costs and create a business advantage. This allows companies to rather focus on their business model than to maintain and invest in the infrastructure (software licenses & hardware).

- **Scalability**
  Cloud Computing resources (i.e. storage or computational power) are allocated in real-time and cloud resources scale the deployed applications automatically on-demand (for example in case of high amounts of requests). This allows cloud users to handle peak loads very efficiently without managing their own infrastructures. For example, load-balancing or developing highly available solutions for their software do not need to be regarded by the cloud users because such solutions are incorporated in the cloud implicitly. By deploying applications and data in the cloud, clients are automatically able to scale up their computational capacities (for example from a few to hundreds of servers) in an instant and on-demand fashion.

- **On-demand**
  Allocating cloud resources on a real-time and on-demand basis helps enterprises to utilize large IT resources instantly and efficiently (see the aspect of efficiency). In contrast to classical long term outsourcing contracts, on-demand usage with pay-per-use revenue models enable cloud users to restructure existing business processes or even to realize the novel business models with little investment (Gartner 2009b). The total cost of ownership (including initial investment in hardware, software licenses, energy, fail-safety and technical engineers) of self-hosted data centers is in contrast to a Cloud Computing approach which minimizes start-up costs and helps enterprises to put new promising business models into the market.

An additional characteristic of Cloud Computing is the support of Service Level Agreements (SLA) defining different service quality guarantees (for example hotline support, web service mean up time or a specific numbers of accessible CPUs) and contractual penalty clauses. Such contracts are of general importance for cost-performance ratio transparency in SOA governance and therefore an essential characteristic for potential future geospatial business models with defined value propositions.

There are still a number of open issues for Cloud Computing. One open issue is the existing barriers of adopting Cloud Computing aspects in existing IT infrastructures, which is exemplified in the so-called "Open Cloud Manifesto"[2]. Especially the absence of cloud interoperability due to vendor specific cloud APIs can be seen as one major obstacle. These specific APIs bind the applications of the cloud users to specific cloud vendors and therefore complicate the migration of applications between different cloud vendors (i.e. vendor lock-in). Standards are needed and will be addressed by the Open Cloud Consortium.

Besides data backup and recovery responsibilities the outsourcing of confidential data from data owners to third party infrastructures is problematic in the context of security. Using public clouds as a deployment platform for applications and services is in most cases not suitable. Private cloud (clouds on private networks) maintained within an entity can help to solve this problem. The identified issues regarding outsourcing of data and reliability of infrastructures are not only specific to cloud infrastructures, but must be addressed for all kinds of distributed architectures.

2.1.2 Cloud Anatomy

The Cloud Computing paradigm replaces the classical multi-tier architecture model of web services and creates a new set of layers (Sun Microsystems Inc. 2009, ANSI 2009) as depicted in Figure 1. *Software as a Service* (SaaS) and *data Storage as a Service* (dSaaS) are the top layers and feature processing and storage facilities through web services. *Platform as a Service* (PaaS) is the middle layer and encapsulates complete development and runtime environments (for example operating systems, databases or web service application frameworks). *Infrastructure as a Service* (IaaS) is the bottom layer and delivers basic computational infrastructures as standardized services over the network. The bottom layer is then based on actual hardware provided to realize a cloud infrastructure.

[2] http://www.opencloudmanifesto.org/

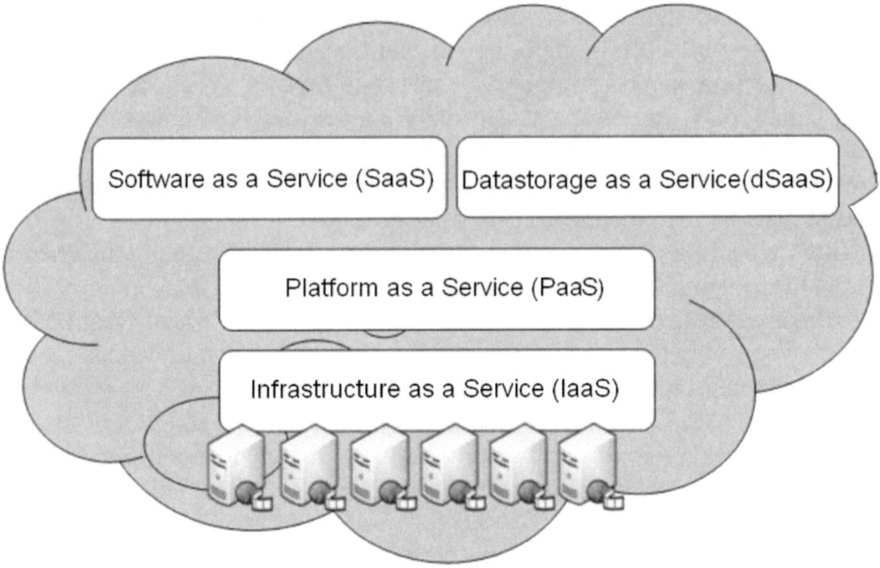

Fig. 1. A short overview about a typical set of Cloud Computing layers

2.2 Spatial Data Infrastructures

Spatial Data Infrastructures (SDIs) are technical, organizational and legal frameworks for geoinformation resources (McLaughlin and Groot 2000). SDIs can be designed differently. For instance, Bernard and Streit (Bernard & Streit 2002) especially focus on the service aspect of SDIs by specifying that an SDI enables the cooperatively use of distributed governmentally or privately held geodata and GI-Service across administrative and system borders. Whereas, McLaughlin and Groot (2000) emphasize the organizational aspect of "[…] delivering spatially resources, from the local level to the global level, in an interoperable way for a variety of uses."

The building blocks of an SDI are the geodata, its technical network, metadata, Web Services and standards (BKG 2002). Specific Web Services provide the geodata and corresponding metadata the Web. To realize communication sufficiently, the services have to be interoperable through standardized interfaces. Onstrud (Onstrud 2007) adds clearinghouses, partnerships, education and communication to this definition. Clearinghouses are used to uniformly search distributed geodata and actually obtain the geodata. Partnerships reduce redundancy and costs. Education and communication enables different entities to communicate knowledge and thereby learn from each other.

Several initiatives are currently in the process of establishing SDIs on multiple levels. From a top-down point of view, on a global level there is i.e. DigitalEarth and on the European level INSPIRE (European Council 2007). Many countries have started to establish their own national SDI, for instance the USA (USGS 2005), Canada (Geoconnection 2004), Germany (GDI-DE 2007), Portugal (Juliao 2009) and Denmark (Jarmbaek et al., 2009).

To actually build SDIs, standards and best practices are required. The Open Geospatial Consortium (OGC) is dedicated to standardize SDI services to enable interoperable communication.

Overall, the main advantages of an SDI for the participating organizations and society are (Bernard et al. 2005):

- Cost effective data production
- Avoidance of duplications
- Efficient data exchange and use over administrative and enterprise borders
- Improvement of decision making on the basis of available high value data.

Based on the presented concepts, Section 3.1 will describe current obstacles in developing SDIs, which can be addressed by integrating the cloud paradigm.

3 Cloud-enabled SDIs

This section provides the design of a cloud-enabled SDI by applying the concepts introduced in Section 2. At first, Section 3.1 analyses obstacles of SDIs. The findings of this analysis are additional input to design a cloud-enabled SDI (Section 3.2).

3.1 Obstacles in SDI Development

SDIs have shown a great potential for enabling the market value of geoinformation as for instance presented in (Micus 2004). However, current SDI development faces different challenges as for example volunteered geoinformation and data harmonization (Craglia 2009). On this basis, the following obstacles can be identified in SDI developments with regards to cloud computing:

- Upfront costs barriers
- Mass market requirements
- Legally binding performance allowances

SDI literature mentions also other obstacles such as organizational aspects (Ollen 2003) which are not considered here due to less relevance to the cloud context.

Volunteered geoinformation is beneficial for SDIs to enhance the availability of real-time data, but appropriate concepts to integrate such geoinformation are not yet available (Craglia 2009). In particular, volunteered geoinformation collected by ordinary users, who in some cases can provide up-to-date data, play an important role especially in risk management scenarios (e.g. geotagged pictures of flooding taken by mobile phones). These geospatial mass market applications, as the name implies, typically yield many concurrent requests which have to be processed. As Scholten et al. show (Scholten et al. 2006) scalability is a problem for SDI services. To meet these requirements of mass market applications for immediate response (i.e. below 5 seconds), scalable solutions are necessary.

Another challenge is the integration of real value-added information, provided by web-based processing. As already mentioned, SDIs are currently in a transition of the focus from data (provider-oriented) to information (user-oriented) (Kiehle 2006, Schaeffer et al. 2009). To generate this information, thorough processing facilities have to be integrated into SDIs. This integration requires large investments in computational and storage resources to handle the intrinsic complexity and huge volumes of geodata (e.g. LIDAR or real-time sensor data) as well as multiple and concurrent requests by mass market applications. Apart from the investments in large-scale computation infrastructure for processing, other investments related to SDI development such as software license costs are typically have to be considered. These investments can be seen as a major obstacle towards the full implementation of SDIs. For instance, to build up the Swedish SDI, more than 150M $ by an annual maintenance cost of 30M $ are reported by (Wigberg 2002). The Italian SDI has already cost over 400M €. Even though most of the money has been spent on data collection, it becomes clear that operating SDIs at a technical level is cost intensive. This shows also the investment of 80M € for infrastructure services for the Italian SDI over a 2 year period (Cappadozzi 2008).

Additionally, SDI initiatives with legal bindings such as INSPIRE (legally binding since 2007) explicitly require guaranteed response times for specific queries (INSPIRE, 2007) (INSPIRE, 2008). For example, the current requirement for processing is a throughput of 1 MB/second and a response time of the service below 1 second. Search queries need to be answered within 3 seconds and services must be able to handle up to 30 of these queries at the same time. Image downloads should have a maximum response time of 5 seconds. To meet the specified performance boundaries in peak times, scalable solutions have to be found which are not yet implemented in SDIs from a technical point of view.

3.2 Design of a Cloud-enabled SDI

This section presents a concept of a cloud-enabled SDI, which integrates the Cloud Computing paradigm with the SDI concept.

In general, there are two options for realizing the integration of Cloud Computing and SDIs:

- Option 1: Adopting Cloud Computing principles and standards to SDIs.
- Option 2: Migrating SDI services on top of a Cloud Computing infrastructure.

Following option 1, SDIs are limited to themselves by creating separate standards and markets and could not benefit from mainstream-IT developments in the future. The authors of this paper favour option 2 which is more beneficial for the GI-domain as it is more open to the mainstream IT world and thereby broadens the opportunities of the GI-domain. Therefore option 2 would in contrast to option 1 allow the combination of SDI and Cloud Computing benefits, while benefiting from new developments in the mainstream IT world at the sae time.

Mapping between SDI and Cloud Computing Components

From an architectural perspective, the integration of SDIs into Cloud Computing infrastructures is shown in Figure 2. In detail, data services (such as WFS) can be considered from a customer perspective as Software as a Service (SaaS), because they offer certain functionality, such as spatio-temporal query for datasets. From a data owner perspective, dSaaS is utilized, because the cloud can store the data served via standardized interfaces over a network. A typical case is a company for remote sensing, storing the large stream of data coming from their satellites and providing these images via data provision services to customers, without dealing with extending memory capacity in their IT infrastructure. SaaS as well as dSaaS rely on PaaS for e.g. the operating system, databases or web service containers, while IaaS describes the hardware level as shown in Section 2.1.2.

Processing instead of data storage aims at deriving information from data, can be seen as a typical SaaS application since customers can use the offered functionality, such as interpolating data on their side. The computation resources are provided via PaaS and IaaS. The same applies for portrayal services such as a WMS or discovery services e.g. Catalog Services (CSW).

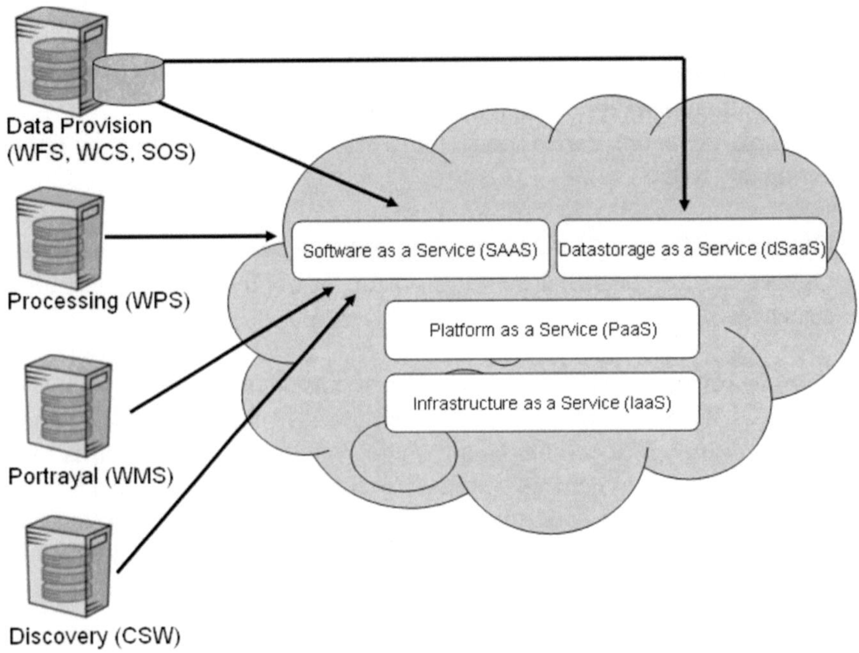

Fig. 2. SDI-Cloud Mapping.

The presented concept addresses the identified obstacles in SDI development (Section 3.1) as explained in detail in the following.

Upfront costs barriers

From a georesource (data/processes) provider perspective, the classic Publish-Find-Bind pattern of SOAs/SDIs (Figure 3) can be applied to the cloud-enabled SDI (Figure 4). According to this classic pattern the georesource providers host their services offering georesources on their own infrastructure and publish these services to a registry. This allows clients to find the georesources and bind (invoke) them. In other words, the georesources are accessed via services based on standardized interfaces over a network. This results in high upfront investments for the georesource owner to cover also peak loads or risk failing of the infrastructure.

Cloud Computing and in particular the aspect of outtasking can be utilized to overcome this high up-front investment for building and maintaining a large in-house IT infrastructure. By delegating computational and storage intensive tasks to third party providers in a cloud and using these tasks via services with standardized interfaces over a network, SDI services can be used in a cloud as shown in Figure 4.

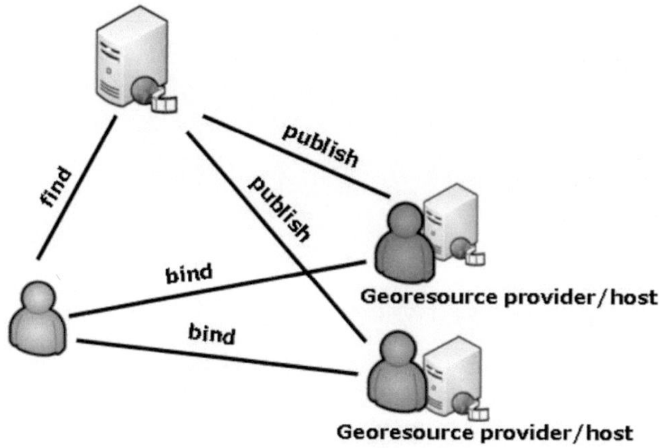

Fig. 3. Classic Publish-Find-Bind SDI pattern

The classic Publish-Find-Bind pattern still applies here, but the georesource provider uses the cloud to host their georesources. Therefore, there is a distinction in this concept between the roles of the georesource provider and georesources host. While the provider still publishes georesources which the customer can discover, the found georesources are bound from the cloud. Therefore, a business relationship has still to be established between the customer and the georesource provider. The revenue model for the georesource provider can be arranged in a flexible way. For instance, on-demand or flat-rate access models may be adequate as they are also the dominated revenue model in public cloud environments such as Google Apps Engine or Amazon Elastic Compute Cloud (Amazon EC2[3]).

Besides, by using standardized service interfaces, cloud infrastructures hosted by different providers can be used interchangeably from a cloud service consumer perspective. In other words, a client application does not need to be aware of whether a service is hosted in a cloud or not and which cloud provider is used. However, different cloud providers still have different internal requirements and capabilities which make it more complicated for the georesource provider to switch clouds for setting up a service in different clouds.

[3] For a complete cost schema, see http://aws.amazon.com/ec2

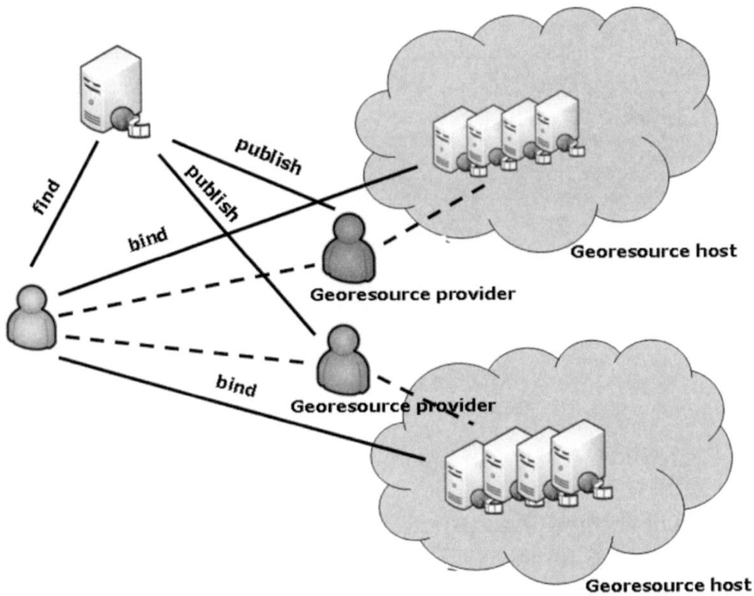

Fig. 4. Publish-Find-Bind in Cloud SDI

Mass market requirements

As identified in Section 3.1, current SDI concepts lack scalability, which is especially crucial for integrating mass-market applications into SDIs. SDIs can benefit from the ability of cloud infrastructures to handle large amount of requests, processes or data. By migrating services into the cloud, the georesources provided by these services are immediately available in a scalable fashion for on-demand use. Figure 5 shows, how the cloud expands from light blue over light violet to blue with the increasing number of user induced requests.

Conceptually, the scalability is automatically available through the cloud without touching the services itself. This implies that existing services can be deployed in a cloud environment without any adjustments to the service implementations.

When deploying SDI components on a cloud infrastructure, they can benefit from the cloud's scalability instantly, but still remain interoperable. Regarding the service interface, there is no difference between a cloud-enabled SDI service and a non-cloud enabled SDI service. In fact they can be used interchangeably and/or sequentially (in a composed workflow of traditional and cloud-enabled SDI components).

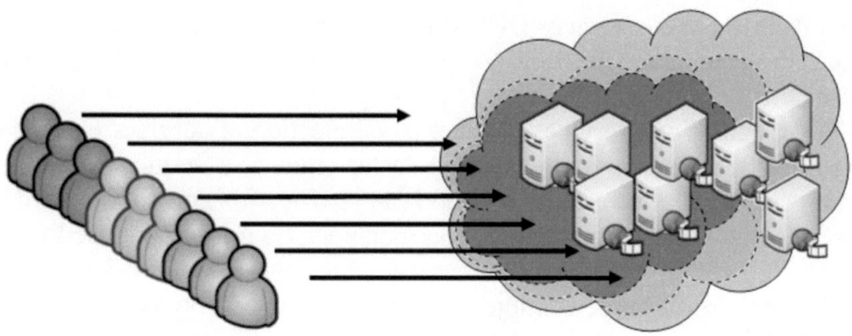

Fig. 5. Scaleable SDIs

Legally binding performance requirements

Existing SDIs which implement a legal framework such as INSPIRE have to meet specific Quality of Service (QoS) parameters as described in Section 3.1. This applies also for private companies providing SDI services which typically have to provide specific QoS levels. Especially the scalability aspect of clouds can help here to process even a large amount of requests in a given time frame. This aspect can be combined with the argument of up-front investments as described before. For instance, for start-up companies the legally binding performance requirements can be a limiting factor to realize innovative ideas if large infrastructure have to be acquired a priory to comply to the performance requirements. The Cloud Computing paradigm can be used to solve this obstacle, because it offers a low-cost way of delegating the performance requirements to a specialized third party georesource host (cloud provider).

For SDIs, this means, that the services of georesources have to be deployed in the cloud as shown in Figure 4.

These theoretic considerations concerning cloud-enabled SDIs will now be evaluated against the background of a real world use case with a special focus on the scalability of cloud-enabled SDI services over existing SDI services.

4 Application of the Use Case for a Cloud-enabled SDI

The scenario is settled in the context of a public risk management use case, in which in-situ-sensor data has to be analyzed for assessing a fictive fire threat in Tasmania. A similar scenario and involved services have been extensively presented in Foerster & Schaeffer (2007) for the area of north-west Spain. This section pushes the scenario idea one step further and leverages cloud enables services to create a highly scaleable solution in Tasmania.

The Amazon Web Services (AWS) together with an OGC Web Processing Service implementation hosting a buffer and intersection process is used in this scenario.

The Amazon Web Services product is a collection of services that are offering Infrastructure as a Service (IaaS), Datastorage as a Service (dSaaS) and some aspects of Platform as a Service (PaaS). The Amazon EC2 provides a web service interface to manage virtual machines (IaaS) that are used to host customer specific applications and can be scaled on-demand to handle peak load. The Amazon Simple Storage Service (Amazon S3) provides a web services interface that can be used to store and retrieve large amounts of data (dSaaS).

To deploy a WPS in Amazon EC2 and thereby to add the SaaS layer, an Amazon Machine Image (AMI) has to be configured. The AMI serves as a template for all instances that have to be setup by the Amazon cloud. Therefore, a WPS has to be installed on the virtual machine following the AMI template on top of a chosen machine setup (IaaS), operating system and servlet container (PaaS). In addition, the whole setup has to be configured to match certain scalability goals (expressed as rules). For this use case, the following rules were applied:

- 1 instance should be running at all times
- a maximum of 12 instances can be created
- if the CPU workload is below 20% in a 30 second interval, the number of instances should be decreased by 1
- if the CPU workload is above 50% in a 30 second interval, the number of instances should be increased by 1.

Once, the AMI is configured, deployed and started, the WPS is accessible via a single URL like any other non cloud-enabled WPS. For instance a standard web client such as OpenLayers or directly as a web service can be used to consume the service. In the given scenario, an expert would discover the URL (find) add the service to the OpenLayers client (bind) and buffer the given wild fire polygons. The resulting layer is then intersected by the given Tasmania road data.

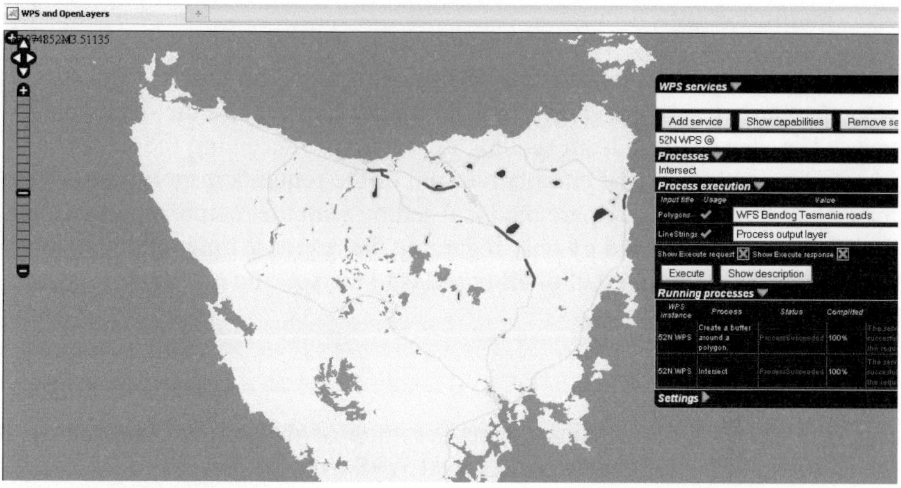

Fig. 6. Result (in red) of cloud enabled WPS intersecting buffered wild fires (violet) and road data (yellow) in Tasmania.

Overall, this allows the user to assess which parts of the road infrastructure are at risk by a fire (see Figure 6, read layer). With an increasing number of requests, the number of WPS instances in the cloud should increase as shown in Figure 5 to meet the scalability goals (i.e. constant response times) (see Section 5 for detailed results). In such risk management situations, in which multiple users with concurrent requests are expected and peak loads on the infrastructure are common, the information about the latest wild fires will still be processed and provided to the user based on real-time processing. The following section examines a stress test simulating such peak loads on the infrastructure and thereby demonstrates the scalability of cloud-enabled SDIs.

5 Stress Test

To demonstrate the scalability of cloud-enabed SDIs, we used a stress test to simulate an increasingly high demand of simultaneous requests (i.e. peak loads). A constant response time by the WPS deployed in the cloud was expected in contrast to a linear rising response time by a non cloud setting. The WPS was stress tested with the simple buffer algorithm, deployed in the Amazon Web Service framework as well as on a local and non cloud-enabled Tomcat installation. The geodata for that process was also delivered via a web service (deployed at the cloud(s) in the first case and deployed on the local and non cloud-enabled machine in the second case).

5.1 Methodology

A cumulative approach was used, starting with 1 and up to 200 requests that were sent nearly simultaneously in a short period of time to the deployed services. The elapsed time from sending the request to receiving the response on its own, as well as for the cumulative sum of the requests/response times was measured. In order to compare the local setting with the remote cloud settings, the results are normalized by only regarding the response time relatively to the maximum/minimum interval of all requests to the specific machine.

5.2 Results

Figure 7 shows the normalized response time of the online (Amazon Web Services) as well as of the local deployed WPS over the number of simultaneously sent requests. Normalization was reached by means of using the interval (min, max) as baseline. The response time of the remote WPS (monotonically increasing line) stays nearly constant up to 200 simultaneous requests whereas the local WPS response time (constant line) grows linearly. For the cloud approach, only one large peak at the beginning can be observed at the beginning and some smaller peaks during the rest of the execution.

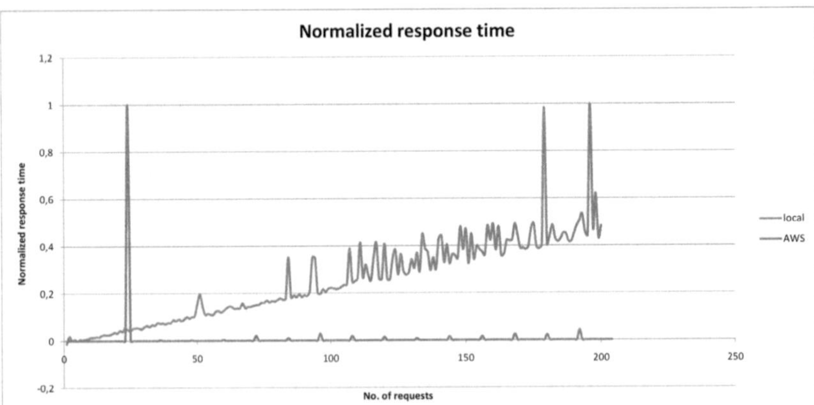

Fig. 7. WPS local vs. WPS in the cloud stress test results

5.3 Evaluation

The performance evaluation shows to some degree that a WPS deployed in the Amazon Web Service scales at high request rates as expected: The response time for many simultaneous requests stays nearly constant in contrast to the non-cloud deployment.

The peak in the beginning of the cloud curve (Figure 7) for the measured response times could be explained by means of managing the (virtual) server instances in the backend. Additionally, the number of instances is increased only by 1 in a 30 second interval, which means, that for the starting period not enough instances are available. We assume that the smaller peaks for the remote WPS are also related to minor background management tasks, such as setting up new instances.

6 Conclusion

This paper presents an approach for integrating SDIs and Cloud Computing technologies to set up a cloud-enabled SDI. A cloud-enabled SDI is identified as beneficial to address the major obstacles of SDI development (Section 2). Different roles in this cloud-enabled SDI are distinguished (Section 3). When integrating Cloud Computing and SDIs the existing publish-find-bind pattern for service interaction can be reused. Therefore, we see a paradigm shift from technological to economical aspects in contrast to a complete paradigm change, because the technical principles stay the same while economical aspects (upfront costs, maintenance, cost-effective production, etc, see section 1) motivate the technological shift.

It also became clear, that the way forward is to bring SDI components into cloud environments instead of adopting mainstream IT techniques such as Cloud Computing for SDIs. This will broaden the business opportunities for SDIs based on the high potential of cloud technologies. Therefore, we can foresee that the components, once deployed in an SDI, could be part of a cloud infrastructure service (belonging to PaaS) as there are already for instance databases or authentication APIs provided in a cloud, for georesources (geoprocessing/geodata).

As discussed, cloud interoperability from a client perspective is given for the geospatial domain because of the well established standards. From a provider perspective, the coupling with each cloud infrastructure is vendor-specific. Therefore, the advance of the SDIs regarding standardization can lead to easy cloud interoperability also for the provider in respect to georesources. Once the standardized SDI services are deployed in a cloud, they can be used by other cloud applications interchangeably. This implies that when a georesource dependent application is migrated from one cloud provider to another, the connection to underlying georesources providing services does not need to be changed due to its standardized access.

Another conclusion we can draw is, that Cloud Computing has the potential to create new business models for SDIs. These business models have to be distinguished from client and provider perspective. From a client perspective, the low up-front investment barrier by using cloud environments for SDI services allows companies to start new business models, which may have not been possible before due to high legally binding requirements. Besides, the outtasking of non-core task of a business process can lead to a modification of exiting business processes, which allows the overall business model to be more flexible to customer needs. This can lead on the provider perspective to specialized SaaS providers which can offer value-added services on-demand in a scalable fashion as for instance shown in the use case. These new SDI business opportunities have to be studied further in the future. Especially, the increasing distribution of smartphones seems to be a promising market, because applications on smartphones typically address the mass market but also lack large processing, storage and battery capacities, which makes it necessary to handle the data in a remote server environment such as the cloud.

The use case and further tests on the scalability part showed that cloud computing keeps its promises in terms of scalability. It could be clearly seen, that the response time stays constant in contrast to a linear increasing response time for a non-cloud approach.

As already discussed in Section 2.1, privacy can be a concern for sensitive data when they are given away to third party public cloud providers. The same problem applies to SDIs and the georesources provided via services in particular, because georesource could e.g. cover sensitive areas such as government buildings or are costly to create. However, certified cloud providers in analogy to certified tax accountants can be one applicable solution to overcome privacy concerns.

References

American National Institut of Standards and Technology (ANSI) (2009) NIST Definition of Cloud Computing. [Online] Available: http://csrc.nist.gov/groups/SNS/cloud-computing/index.html

Bernard, L. and Streit, U. (2002) Geodateninfrastrukturen und GI- Dienste –

Aktueller Forschungsstand und Forschungsprobleme. In: Seyfert, E. [Ed..] 22. Wissenschaftlich-Technische Jahrestagung der DGPF: 11-20, Neubrandenburg.

Bernard L, Fitzke J, Wagner R (2005) Geodateninfrastruktur – Grundlagen und Anwendungen. Wichmann, Heidelberg.

Bundesamt für Kartographie und Geodäsie (BKG) (2003) Geoinformation und moderner Staat. Interministerieller Ausschuss für Geoinformation (IMAGI), Bundesamt für Kartographie und Geodäsie, Frankfurt

Cappadozzi E (2008) The cost of a national SDI for Italy. [Online] Available: http://www.gistandards.eu/MWS/images/stories/presentations/Cappadozzi.ppt

Craglia M (2009) From INSPIRE to Digital Earth: Research Challenges for the Next Generation Spatial Data Infrastructures. In: Proceedings of the 12th AGILE Conference - Hannover, Germany.

Dean J, Ghemawat S (2004) Mapreduce: Simplified data processing on large clusters. [Online]. Available: http://www.usenix.org/events/osdi04/tech/dean.html

European Council (2007) Directive 2007/2/EC of the European Parliament and of the Council of 14 March 2007 establishing an Infrastructure for Spatial Information in the European Community (INSPIRE). In: Official Journal of the European Union, Vol 50, Brussels.

Foster I, Zhao Y, Raicu I, Lu S (2008) Cloud computing and grid computing 360-degree compared. [Online]. Available: http://arxiv.org/abs/0901.0131

Gartner (2008) Gartner Says Cloud Computing Will Be As Influential As E-business. Gartner Press Release. [Online] Available http://www.gartner.com/it/page.jsp?id=707508

Gartner (2009a) Gartner Says Cloud Application Infrastructure Technologies Need Seven Years to Mature. Gartner Press Release. [Online]. Available: http://www.gartner.com/it/page.jsp?id=871113

Gartner (2009b) Outsourcing trip report. Gartner Press Release. Gartner Press Release. [Online]. Available: http://www.gartner.com/it/content/754100/754124/outsourcing_2009_trip_report.pdf

GDI-DE (2007) Architektur der Geodateninfrastruktur Deutschland. [Online]. Available: http://www.gdi-de.de/de_neu/download/AK/GDI_ArchitekturKonzept_V1.pdf

Geoconnections (2004) Guide to the Canadian Spatial Data Infrastructure. [Online]. Available: http://www.geoconnections.org/publications/Technical_Manual/html_e/cgdiindex.html

Hartig K (2008) What is Cloud Computing? The cloud is a virtualization of resources that maintains and manages itself. .In: NET Developers Journal, SYS-CON Media.

INSPIRE (2007) INSPIRE Network Services Performance Guidelines, INSPIRE Consolidation Team, European Commission.

INSPIRE (2008) INSPIRE Network Services Architecture. Network Services Drafting Team, European Commission

Jarmbaek J (2009) The Danish Way, Development of the Danish Infrastructure for Spatial Information Through Binding Collaboration. In: Proceedings of GSDI 11 World Conference, 2009, Rotterdam, The Netherlands

Juliao R (2009) Portugal and Spain Twin SDI's - From National Projects to an Iberian SDI. In: Proceedings of GSDI 11 World Conference, 2009, Rotterdam, The Netherlands

Kiehle C, Greve K, Heier C (2007) Requirements for Next Generation Spatial Data Infrastructures-Standardized Web Based Geoprocessing and Web Service Orchestration. In: Transactions in GIS 11(6):819–834, doi:10.1111/j.1467-9671.2007.01076.x

McLaughlin J, Groot R (2000) Geospatial data infrastructure: concepts, cases and good practice. In: Spatial Information Systems and Geostatistics Series, University Press, Oxford

Micus (2004) The market for geospatial information - potentials for employment, innovation and value added. [Online]. Available: http://www.micus.de/pdf/micus_geoinfo_germany.pdf

Myerson J (2008) Cloud computing versus grid computing - Service types, similarities and differences, and things to consider, IBM Corporation. [Online] Available: http://www.ibm.com/developerworks/web/library/wa-cloudgrid/

Ollén J (2003) Spatial Data Infrastructure - A Tool for Growth and Suistanable Development in Europe. From OEEPE to EuroSDR: 50 years of European Spatial Data Research and beyond - Seminar of Honour, (eds.) C. Heipke, EuroSDR, Munich, Germany, Nr. 46,. 23-29.

Onsrud H (2007) Research and Theory in Advancing Spatial Data Infrastructure Concepts.ESRI Press, Redlands.

Schäffer B, Baranski B, Foerster T, Brauner J (2009) A Service-Oriented Framework for Real-time and Distributed Geoprocessing. In: Proceedings of the International Opensource Geospatial Research Symposium, 2009 (not published yet)

Scholten M, Klamma R, Kiehle C (2006) Evaluating performance in spatial data infrastructures for geoprocessing. In: IEEE InternetComputing 10, no. 5 (October 2006): 34-41, doi:10.1109/MIC.2006.97

Sun Microsystems Inc. (2009) Cloud Computing at a higher level. [Online] Available: https://slx.sun.com/files/Cloud_Computing_Brochure_2009.pdf

USGS (2005). The National Spatial Data Infrastructure. [Online] Available: http://www.fgdc.gov/nsdi/library/factsheets/documents/nsdi.pdf

Wigberg, H.-E. (2002) The Swedish National Spatial Data Infrastructure [Online] Available: http://www.fig.net/pub/fig_2002/Ts3-5/TS3_5_wiberg.pdf

MIX
Papier aus verantwortungsvollen Quellen
Paper from responsible sources
FSC® C105338

If you have any concerns about our products,
you can contact us on
ProductSafety@springernature.com

In case Publisher is established outside the EU,
the EU authorized representative is:
Springer Nature Customer Service Center GmbH
Europaplatz 3, 69115 Heidelberg, Germany

Printed by Libri Plureos GmbH
in Hamburg, Germany